Consumer Perception of Food Attributes

Editors

Shigeru Matsumoto
Department of Economics
Aoyama University
Tokyo, Japan

Tsunehiro Otsuki
Osaka School of International Public Policy
Osaka University
Osaka, Japan

The opinions expressed and arguments employed in this publication are the sole responsibility of the authors and do not necessarily reflect those of the OECD or of the governments of its Member countries.

This Conference was sponsored by the OECD Co-operative Research Programme: Biological Resource Management for Sustainable Agricultural Systems, whose financial support made it possible for the invited speakers to participate in the Conference.

CRC Press
Taylor & Francis Group
Boca Raton London New York

CRC Press is an imprint of the
Taylor & Francis Group, an **informa** business

A SCIENCE PUBLISHERS BOOK

CRC Press
Taylor & Francis Group
6000 Broken Sound Parkway NW, Suite 300
Boca Raton, FL 33487-2742

First issued in paperback 2021

© 2018 by Taylor & Francis Group, LLC
CRC Press is an imprint of Taylor & Francis Group, an Informa business

No claim to original U.S. Government works

Version Date:20180521

ISBN-13: 978-0-367-78109-5 (pbk)
ISBN-13: 978-1-138-19684-1 (hbk)

**Visit the Taylor & Francis Web site at
http://www.taylorandfrancis.com**

**and the CRC Press Web site at
http://www.crcpress.com**

Preface

A rich diet is prerequisite for a rich life. Governments have made enormous efforts, sustained over a long period, to provide their citizens with necessary nutrition and caloric intake. In addition to such efforts, technological innovations in the agricultural sector and development of distribution networks have now enabled people in developed countries to obtain sufficient nutrition and caloric intake to support their daily lives. The progress in technology and distribution has further enabled us to enjoy a wide variety of foods. Meanwhile, consumers' desire for foods has evolved from the simple need to fill their stomach with tasty foods to more complicated ones. Producers, distributors, and governments must respond to these complex consumer demands.

Due to the development of distribution networks, consumers are now able to obtain food products that are produced far away. Nonetheless, they want to understand food attributes and how the foods were produced before purchase. On the other hand, producers want their good production practices and desirable food properties to be adequately evaluated. Thus, both consumers and producers desire more effective communication.

Food safety management is an important part of food production. Expanded distribution networks also mean that when food contamination and animal diseases outbreaks occur, they spread instantaneously worldwide, throwing food markets into great turmoil. Although consumers expect their own governments to ensure food safety, food safety management can no longer be dealt with by a single country alone. International cooperation is becoming increasingly important to maintain a safe and rich diet.

Moreover, diet is closely tied to cultures and customs. Therefore, the perception of food attributes differs greatly across countries and ethnic groups. Since neglecting the eating habits of others may amount to denying their individual values, very careful consideration is necessary. Past food disputes teach us that it is not sensible to evaluate food management solely from a scientific perspective.

Against this backdrop, we invited researchers with diverse backgrounds from various countries to participate in this book project. We also held an international symposium from 18–20 May 2018, sponsored by the OECD Co-operative Research Programme on Biological Resource Management for Sustainable Agricultural Systems and the Institute of Economic Research of Aoyama Gakuin University. The intensive discussions arising from the symposium facilitated communication exchange among researchers of various backgrounds. Consequently, the problems associated with food attribute management and corresponding necessary measures were reconfirmed. In this book, consumer attitudes toward foods and corresponding policies are discussed from various aspects such as safety, use of new technology, health, sustainability, animal protection, and trade restrictions.

Derived from the experiences of comparative studies across countries and from the rich literature review of various research fields, the findings of this book offer insights for the management of food attributes. This book will appeal to those who intend to develop better food policies by understanding consumer perception of food attributes.

Shigeru Matsumoto
Tokyo, Japan

Tsunehiro Otsuki
Osaka, Japan

Acknowledgment

This book emerged from the International Symposium on "Food Credence Attributes: How Can We Design Policies to Meet Consumer Demand?" held at Aoyama Gakuin University, Japan in May 2017. The symposium was sponsored by the OECD Co-operative Research Programme on Biological Resource Management for Sustainable Agricultural Systems and the Institute of Economic Research of Aoyama Gakuin University. Their financial support facilitated the participation of eminent speakers.

In addition to the contributors to this book, many researchers, including Dr Harunobu Amagase, Dr Rafael Blasco, Prof. Lynn Frewer, Dr Viet Ngu Hoang, Prof Naoto Jini, Mr Yasushi Kuriyama, Dr Kaoru Nabeshima, and Dr Setsuko Todoriki, participated in the symposium. Their participation greatly facilitated the discussion at the symposium and subsequently enriched the content of this book. Tokyo Center of Economic Research sponsored Dr Viet Ngu Hong. Mrs Emiko Kuramochi of the Institute of Economic Research of Aoyama Gakuin University supported the coordination of the symposium, which would not have been possible without her support. A family excursion was arranged as well. We also thank Mrs Yoko Matsumoto for her generous help.

Contents

1

Overview

Céline Giner,[3,]* *Shigeru Matsumoto*[1] and
Tsunehiro Otsuki[2]

Introduction

Our lifespan is limited and so also the number of meals we can partake. Even if we have three meals a day for 90 years regularly, we cannot eat more than 100,000 meals in our entire lifetime. Given this limitation, people tend to seek more satisfying foods. This is not a new phenomenon. In earlier times, our ancestors sought just the nutrients necessary for survival. They then traveled around the world and looked for more satisfying foods.

Modernization in agriculture and expansion of international trade has rapidly increased food variety. Compared to our ancestors, we can eat a wide variety of foods. However, deciding what food to eat has become a more complex task. Suppose we find two types of garlic in a grocery store: one imported from Spain and the other from China. Which garlic are we going to choose? Should we assume that difference in quality is fully reflected by the difference in price? Or should we purchase Spanish garlic to make Spanish gambas al ajillo, but purchase Chinese garlic to make Chinese garlic shrimp? Diversification of production methods and the complexity of the food supply chain further add to consumer confusion with regard to food choices. Can we assume that the social and environmental impacts of garlic production are about the

---

[1] Department of Economics, Aoyama Gakuin University, Room 828, Building 8, 4-4-25 Shibuya, Shibuya, Tokyo, Japan, 150-8366.
E-mail: shmatsumoto@aoyamagakuin.jp
[2] Department of Economics, Osaka School of International Public Policy, Osaka University, 1-31 Machikaneyama, Toyonaka, Osaka, 560-0043 Japan.
E-mail: otsuki@osipp.osaka-u.ac.jp
[3] Agricultural Policy Analyst, Trade and Agriculture Directorate, OECD, 2, rue André-Pascal, 75775 Paris CODEX 16, France.
* Corresponding author: Celine.GINER@oecd.org

same between the countries? Or should we be concerned by the difference in production and delivery methods at the time of purchase?

Consumers seek food attribute information at the time of purchase and buy food with certain properties. They care about food production methods, its geographic origin, and its potential impacts on human health. Consumers are not able to check the accuracy of food attributes before or even after consumption. For the past several decades, many scholars have analyzed the value that consumers place on credence attributes and reported that consumers may be willing to pay a premium for foods with these desirable properties. In addition, the studies reveal that individuals place greater importance on some credence attributes than others. For example, some are seriously concerned about animal welfare, while others are solely concerned about food safety. This book summarizes the recent empirical findings from scholarly works about consumers' valuation of food credence attributes. We believe that such knowledge benefits producers, processors, retailers, and policy makers.

Appearance provides little or no indication of how food is produced. This information asymmetry harms transaction efficiency between consumers and producers. Since consumers cannot verify food credence attributes, they rely on information provided by consumer organizations, distributors, governments, and food producers.

Many programs have been developed to more effectively inform consumers regarding food production processes. For instance, organic and country-of-origin labels are commonly used in developed countries. The question then arises whether such labeling programs have an impact on consumers' food choices and improve social welfare. This book discusses the effectiveness of the programs that have been introduced to strengthen the relationship and to resolve information asymmetry between producers and consumers.

The book consists of four parts.

Part I addresses food safety—consumer's primary concern about food:

- Chapter 2 overviews consumer perspectives about the benefits and risks of novel technologies such as irradiation, genetic modification, and nanotechnology. The chapter also studies pesticides and production-enhancing hormones used in agriculture.
- Chapter 3 analyzes consumer response to animal diseases and evaluates the effectiveness of the countermeasures undertaken in affected countries.
- Chapter 4 studies the regulation of genetically modified (GM) foods in the United States and reports that most Americans know little about GM foods.
- The nuclear accident at Fukushima Power Plant in Japan has had a tremendous impact on the Japanese agricultural market. Chapter 5 analyzes the sales data in the Tokyo wholesale market to examine how radioactive contamination has changed consumers' willingness to pay for vegetables from affected regions.

Food credence attributes are also of great concern in the context of international trade. They may have heterogeneous, or sometimes opposite effects depending on their net trade position or on their economic development. Some countries demand that food or agricultural products meet certain desirable properties. This implies substantial commitment and constraints for producers in exporting countries and may also constitute trade barriers.

Part II addresses the relationships between international trade and food credence attributes:

- Chapter 6 gives an overview of the role of food credence attributes in international trade. The chapter demonstrates the impact of regulation of credence goods and attributes on trade and welfare. It observes the potential protectionist implications of the regulations in place.
- Chapter 7 examines the effect of food safety standards on the international food trade. The chapter demonstrates the country-, market-, and consumer-level impacts of food safety standards and attempts to resolve the current debates on whether food safety standards promote or impede trade.
- Chapter 8 discusses how food credence attributes such as organic farming and geographic indication can be mobilized to enhance economic development toward a win–win outcome through case studies on cross-border value-chain management in the Mekong region.
- Chapter 9 addresses agricultural biodiversity as a diversification of farm production while providing environmental, economic, and sociocultural benefits to the local and global society. The chapter attempts to empirically identify farm- and market-level factors that promote agricultural biodiversity.

A rich dietary life not only satiates our stomach but also enriches our life. As we can infer from the fact that many festivals and events are linked to the harvesting of agricultural produce and also our dietary habits have been closely related to culture and/or religion.

Part III discusses cultural and ethical aspects of food credence attributes. If consumers have cultural and ethical motivations for their diet and are trying to express their own identity through food consumption, then producers and governments cannot obtain consumer support simply by providing food attribute information.

- Chapter 10 summarizes the power and potential hazards associated with the promotion of food credence attributes. Typical consumers do not have the ability to gather and process all relevant information required for an informed food choice.
- Using the complexity theory and multi-criteria analysis, Chapter 11 discusses how ethical choices of consumers are to play a key role in a sustainable food system.
- We use animals for crop production and slaughter animals for food. The ethical aspects of agricultural production are emphasized in relation to animals. Chapter 12 focuses on consumers' attitudes toward farm animal welfare.
- It is relatively easy to share culture and ethics with neighbors but difficult to share with those far away. Many countries have introduced country-of-origin labeling (COOL) for those across in the past two decades. Chapter 13 assesses the impact of such COOLs. Specifically, the chapter examines how consumers' trust toward their own government determines the acceptance of imported food.

- Regional origin labels are increasingly popular in recent years. Chapter 14 studies regional origin logos/labels introduced in the Czech Republic and examines whether logos/labels are used at the point of purchase.
- Due to the recent large-scale commercialization in the organic food development, a few US consumers became skeptical about organic production labeled food and started to regard local food and organic food as substitute, rather than complementary. Chapter 15 focuses on this structure change and discusses its potential impact on agricultural markets.

Part IV discusses marketing and regulation for food credence attributes.

- Chapter 16 reviews the regulations of food marketing and examines the mechanics of consumer inference and choice, and the means by which food marketers seek to influence them.
- Although consumers often describe that health considerations are important drivers of their food choices, health, and nutrition information is not always available at the time of purchase. Chapter 17 overviews the literature on health-related food labeling and then reports the findings from a Norwegian study of consumer perceptions of food warnings.
- Consumers with different dietary habits may utilize the same food attribute information differently. Chapter 18 examines how dietary restraint interacts with GM/non-GM food labeling to impact consumers' healthfulness perceptions and subsequent consumption.
- Chapter 19 discusses the difference and competition between local and the existing global certification programs. The chapter specifically analyzes why and how the Japanese local certification programs have been promoted in the context of the Tokyo 2020 Olympics, while global certification programs, such as GLOBALG.A.P., MSC, and ASC, were widely used at the last two Olympic games in London and Rio de Janeiro.

The unique contribution of this book is to provide analyses and discussions from various academic disciplines, including marketing science that examines consumer food choice behavior, economics that quantifies the valuation of food attributes, and international law that analyzes standards prepared for regulating food credence attributes. Since food dietary problem is not a simple one that can be resolved by pursuing one approach, we diversify research methods. However, regardless of the differences in such research methodologies, many common problems are confirmed. Based on these findings, we make several practical recommendations for management of food credence attributes in the conclusion section. We believe that these recommendations are valuable for academic researchers, administrators, business persons, consumers, as well as agricultural farmers.

PART I
Food Safety

2

Consumer Perspectives on Food Safety Issues
Novel Technologies, Chemical Contaminants, Organic Food and Deceptive Practices

Anne Wilcock,[1,*] *Brita Ball*[2] *and Jana Gorveatt*[3]

Introduction

What qualities do consumers look for in their food? The list is a long one, and includes aesthetics, sensory characteristics, brand image, packaging and food safety (Lobb 2005, Verbeke et al. 2007, Rijswijk and Frewer 2008). Which of these is most important varies widely from one culture to the next and one consumer to the next (Wilcock et al. 2004). In developed countries, for example, food safety is expected whereas in developing countries it is a quality that consumers hope for.

There have been numerous reports of consumer concern about various food credence attributes such as irradiation, genetic modification, nanotechnology, use of hormones in food animals, and organic farming. Scholars from various parts of the world have analysed consumers' assessments of food credence attributes and reported that they are willing to pay premiums for the foods that have what they consider to be desirable properties. Such assessments have been shown to differ among countries.

[1] University Professor Emerita, University of Guelph, Guelph, Ontario, Canada N1G 2W1.
[2] Principal Consultant, Brita Ball & Associates, Guelph, Ontario, Canada N1G 4H6.
  E-mail: brita@fsculture.com
[3] Department of Food Science, University of Guelph, Guelph, Ontario, Canada N1G 2W1.
  E-mail: gorveatj@uoguelph.ca
* Corresponding author: wilcock@uoguelph.ca

Consumer views of food credence attributes often differ from food industry and public health perspectives. Some novel technologies, like irradiation, can significantly improve product safety and shelf life, while genetic modification and nanotechnology result in desirable properties that often benefit industry and promote innovation. These industrial benefits are often viewed unfavorably by consumers. An understanding of consumer views is important to the promotion and success of products in the market.

This chapter provides an overview of consumer perspectives about the benefits and risks of novel technologies such as irradiation, genetic modification and nanotechnology. The literature on consumer perspectives about chemical contaminants such as pesticides as well as the use of hormones in food animals is also reviewed. The term 'organic' has different meanings; we have made an effort to provide an overview of consumer perspectives of food grown organically in contrast to conventionally grown food. The chapter ends with a discussion about deceptive practices such as food fraud, an issue that unfortunately is becoming widespread due to the prevalence of global trade.

Novel Technologies

Consumer knowledge of novel food technologies is generally limited. A lack of information leads to consumers' misconceptions about, and apprehensions of, new food technologies. Safety is a concern for consumers, and misconceptions often lead to the belief that food is unsafe or to uncertainty about long-term health effects. Educating consumers has been shown to improve their acceptance of food technologies such as genetic engineering, irradiation and nanotechnology. Improving consumer attitudes is essential to ensuring the sale of foods processed with these technologies.

The lack of consumer acceptance of novel technologies has contributed to their limited application in some jurisdictions (Roberts 2016). Governments vary in their management and approval of these products, and regulations are still developing (Roberts 2016). Consumer insistence on regulation of these technologies has led to some governments taking action and developing labelling regulations (Schilling et al. 2003, FAO-WHO 2010, WHO-FAO 2011).

Irradiation

Food irradiation is a promising technology that uses a carefully controlled process to expose food to ionizing energy primarily to reduce pathogen load and spoilage organisms, and prevent the spread of insects (O'Bryan et al. 2008, Eustice and Bruhn 2013). The process is not the same as radioactive contamination of foods (WHO 1994, European Commission 2017). Based on more than three decades of research, the World Health Organization (1994) determined that treating foods with doses of up to 10 kGy[1] produces no measurable radioactivity. Scientific reports continue to emphasize the effectiveness of irradiation for food safety while maintaining wholesomeness and nutritional value (Farkas 1998, FAO/IAEA/WHO 1999, Tauxe 2001, Acheson and Steele 2001, Eustice and Bruhn 2013). Despite its endorsement

---

[1] 1000 Grays; 1 Gray = 1 Joule of ionizing radiation absorbed per kilogram of matter.

by numerous organizations in support of food safety and public health, the technology has been regarded with suspicion by consumers (Tauxe 2001, Cardello et al. 2007, van Dijk et al. 2011, Eustice and Bruhn 2013). Public concern has remained about the healthfulness and safety of the application of technology in food (Tauxe 2001, Eustice and Bruhn 2013).

Proponents of food irradiation focus on its public health value as well as the safety of the process, weighed against negligible losses in nutritional value when used at approved levels. They compare the potential impact of the technology to the public health improvements gained from thermal pasteurization of milk and retort canning standards. According to Tauxe (2001), if half the poultry, pork, ground beef and processed meats in the United States were irradiated, illnesses and deaths associated with infections from five common microbial pathogens would be reduced by 25%. The irradiation of poultry products alone has the potential to reduce public health issues and their economic impact by reducing the burden of Campylobacter and Salmonella, the top two bacterial agents of foodborne illness in the United States, Canada and the United Kingdom (Thomas et al. 2013). In fact, nutritional losses are less than losses produced from pasteurization or canning (Tauxe 2001). Furthermore, sensory evaluation of irradiated products has confirmed that acceptable quality can be maintained. Foods irradiated at 10 kGy or less may not be distinguishable from non-irradiated products (Andrews et al. 2003, O'Bryan et al. 2008, Maherani et al. 2016).

Lack of awareness about irradiation technology and its benefits is a major reason cited for its lack of acceptance (Tauxe 2001, Nayga et al. 2005, Frewer et al. 2011, Eustice and Bruhn 2013, Maherani et al. 2016). Studies that provide respondents with information about the benefits of irradiation technology have demonstrated the effectiveness of various types of educational material (Bruhn 1995, Bruhn 1998, Nayga et al. 2005).

The evidence overwhelmingly supports irradiation at approved levels as a way to improve the safety of the food supply with limited compromise to quality and nutrition. Further research is needed to determine the optimal approaches to increase public awareness of, and knowledge about, the benefits of food irradiation. As with other innovations, we can expect a core group of consumers to reject the technology for philosophical or other reasons.

Genetic modification

Genetic engineering (GE) of plants and animals using biotechnology techniques to achieve desired genetic traits has created controversy and confusion among consumers. Genetic changes occur naturally in plants and animals through breeding and spontaneous mutation. Humans have used traditional breeding techniques to achieve desired traits for thousands of years. Natural genetic mutations have also resulted in offspring with desirable traits, such as the colour of 'Ruby Red' grapefruit (Piotrowski 2015) and double muscling in cattle (Fiems 2012). The creation of desired traits can be hastened by intentional manipulation of genetic material using natural or traditional approaches, or biotechnology techniques.

The use of molecular biology techniques for genetic modification has led to consumer concerns about potential risks to human health, the environment and local economies, particularly with the insertion of genetic material from vastly different organisms. Some countries, including the United States and Canada, consider any change in the genotype of an organism to be genetic modification, while the use of biotechnology techniques is deemed genetic engineering. In contrast, the European Union, Australia and others exclude traditional and naturally occurring changes from their definition of genetically modified organisms (GMOs) (König et al. 2004, D'Ambrosio et al. 2016). Despite official definitions, popular use of the GMO term refers to genetically engineered (GE) organisms.

To ensure their safety, GE organisms and foods are required to undergo a comprehensive testing and approval process if the product is not substantially equivalent to its non-modified counterpart (König et al. 2004). Safety testing has led to the rejection of some GE organisms. For example, there have been reports of allergic reactions among subjects exposed to transgenic methionine-enhanced soybeans that contained brazil nut protein (Nordlee et al. 1996). Reports such as these may concern consumers who are not aware of safety testing protocols for GE foods.

Although GE foods have been approved for human consumption in many countries, product marketing and consumer acceptance have played an important role in their eventual success or failure in the market place (Schurman and Munro 2009). Tinned GE tomatoes were overtly introduced to and accepted in Britain during the 1990s (Miles et al. 2005, Schurman and Munro 2009). However, when GE soy was quietly introduced without regard to consumers' right to know what they were eating, the anti-biotechnology backlash led to the British and EU moratorium on allowing bioengineered products until strict legislation and systems to segregate and trace GE crops were in place (Costa-Font et al. 2008, Schurman and Munro 2009). Effective segregation of GE and non-GE organisms and foods as well as mandatory labelling enable European consumers to make a choice about what they purchase.

Nanotechnology

Nanotechnology is a rapidly growing field of science that is transforming the food industry. The technology is currently used in pharmaceuticals, cosmetics and other consumer goods. However, the application of engineered nanomaterials in foods is in its infancy.

Nanotechnology can be applied in the food industry at various points in the farm-to-fork continuum including agriculture, food production, food processing, food preservation and food packaging (FAO-WHO 2010). Many products that are available in consumer markets are in the form of nanotechnology-based food, health food products and, most commonly, food packaging (FAO-WHO 2010).

Nanosized particles can be found naturally in foods or deliberately synthesized by creating particles between 1 and 100 nanometers (nm). Nanomaterials can improve shelf life and contribute to the wholesomeness of foods by improving the availability and delivery of essential nutrients and by reducing salt, sugar and fat content with limited effect on flavour. It can also be used to improve desirable attributes such as taste, texture and/or colour. The primary use is currently nanoencapsulation of nutrients and probiotics; however, the potential market for engineered nanofoods is huge.

Many consumers are unaware of the technology and its potential benefits and risks. The minute size of engineered nanomaterials increases the surface area relative to their mass, impacts the reactivity of their surfaces, and provides the ability to pass through cell membranes and accumulate in vital organs (van Calster et al. 2011). There is considerable uncertainty about the safety of using nanomaterials in foods because of their potential toxicity. As a result, it is creating controversy in some countries where nanoparticles were reported in baby foods despite their potential toxicity (Miller and Senjen 2008).

Nanomaterials are also being used in packaging and on food contact surfaces. Passive food packaging films produced using nanoclay are able to reduce gas permeability, slowing the flow of oxygen and increasing shelf life. Nanosensors in packaging are used to create interactive 'smart' packaging for different purposes. Bioactive packaging can detect product spoilage or ripeness, temperature and humidity changes during storage, and the presence of pathogens or their toxins. Nanoengineered food contact surfaces containing silver or other antimicrobial nanoparticles can reduce the food safety risk by preventing bacteria from adhering to surfaces, thereby reducing biofilm formation. The drawback to nanomaterial use in these applications is the uncertainty about the health impact from potential migration of the nanomaterials into the food.

The food industry has an opportunity to create and maintain consumer trust by providing relevant information and disclosing nanoengineered ingredients. Large food manufacturers are undertaking research and development using nanotechnology applications in food; few have revealed which, if any, products are on the market. Friends of the Earth groups report the presence of nanoparticles in hundreds of foods (Miller and Senjen 2008) although it is not clear the extent to which they are naturally occurring or the result of nanoengineering.

Lack of disclosure about food companies' research on and use of nanotechnologies is fueling concern among consumer groups about the potential safety of products, especially where labelling is not required. Lack of information about the benefits to consumers and product safety related to nanotechnology serve to heighten consumer concern. When other novel technologies were introduced, consumers accepted them more readily when they perceived that the benefits to themselves or society outweighed the risks (Giles et al. 2015). Will companies remember these lessons when they bring foods made with nanotechnology to the market?

Chemical Contaminants

According to the International Food Information Council (IFIC) 2015 survey (2015), American consumers were more concerned about 'chemicals' in their food than they were in 2014. IFIC showed an increase from 23% of respondents with concern in 2014 to 36% in 2015. Unintentional chemical contamination can stem from a range of sources including the improper use of pesticides and other compounds used in the food and agricultural industries. Chemicals responsible for such contamination include fertilizers and pesticides for crop protection, improperly used growth hormones designed to increase market yield in livestock, and food additives. While such chemicals are regulated in many countries, consumer concern about the use of such chemicals has been widely reported in the literature (Williams and Hammitt 2001, Krishna and Qaim 2008). Contamination of the food supply may also occur from production that occurs under conditions of general environmental pollution (Yeung and Morris 2001).

Slovic and colleagues (1987) suggested the following list of characteristics to explain consumers' risk perception: severity of consequences, control of risk, immediacy of effect, voluntariness of risk, knowledge about risk, newness, chronic-catastrophic, and common-dread. They showed that some of these characteristics were correlated with one another, and suggested that three antecedent factors (previously identified in several studies) influence risk perception. These are dread, fear of the unknown, and the number of people exposed to the risk. Chemical hazards, including pesticides and antibiotics, have been rated high on the 'dread' factor (Raats and Shepherd 1996) and relatively high on the unknown factor because they are perceived as unnatural and unfamiliar by consumers (Frewer et al. 1998).

Pesticides

The use of pesticides has contributed to unprecedented growth in agricultural output. The majority of pesticides used throughout the world are used on fruit and vegetable crops to control insects or disease and thereby improve crop productivity. If their use were prohibited, productivity would plummet and food prices would increase substantially. Pesticides approved for use on food crops are highly toxic to specific organisms and less so to non-targeted organisms and the environment (U.S. Environmental Protection Agency 2017).

The use of pesticides, however, is not without risks. Pesticide residues may be present in food, drinking water, or in the residential or occupational environment. Consumers may become exposed by ingesting them, inhaling them, or coming into contact with them through the skin. It is the totality of exposure, not exposure to a single substance, that creates a health risk to humans.

Consumer concern over pesticide residues in conventionally-grown food is extremely high. In one survey, for example, consumers estimated the median fatality rate to be between 50 and 200 per million (Williams and Hammitt 2001). Concern about pesticide residues has been reported to be higher among females, households with larger numbers of children, and consumers over 35 years of age (Govindasamy and Italia 1998). Concern was lower in households with higher levels of education and higher incomes (Byrne et al. 1991, Govindasamy and Italia 1998). Florax and colleagues (2005) reported a significant willingness by consumers to avoid pesticide-related risks.

There have been several studies on pesticide residues on produce in both developed and developing countries. A report published by the European Food Safety Authority (2017) stated that 53% of conventionally-grown samples that were collected across the European Union and then tested were free of quantifiable residues and 97% were within legal limits. A total of 99.3% of organic foods were residue-free or within legal limits.

A study of pesticide contamination of seasonal vegetables in India (Kumari et al. 2004) showed the presence of organophosphates followed by synthetic pyrethroids and organochlorides. Of the organophosphates, residues of monocrotophos, quinalphos, and chlorpyriphos exceeded the maximum residue limit (MRL) in almost one-quarter of the samples tested. In a more recent study, also in India (Mandal and Singh 2010), 42% of samples of cauliflower tested for levels of pesticide residues were contaminated with measurable levels of pesticides. The same three classes of chemicals were identified, this time with the organophosphates being the dominant group, followed by the organochlorides and the pyrethroids; none, however, exceeded the MRLs. Suresh et al. (2015) reported that 85% of Indian consumers were willing to pay more for residue-safe vegetables. A pesticide contamination study in China (Chen et al. 2011) showed that insecticide and fungicide residues exceeded the MRLs in pakchoi cabbage, legumes, and leaf mustard.

Exposure to higher residues of pesticides can lead to numerous health issues as well as environmental pollution. It has been suggested that farmers in highly populated regions of the world will continue to use pesticides, at least in the short term, and that pesticide residues will continue to be a food safety concern (Wilson and Tisdell 2001). The food safety risks associated with pesticide residues are variable and appear to depend on the jurisdiction in which they are used.

Production enhancing hormones

Consumers have expressed concern about the safety of food containing 'chemicals'. This includes concern about residues of production-enhancing products that farmers use with their livestock. The most well-known types of products are hormonal active growth promoters ('hormones'). Some, but not all, countries permit the use of some hormones and drugs to improve production efficiency, thereby reducing costs to the farmer. Countries that permit hormones or other chemicals as production aids may limit their use to certain species and for specific purposes. The United States, for example, restricts steroid hormones to beef cattle and sheep (Center for Veterinary Medicine 2015) while Canada restricts usage to beef cattle (Veterinary Drugs Directorate 2012a).

Growth promoting hormones – beef cattle

Growth promoting hormones are found naturally in livestock at various ages and physiological conditions (Waltner-Toews and McEwen 1994, Raun and Preston 1997). When used as production aids, they are typically implanted in an ear or added to feed (Stephany 2009). Testosterone, estradiol-17β (estrogen) and progesterone are natural hormones which are identical to those produced naturally by humans (Hobbs 2014). No maximum residue limits are established for natural hormones because they are used at levels that are within the normal ranges found in livestock. Various synthetic hormones that mimic these have been evaluated for safety (Doyle 2000, Veterinary Drugs Directorate 2012b). Several international and European scientific organizations declared implants to be safe if used as directed; six of eight 'significant reviews' found no evidence that residues would result in adverse human health effects (Doyle 2000).

The United States, Canada, Australia, New Zealand and some other countries permit the restricted use of a limited number of hormones to promote growth (Stephany 2009). Waltner-Toews and McEwen (1994) identify risks if farmers use hormonal growth promotants improperly by incorrect implantation, excessive doses or having an insufficient withdrawal period. Countries with effective inspection, monitoring and enforcement systems may find low levels of residues and undetected or rare instances of illegal use, such as Canada (Waltner-Toews and McEwen 1994).

The European Union has cited the precautionary principle and banned the use of growth promoting hormones in slaughter animals. Illegal use of hormones has been an issue in some countries when legal alternatives were not available (Waltner-Toews and McEwen 1994). According to Stephany (2009), there has been evidence of illegal use of hormones in the European Union by intra-muscular injection, application on the skin of animals and in feed. Stephany (2009) indicated the use of 'dozens' of illegal compounds in the EU available in the black market and estimates the range of misuse at 5–15% across Member States. Does the ban on hormone use for growth promotion actually increase the risk to consumers?

Consumers are often unaware of the various restrictions so they generalize concerns across animal production systems. For example, Brewer and Rojas (2008) identified nearly one-half of U.S. consumers were concerned about hormone residues in meat, poultry and milk in the early part of the century. Growth promoting hormone treatments are not permitted for pigs or poultry in the United States (Center for Veterinary Medicine 2015).

Hartman et al. (1998) reported that a German nutritional study estimated that milk products contribute 60–80% of hormones, hormone precursors and intermediates normally consumed in a human diet. The remaining dietary contribution of these compounds is divided between eggs and plant products (e.g., haricot beans, wheat, potatoes, soybeans and corn oil) equally with meats and meat products (including muscle meats, liver, sausages, poultry and fish). Stephany (2009) indicated that a major source of estradiol in a western diet is hens' eggs which are known to be high in cholesterol. Naturally occurring hormones and related chemicals are consumed daily, and there is international scientific consensus about the safety of approved compounds. Are consumer concerns about hormone residues in meat misplaced in jurisdictions in which effective government oversight exists?

Bovine growth hormone – dairy cattle

Bovine somatotropin (BST) is a growth hormone that is found naturally in bovine cow milk. When used to treat lactating dairy cattle, BST increases milk production by 10–20% (Dobson 1996, Crooker et al. 2004). Biotechnology techniques have made it practical to commercially produce BST that can be used by regular injection during lactation periods. The recombinant BST (rBST) molecule varies slightly from BST in chemical structure and biological activity. As a result, rBST-treated cows produce milk that is considered to have essentially the same composition of nutrients and hormones as milk produced by untreated cows (Crooker et al. 2004, Vicini et al. 2008, WHO-FAO 2014).

Studies on the safety of rBST indicate that there is little risk to human health from milk or meat from treated cows. The large molecular structure of rBST means it would be digested in the human gut when consumed. It is also denatured through pasteurization or cooking at high temperatures. If any intact molecules were to get through, the bovine somatotropin molecule would be biologically inactive as it cannot bind to human somatotropin receptor sites (Waltner-Toews and McEwen 1994, Crooker et al. 2004, WHO-FAO 2014). JEFCA (2014) recently reaffirmed its decision to not specify acceptable daily intake and maximum residue limits for rBST due to its safety.

While there is no difference in BST levels between treated and untreated cows, some concern has been expressed about the increase in the insulin-like growth factor 1 (IGF-1). Bovine IGF-1 is identical to human IGF-1 which is present in human breast milk at levels as high or higher than bovine milk from rBST-treated cows. According to Waltner-Toews and McEwen (1994), the amount of IGF-1 in the milk and meat of treated cows is about twice the normal amount. Even at increased levels, the IGF-1 concentration remains within a normal range and, if ingested, is degraded in the human gut. WHO and FAO (2014) affirmed the safety of milk from rBST-treated cows, saying that any contribution of IGF-1 to humans through the milk would be extremely low relative to naturally produced human IGF-1.

The United States, Brazil, Mexico and other countries have approved rBST use (Dobson 1996), while Canada (Veterinary Drugs Directorate 2012a), the EU (Wiener and Rogers 2002) and others considered non-food outcomes related to animal health and welfare in their decisions to ban rBST use. Still, concerns about increased mastitis resulting in increased use of antimicrobials, potentially leading to antimicrobial residues or bacterial resistance, have been proven false (Crooker et al. 2004).

Some consumers prefer to avoid milk from rBST-treated cows (Vicini et al. 2008, Kolodinsky 2008); however, labelling milk with the rBST-free attribute has been controversial because of the potential to stigmatize milk that may come from treated cows (Kanter et al. 2009). The US Food and Drug Administration had permitted labelling based on production claims as no material differences had been found in nutritional quality and safety (Wheeler 2011). A recent court case confirmed processors' rights to label milk rBST-free based on composition throughout the United States (Wheeler 2011). More recently, researchers (Ludwig et al. 2013) identified several biomarkers and a rapid test that has the potential to detect the use of rBST. The biomarkers include IGF-1, as well as osteocalcin, the insulin-like IGF binding protein 2 (IGFBP2) and antibodies to rBST treatment. Effective risk communication may help mitigate some consumer concern without stigmatizing conventional milk which may contain milk from both treated and non-treated cows.

Organic Food

Consumer perception of the 'organic' attribute as applied to food is complex. The term has various interpretations—what the food is or isn't; what it has or doesn't have in it or on it; and what practices, chemicals or aids were or were not used during production or processing. To ensure specific standards for the term, its use is restricted in countries such as Canada, the United States (Campbell et al. 2014), and members of the EU (Lairon 2010) where it can be used only for food produced under defined conditions. Organic standards may specifically exclude, for example, the use of the novel technologies which we have described above, the use of hormone treatments for production enhancement and the use of synthetic pesticides. Most of the defined conditions for organic production are not directly related to food safety.

Because of the restrictions, some consumers believe organic food is better for the environment and livestock, better tasting, more nutritious and/or safer than conventional foods (Bourn and Prescott 2002, Kijlstra et al. 2009, Lairon 2010, Van Loo et al. 2012, Campbell et al. 2014). Nevertheless, claims that organic foods are safer than conventional are questionable (Magkos et al. 2006, Kijlstra et al. 2009, Lairon 2010, Van Loo et al. 2012, Barański et al. 2014). Several microbiological and chemical food safety risks may increase due to organic production systems while others may be lower (Kijlstra et al. 2009, Van Loo et al. 2012). Reviews have shown conflicting results in comparisons of *Salmonella, Campylobacter, E. coli* and parasites in organic versus conventional foods (Kijlstra et al. 2009, Van Loo et al. 2012). Kijlstra et al. (2009) found evidence of increased risk in organic eggs versus conventional ones due to chemical residues, specifically dioxins from laying hens permitted to forage outdoors in the EU, and lead contamination from free-foraging hens in the UK. Nitrate content, however, which is thought by some to be an issue, is lower in some organic crops (Magkos et al. 2006, Lairon 2010, Lima and Vianello 2011).

Specific concern about pesticide residues is a common reason consumers have given for choosing organic over conventional (Bourn and Prescott 2002). However, the use of natural pesticides and fertilizers is permitted for organic crop production. Furthermore, in some jurisdictions hormones and antimicrobials can be used for organic livestock. Van Loo et al. (2012) and Magkos et al. (2006) report that pesticide residues in both conventional and organic products are well below maximum residue limits so consumers should not be concerned. Baranski et al. (2014), in contrast, reported pesticide residues at levels above the maximum residue limits and higher levels of cadmium in organic crops than in conventional crops. Van Loo et al. (2012) indicate that the risk to children from exposure to pesticide residues may have been underestimated, adding uncertainty about the safety of organic foods for this population. Is food safety a legitimate reason for consumers to purchase organic products? The evidence suggests it is not but leaves room for consumers to doubt.

Deceptive Practices

The food supply must be protected against contamination because such contamination can be serious or even fatal to consumers. While most contamination is unintentional, some is deliberate. Deliberate contamination, or 'food fraud' has been defined as the "deliberate and intentional substitution, addition, tampering, or misrepresentation of food, food ingredients, or food packaging; or false or misleading statements made about a product, for economic gain" (Spink and Moyer 2011). There has been a steady increase in the number of food fraud cases worldwide, and this has been attributed to a combination of the increase in international trade, emerging markets, and increases in food prices (Holbrook 2013, Di Pinto et al. 2015).

Deliberate contamination of food can be accomplished by adulteration, counterfeiting, comingling, and substitution. Examples include "watering down or adding inert ingredients to products such as infant formula and drugs, relabeling products that have passed their expiration dates, relabeling to change country of origin and substituting cheaper species of fish for more expensive ones" (Zach et al. 2012). Undeclared allergens are an obvious food safety risk from substitution. Other health risks can come from using deliberately inferior, deteriorated or contaminated foods.

Food safety became a major issue in domestic food markets in China between 2003 and 2006 because of incidents involving food poisoning, discovery of dangerous dyes and additives to food products, fraudulent products, and sale of food that had passed its expiration date (Wang et al. 2008). In 2008, deliberate contamination of infant formula with melamine killed young children. Imports from China tend to be highlighted, but those from other countries can also be problematic. According to a recent ABC (2013) News report on counterfeit foods, 7% of the U.S. food supply is estimated to be counterfeit. Products such as olive oil, spices, tea, pomegranate juice and lemon juice were highlighted; in fact, counterfeiters are interested only in financial gain, so all food products are potential targets. Governance in the form of anti-counterfeit measures by food manufacturers, ongoing diligence, training by trade associations, and government initiatives and regulations all support the battle against counterfeit products.

In 2013, a large seafood fraud investigation by Oceana revealed that almost "nine in 10 samples sold as snapper were mislabeled, while 59% of samples sold as tuna were mislabeled" (Holbrook 2013). The investigation went on to reveal that seafood purchased in grocery stores, traditional restaurants, and sushi restaurants was mislabeled between 18% and 74% of the time (Holbrook 2013). Species substitution and other fraudulent activities such as short-weighting and falsification of import/export documents put "consumers and restaurants trying to make honest, eco-friendly choices at a disadvantage" (Anonymous 2011).

The substitution of horse meat for beef and other meats in Europe in 2013 led to a call for greater transparency in the meat industry and the implementation of strict regulations on food adulteration (Premanandh 2013). A recent study of processed chicken sausages, pork sausages, pate samples, and meat patties containing chicken, pork and/or bovine meat purchased in grocery stores in Italy confirmed the widespread problem of incorrect identification of the type of meat in more than half of the products tested (Di Pinto et al. 2015). While the authors of the study acknowledge that such errors could be either unintentional or deliberate, they state that the outcomes nonetheless include "consumer deception, potential health risks and the inability of individuals to choose products on the basis of their religious and ethical beliefs" (Di Pinto et al. 2015).

Even organic products are not immune. An article in *the Washington Post* (*The Washington Post* 2017) highlighted deficiencies in the organic market in the US, which currently accounts for $40 billion in sales on an annual basis. Milk, in particular, that is certified 'USDA Organic' has not necessarily been inspected according to the rules. Inspectors are hired by the farmers themselves (from a list of USDA-approved inspectors) for annual audits; this alone is a weakness in that the inspectors may wish to please 'their employers'. A requirement of the certification is that the cows spend a specific amount of time on pasture that is itself organic during the grazing season; if the audit is performed outside normal grazing season, what evidence is there that this is happening? Milk that complies with the 'USDA Organic' standard contains two fats (conjugated linoleic acid and Omega-3) with health benefits for humans, even though the amounts present in the organic milk are very small. Furthermore, organic milk contains less linoleic acid, an Omega-6 fat (*The Washington Post* 2017). The steady increase in the consumption of organic food in the US suggests that consumers trust the USDA organic label, but are they being misled?

The issue of authenticity has also infiltrated the halal food and beverage market, which is expected to account for as much as 17.4% of the food market worldwide by 2018 (Lubis et al. 2016). The term 'halal' means 'permissible'; with reference to food, the term refers to food that has been prepared according to "Islamic ideals that emphasize purity and cleanliness to promote one's health and wholesomeness" (Lubis et al. 2016). There are many types of food considered halal; descriptions of them are beyond the scope of this paper. There have been numerous discoveries of non-halal ingredients being labelled as halal. For this reason, consumers of halal food have been reported to be concerned about the authenticity of their food. The various methods currently used to investigate adulteration of halal foods as well as potential methods that may be used in the future have been discussed elsewhere (Lubis et al. 2016).

Issues with unintentional and deliberate contamination of food have led to increased consumer mistrust in the food supply, and hence the importance of tools such as traceability. An exploratory study of traceability of dairy products in Canada—most of which are produced within the country—concluded that traceability could "not only add value to current products available to consumers, but it could also empower consumers to protect themselves" (Charlebois and Haratifar 2015).

Importation of foods and ingredients that are used to produce processed foods presents a particular challenge. In the latter case, the opportunity for deliberate contamination is increased because of the potential for a raw material to contaminate numerous products and also because of the greater complexity of the food chain.

Accidental contamination of food may also occur. This may occur in imported foods because members of the food chain, i.e., farmers through food processors, are not familiar with specific safe practices or with additives that are approved for use.

Protecting consumers from both deliberate and accidental contamination has been, and will continue to be, an ongoing challenge. Traditional food safety initiatives are no longer considered adequate to protect the food supply from deliberate attack. A variety of measures that span the entire food production cycle are used. At the farm level, security cameras may be used to enhance the surveillance of facilities, including fields and parking facilities. At packing plants and processing facilities, Hazard Analysis and Critical Control Points (HACCP) programs can and should be carefully examined to identify and strengthen the most vulnerable points. During shipping, the opportunity for tampering and terrorism can be reduced with secure locks, security systems and seals. At retail, labels on food should provide accurate and adequate information to allow consumers to make informed choices about the food they purchase. Testing to ensure that the content of a package matches the information on the label has become a routine practice. It has been suggested that a description of issues of authenticity on products would help to increase consumer confidence in those products (Premanandh 2013).

While consumers want to be assured that food is well protected against both unintentional and deliberate contamination, this protection comes with a cost. All security enhancements must be balanced against the tight profit margins on food. Even with these safeguards, consumers may not understand the crucial role of food safety regulations. In order to protect consumers, it is important to understand their attitudes towards food safety.

Summary

In this chapter we discussed consumer concerns about three novel technologies and 'chemicals', specifically pesticides and production enhancing hormones used in agriculture. We provided a scientific perspective on the safety of these technologies as contrasted with concerns voiced about them. We also discussed evidence of chemical residues in the food supply, which appear to correspond to regions where food safety legislation and/or regulatory enforcement may be inadequate. We discussed organic foods and the food safety implication of deceptive practices known as 'food fraud', concluding that organic foods may not be the panacea for consumers concerned about food safety. Consumers' attitudes are complex, and are formed by the interaction of many factors (demographics, social, cultural, economic conditions, etc.). They vary from one consumer to the next, and change over time. That makes any discussion of consumer perspectives of food safety attributes a challenge.

The food supply in developed countries with robust enforcement systems is generally considered to be safe. However, consumers' perceptions of technology, industry practices and the effectiveness of regulatory oversight affect confidence in their country's food supply. Some concerned consumers are willing to pay more for food without attributes they want to avoid (e.g., irradiated, genetically modified, nanoengineered, with potential for chemical residues) than foods that might have these attributes. These consumers count on accurate labelling to make informed product selections. In jurisdictions with voluntary food labelling regulations, however, consumers may opt to purchase certified organic products. Yet with improper use of agricultural chemicals and deceptive practices, even foods with mandatory labelling and certified organic products may not be what they seem. What realistic options do concerned consumers have? What can people who are food-insecure do? What more do regulatory agencies around the world need to do to ensure the safety of their food supply and build consumer trust?

References

ABC News. 2013. How to spot counterfeit foods. Telev Broadcast. <http://abcnews.go.com/WNT/video/spot-counterfeit-food-18288837> (accessed on 15 March 2017).

Acheson, D. and J.H. Steele. 2001. Food irradiation: A public health challenge for the 21st century. Clin. Infect. Dis. 33: 376–377.

Andrews, L., M. Jahncke and K. Mallikarjunan. 2003. Low dose gamma irradiation to reduce pathogenic vibrios in live oysters (Crassostrea virginica). J. Aquat. Food Prod. Technol. 12: 71–82.

Anonymous. 2011. Oceana launches stop seafood fraud campaign. Supermark News. <http://www.supermarketnews.com/seafood/oceana-launches-stop-seafood-fraud-campaign> (accessed on 17 May 2017).

Barański, M., D. Średnicka-Tober, N. Volakakis, C. Seal, R. Sanderson, G.B. Stewart, C. Benbrook, B. Biavati, E. Markellou, C. Giotis et al. 2014. Higher antioxidant and lower cadmium concentrations and lower incidence of pesticide residues in organically grown crops: a systematic literature review and meta-analyses. Br. J. Nutr. 112: 794–811.

Bourn, D. and J. Prescott. 2002. A comparison of the nutritional value, sensory qualities, and food safety of organically and conventionally produced foods. Crit. Rev. Food Sci. Nutr. 42: 1–34.

Brewer, M.S. and M. Rojas. 2008. Consumer attitudes toward issues in food safety. J. Food Saf. 28: 1–22.

Bruhn, C.M. 1995. Consumer attitudes and market response to irradiated food. J. Food Prot. 58: 175–181.

Bruhn, C.M. 1998. Consumer acceptance of irradiated food: theory and reality. Radiat. Phys. Chem. 52: 129–133.

Byrne, P.J., G.M. Gempesaw and U.C. Toensmeyer. 1991. An evaluation of consumer pesticide residue concerns and risk information sources. South. J. Agr. Econ. 23: 167–174.

van Calster, G., D.M. Bowman and J. D'Silva. 2011. Protecting consumers or failing them? The regulation of nanotechnologies in the EU. Eur. J. Consum. Law. 1: 85–113. <https://papers.ssrn.com/sol3/papers.cfm?abstract_id=1911902> (accessed on 28 June 2017).

Campbell, B.L., H. Khachatryan, B.K. Behe, J. Dennis and C. Hall. 2014. U.S. and Canadian consumer perception of local and organic terminology. Int. Food Agribus. Manag. Rev. 17: 21–40.

Cardello, A.V., H.G. Schutz and L.L. Lesher. 2007. Consumer perceptions of foods processed by innovative and emerging technologies: A conjoint analytic study. Innov. Food Sci. Emerg. Technol. 8: 73–83.

Center for Veterinary Medicine. 2015. Steroid hormone implants used for growth in food-producing animals. <https://www.fda.gov/animalveterinary/safetyhealth/ productsafetyinformation/ucm055436.htm> (accessed on 10 August 2017).

Charlebois, S. and S. Haratifar. 2015. The perceived value of dairy product traceability in modern society: An exploratory study. J. Dairy Sci. 98: 3514–3525.

Chen, C., Y. Qian, Q. Chen, C. Tao, C. Li and Y. Li. 2011. Evaluation of pesticide residues in fruits and vegetables from Xiamen, China. Food Control 22: 1114–1120.

Costa-Font, M.J.M. Gil and W.B. Traill. 2008. Consumer acceptance, valuation of and attitudes towards genetically modified food: Review and implications for food policy. Food Policy 33: 99–111.

Crooker, B.A., D.E. Otterby, J.G. Linn, B.J. Conlin, H. Chester-Jones, L.B. Hansen, W.P. Hansen, D.G. Johnson, G.D. Marx, J.K. Reneau et al. 2004. Dairy research and bovine somatotropin. Minneapolis, MN. <blog.wsd.net/jegelund/files/2012/02/Dairy-Research-and-Bovine-Somatotropin.pdf> (accessed on 14 August 2017).

D'Ambrosio, C., A.L. Stigliani and G. Giorio. 2016. Food from genetically engineered plants: Tomato with increased β-carotene, lutein, and xanthophylls contents. *In*: Watson, R.R. and V.R. Preedy (eds.). Genet. Modif. Org. Food – Prod. Safety, Regul. Public Heal. Elsevier Online. <http://dx.doi.org/10.1016/B978-0-12-802259-7.00033-6> (accessed on 6 May 2017).

van Dijk, H., A.R.H. Fischer and L.J. Frewer. 2011. Consumer responses to integrated risk-benefit information associated with the consumption of food. Risk Anal. 31: 429–439.

Dobson, W.D. 1996. The BST Case. Madison. <https://aae.wisc.edu/pubs/sps/the-bst-case/> (accessed on 14 August 2017).

Doyle, E. 2000. Human safety of hormone implants used to promote growth in cattle. Madison, WI. <https://fri.wisc.edu/files/Briefs_File/hormone.pdf> (accessed on 13 August 2017).

European Commission. 2017. Food Irradiation - European Commission. <http://ec.europa.eu/ food/safety/biosafety/irradiation_en> (accessed on 17 April 2017).

European Food Safety Authority. 2017. The 2015 European Union report on pesticide residues in food. EFSA J. 15: 1–134.

Eustice, R.F. and C.M. Bruhn. 2013. Consumer acceptance and marketing of irradiated foods. pp. 173–195. *In*: Xuetong Fan and Christopher H. Sommers (eds.). Food Irradiat. Res. Technol. 2nd edn., Blackwell Publishing and the Institute of Food Technologies. <http://ucce.ucdavis.edu/files/datastore/234-2467.pdf> (accessed on 9 June 2017).

FAO/IAEA/WHO. 1999. High-dose irradiation: Wholesomeness of food irradiated with doses above 10 kGy. Geneva: World Health Organization. <http://www.who.int/foodsafety/ publications/food-irradiated/en/> (accessed on 17 April 2017).

FAO-WHO. 2010. FAO/WHO Expert meeting on the application of nanotechnologies in the food and agriculture sectors: potential food safety implications. Rome, Italy. <http://www.fao.org/docrep/012/i1434e/i1434e00.pdf> (accessed on 9 June 2017).

Farkas, J. 1998. Irradiation as a method for decontaminating food: A review. Int. J. Food Microbiol. 44: 189–204.

Fiems, L.O. 2012. Double muscling in cattle: Genes, husbandry, carcasses and meat. Animals. 2: 472–506.

Florax, R.J.G.M., C.M. Travisi and P. Nijkamp. 2005. A meta-analysis of the willingness to pay for reductions in pesticide risk exposure. Eur. Rev. Agr. Econ. 32: 441–467.

Frewer, L.J., K. Bergmann, M. Brennan, R. Lion, R. Meertens, G. Rowe, M. Siegrist and C. Vereijken. 2011. Consumer response to novel agri-food technologies: Implications for predicting consumer acceptance of emerging food technologies. Trends Food Sci. Technol. 22: 442–456.

Frewer, L.J., C. Howard and R. Shepherd. 1998. Understanding public attitudes to technology. J. Risk Res. 1: 221–235.

Giles, E.L., S. Kuznesof, B. Clark, C. Hubbard and L.J. Frewer. 2015. Consumer acceptance of and willingness to pay for food nanotechnology: a systematic review. J. Nanopart. Res. 17: 467.

Govindasamy, R. and J. Italia. 1998. Predicting consumer risk perceptions towards pesticide residue: a logistic analysis. Appl. Econ. Lett. 5: 793–796.

Hartmann, S., M. Lacorn and H. Steinhart. 1998. Natural occurrence of steroid hormones in food. Food Chem. 62: 7–20.

Hobbs, J.E. 2014. Canada, US-EU beef hormone dispute. pp. 273–279. *In*: Paul B. Thompson and David M. Kaplan (eds.). Encycl. Food Agr. Ethics, Dordrecht: Springer Netherlands.

Holbrook, E. 2013. Dining on deception: the rising risk of food fraud and what is being done about it. Risk Manag. 60: 29–33.

International Food Information Council. 2015. What's your health worth? Food & Health Survey 2015. <www.foodinsight.org/sites/default/files/2015-Food-and-Health-Survey-Full-Report.pdf> (accessed on 16 August 2017).

Kanter, C., K.D. Messer and H.M. Kaiser. 2009. Does production labeling stigmatize conventional milk? Am. J. Agr. Econ. 91: 1097–1109.

Kijlstra, A., B.G. Meerburg and A.P. Bos. 2009. Food safety in free-range and organic livestock systems: risk management and responsibility. J. Food Prot. 72: 2629–2637.

Kolodinsky, J. 2008. Affect or information? Labeling policy and consumer valuation of rBST free and organic characteristics of milk. Food Policy 33: 616–623.

König, A., A. Cockburn, R.W.R. Crevel, E. Debruyne, R. Grafstroem, U. Hammerling, I. Kimber, I. Knudsen, H.A. Kuiper, A.A.C.M. Peijnenburg et al. 2004. Assessment of the safety of foods derived from genetically modified (GM) crops. Food Chem. Toxicol. 42: 1047–1088.

Krishna, V.V. and M. Qaim. 2008. Consumer attitudes toward GM food and pesticide residues in India. Rev. Agr. Econ. 30: 233–251.

Kumari, B., V.K. Madan, J. Singh, S. Singh and T.S. Kathpal. 2004. Monitoring of pesticidal contamination of farmgate vegetables from Hisar. Environ. Monit. Assess. 90: 65–71.

Lairon, D. 2010. Nutritional quality and safety of organic food. A review. Agron Sustain. Dev. 30: 33–41.

Lima, G.P.P. and F. Vianello. 2011. Review on the main differences between organic and conventional plant-based foods. Int. J. Food Sci. Technol. 46: 1–13.

Lobb, A. 2005. Consumer trust, risk and food safety: a review. Food Econ. 2: 3–12.

Van Loo, E.J., W. Alali and S.C. Ricke. 2012. Food safety and organic meats. Annu. Rev. Food Sci. Technol. 3: 203–225.

Lubis, H.N., N.F. Mohd-Naim, N.N. Alizul and M.U. Ahmed. 2016. From market to food plate: Current trusted technology and innovations in halal food analysis. Trends Food Sci. Technol. 58: 55–68.

Ludwig, S.K.J., N.G.E. Smits, F.T. Cannizzo and M.W.F. Nielen. 2013. Potential of treatment-specific protein biomarker profiles for detection of hormone abuse in cattle. J. Agr. Food Chem. 61: 4514–4519.

Magkos, F., F. Arvaniti and A. Zampelas. 2006. Organic food: Buying more safety or just peace of mind? A critical review of the literature. Crit. Rev. Food Sci. Nutr. 46: 23–56.

Maherani, B., F. Hossain, P. Criado, Y. Ben-Fadhel, S. Salmieri and M. Lacroix. 2016. World market development and consumer acceptance of irradiation technology. Foods. 5: 79.

Mandal, K. and B. Singh. 2010. Magnitude and frequency of pesticide residues in farmgate samples of cauliflower in Punjab, India. Bull. Environ. Contam. Toxicol. 85: 423–426.

Miles, S., Ø. Ueland and L.J. Frewer. 2005. Public attitudes towards genetically-modified food. Br. Food J. 107: 246–262.

Miller, G. and R. Senjen. 2008. Out of the laboratory and on to our plates: Nanotechnology in food & agriculture. Friends of the Earth. <http://www.foeeurope.org/activities/ nanotechnology/Documents/Nano_food_report.pdf> (accessed on 19 May 2017).

Nayga, R.M., W. Aiew and J.P. Nichols. 2005. Information effects on consumers' willingness to purchase irradiated food products. Rev. Agr. Econ. 27: 37–48.

Nordlee, J.A., S.L. Taylor, J.A. Townsend, L.A. Thomas and R.K. Bush. 1996. Identification of a Brazil-Nut allergen in transgenic soybeans. N. Engl. J. Med. 334: 688–692.

O'Bryan, C.A., P.G. Crandall, S.C. Ricke and D.G. Olson. 2008. Impact of irradiation on the safety and quality of poultry and meat products: A review. Crit. Rev. Food Sci. Nutr. 48: 442–457.

Di Pinto, A., M. Bottaro, E. Bonerba, G. Bozzo, E. Ceci, P. Marchetti, A. Mottola and G. Tantillo. 2015. Occurrence of mislabeling in meat products using DNA-based assay. J. Food Sci. Technol. 52: 2479–2484.

Piotrowski, M. 2015. Biotechnology of higher plants. pp. 219–266. *In*: Kück, H.-U. and N. Frankenberg-Dinkel (eds.). Biotechnology. München, Boston: Walter de Gruyter GmbH & Co. KG.

Premanandh, J. 2013. Horse meat scandal—A wake-up call for regulatory authorities. Food Control. 34: 568–569.

Raats, M. and R. Shepherd. 1996. Developing a subject-derived terminology to describe perceptions of chemicals in foods. Risk Anal. 16: 133–146.

Raun, A.P. and R.L. Preston. 1997. History of hormonal modifier use. pp. 1–9. *In*: 1997 OSU Implant Symp. Stillwater, OK: OSU Beef Extension. <www.beefextension.com/proceedings/ implant_97/97-2.pdf> (accessed on 13 August 2017).

Rijswijk, W. Van and L.J. Frewer. 2008. Consumer perceptions of food quality and safety and their relation to traceability. Br. Food J. 110: 1034–1046.

Roberts, P. 2016. Food irradiation: Standards, regulations and world-wide trade. Radiat. Phys. Chem. 129: 30–34.

Schilling, B.J., W.K. Hallman, F. Hossain and A.O. Adelaja. 2003. Consumer perceptions of food biotechnology: Evidence from a survey of U.S. consumers. J. Food Distrib. Res. 34: 30–35.

Schurman, R. and W. Munro. 2009. Targeting capital: A cultural economy approach to understanding the efficacy of two anti-genetic engineering movements. Am. J. Sociol. 115: 155–202.

Slovic, P. 1987. Perception of risk. Science. 236: 280–285.

Spink, J. and D.C. Moyer. 2011. Background: Defining the public health threat of food fraud. J. Food Sci. 76: 1–7.

Stephany, R.W. 2009. Hormonal growth promoting agents in food producing animals. pp. 355–367. *In*: Thieme, D. and P. Hemmersbach (eds.). Doping Sport Biochem. Princ. Eff. Anal. 258. HEP Volume. Springer, Berlin, Heidelberg.

Suresh, A., G.K. Jha, S. Raghav, P. Supriya, A. Lama, B. Punera, R. Kumar, A.R. Handral, J.P. Sethy, R.G. Gidey et al. 2015. Food safety concerns of consumers: A case study of pesticide residues on vegetables in Delhi. Agr. Econ. Res. Rev. 28: 229–236.

Tauxe, R.V. 2001. Food safety and irradiation: Protecting the public from foodborne infections 1. Emerg. Infect. Dis. 7: 516–521.

The Washington Post. 2017. Milk from large dairies, including Colorado's Aurora Organic Dairy, may not be as organic as you think. Print Denver Post. <http://www.denverpost.com/ 2017/05/02/aurora-organic-milk-may-not-be-organic/> (accessed on 13 August 2017).

Thomas, M.K., R. Murray, L. Flockhart, K. Pintar, F. Pollari, A. Fazil, A. Nesbitt and B. Marshall. 2013. Estimates of the burden of foodborne illness in Canada for 30 specified pathogens and unspecified agents, circa 2006. Foodborne Pathog. Dis. 10: 639–648.

U.S. Environmental Protection Agency. 2017. Pesticides and Public Health.

Verbeke, W., L.J. Frewer, J. Scholderer and H.F. De Brabander. 2007. Why consumers behave as they do with respect to food safety and risk information. Anal. Chim. Acta 586: 2–7.

Veterinary Drugs Directorate. 2012a. Questions and Answers—Hormonal Growth Promoters. <https:// www.canada.ca/en/health-canada/services/drugs-health-products/veterinary-drugs/ factsheets-faq/ hormonal-growth-promoters.html> (accessed on 13 August 2017).

Veterinary Drugs Directorate. 2012b. Setting Standards for Maximum Residue Limits (MRLs) of Veterinary Drugs Used in Food-Producing Animals. <https://www.canada.ca/en/health-canada/services/drugs-health-products/veterinary-drugs/maximum-residue-limits-mrls/setting-standards-maximum-residue-limits-mrls-veterinary-drugs-used-food-producing-animals.html> (accessed on 13 August 2017).

Vicini, J., T. Etherton, P. Kris-Etherton, J. Ballam, S. Denham, R. Staub, D. Goldstein, R. Cady, M. McGrath and M. Lucy. 2008. Survey of retail milk composition as affected by label claims regarding farm-management practices. J. Am. Diet. Assoc. 108: 1198–1203.

Waltner-Toews, D. and S.A. McEwen. 1994. Residues of hormonal substances in foods of animal origin: a risk assessment. Prev. Vet. Med. 20: 235–247.

Wang, Z., Y. Mao and F. Gale. 2008. Chinese consumer demand for food safety attributes in milk products. Food Policy. 33: 27–36.

Wheeler, S. 2011. Got (rbST-Free) milk—The sixth circuit overturns Ohio's milk labeling Restrictions. Ecol. Law Q. 38: 571–578.

Wiener, J.B. and M.D. Rogers. 2002. Comparing precaution in the United States and Europe. J. Risk Res. 5: 317–349.

Wilcock, A., M. Pun, J. Khanona and M. Aung. 2004. Consumer attitudes, knowledge and behaviour: A review of food safety issues. Trends Food Sci. Technol. 15: 56–66.

Williams, P.R.D. and J.K. Hammitt. 2001. Perceived risks of conventional and organic produce: pesticides, pathogens, and natural toxins. Risk Anal. 21: 319–30.

Wilson, C. and C. Tisdell. 2001. Why farmers continue to use pesticides despite environmental, health and sustainability costs. Ecol. Econ. 39: 449–462.

WHO. 1994. Safety and nutritional adequacy of irradiated food. Geneva. <http://apps.who.int/iris/bitstre am/10665/39463/4/9241561629-eng.pdf> (accessed on 17 April 2017).

WHO-FAO. 2011. Labelling of foods and food ingredients obtained through certain techniques of genetic modification/genetic engineering. (CL 2010/19-FL). Codex Aliment Comm CX/FL. 39: 1–24. <http:// www.fao.org/tempref/codex/Meetings/CCFL/ccf139/f139_12e.pdf> (accessed on 10 May 2017).

WHO-FAO. 2014. Evaluation of certain veterinary drug residues in food. World Health Organ Tech. Rep. Ser. 988: 1–123.

Yeung, R.M.W. and J. Morris. 2001. Food safety risk: Consumer perception and purchase behaviour. Br. Food J. 103: 170–187.

Zach, L., M.E. Doyle, V. Bier and C. Czuprynsk. 2012. Systems and governance in food import safety: A U.S. perspective. Food Control. 27: 153–162.

3

Consumers' Food Safety Concern over Animal Diseases

Doo Bong Han[1,]* and *Jung Yun Choi*[2]

Introduction

Recent outbreaks of contagious animal diseases such as bovine spongiform encephalopathy (BSE), often called 'mad cow disease' and foot-and-mouth disease (FMD) have augmented public awareness and concerns about the potentially devastating impacts of these animals and other animal diseases. Indeed, the occurrence and the appearance of FMD, BSE, and other animal diseases have had a significant adverse impact not only on the relevant markets, industries, and trade, but also human health, which have resulted in considerable economic and social costs (Beach et al. 2007). In addition, the recent rapid expansion in global trade and travel increases the likelihood of the spread of animal disease and threat to food safety (Davis 2002). According to the World Organization for Animal Heath (OIE), 60% of the pathogens that affect humans are known to originate from animals. As animal health issues are closely linked to food safety as well as human health, animal diseases have raised consumers' concerns about food hazards. This has led to significant changes in consumers' perceptions of food-related risks and their purchasing behaviors. In response to growing consumers' concerns about food safety related to animal diseases, governments and industries have developed policy measures to prevent and control the spread of animal diseases, and to secure food safety. Nonetheless, there have been doubts about scientific knowledge

[1] Department of Food and Resource Economics, Korea University, 145 Anamro, Seongbukgu, Seoul, 02541, Republic of Korea.
[2] Research Professor, Department of Food and Resource Economics, Korea University.
E-mail: jychoi1202@korea.ac.kr
* Corresponding author: han@korea.ac.kr

and appropriate preparedness of the potential threats to animal health and food safety. In recognition of the need for a better understanding of the diverse impacts of animal disease outbreaks on industries, markets, and consumers, this chapter reviews the recent findings of empirical studies related to three major animal diseases (e.g., bovine spongiform encephalopathy (BSE), foot and mouth disease (FMD), and avian influenza (AI)), and the current policy strategies to prevent animal diseases and secure food safety, which will provide useful implications for future food policy.

An Overview of Literature on Animal Disease

This section reviews the vast empirical literature on the major streams in the study of animal diseases (BSE, FMD, AI). Studies related to animal diseases can be broadly classified into the following categories: (1) those that examine the economic welfare impact of animal diseases; (2) those that evaluate regulation and policy measures for preventing animal diseases and ensuring food safety; (3) those that study consumers' responses such as consumers' willingness to pay for food safety; and (4) those that investigate consumers' purchasing behaviors associated with perceived risk.

Economic welfare impacts

Many scholars have investigated economic welfare impacts of animal disease outbreak in countries with export potential (Cairns et al. 2016, Tozer et al. 2015, Tozer and Marsh 2012, Twine et al. 2016, Wigle et al. 2007, McDonald and Roberts 1998, Mainland and Ashworth 1992). These studies generally reported that animal disease outbreaks would impose significant economic welfare losses to the relevant markets, and industries, but overall insignificant welfare losses to consumers. For example, a recent study of Cairns et al. (2016) analyzed the economic impact of an FMD outbreak on Ontario's beef sector using a partial equilibrium model. They found that economic losses would be large with no significant difference between diseased (a total loss of $235 million) and non-diseased areas (a total loss of $217 million). Despite the economic losses, they also found that consumers and retailers would incur almost no welfare losses due to the increase in imports and prices. Tozer et al. (2015) assessed economic welfare impacts (the effects on the cattle industry, and producer and consumer welfare) of a hypothetical FMD outbreak in the Canadian beef cattle sector by using an intertemporal economic model. Their results showed that total welfare losses could decrease as the level of traceability or surveillance of susceptible animals increase while changes in consumer surplus are positive but producers remain losers. Tozer and Marsh (2012) investigated the domestic and trade impacts on the Australian beef industry when a hypothetical FMD outbreak occurs. Using a bio-economic optimization model, they found that the changes in economic surplus due to FMD outbreak ranged from a positive net gain of $57 million to a net loss of $1.7 billion. They also found that the impacts on producers and consumers could vary with the location of the outbreak, control levels, and the nature of any trade restriction over time. Twine et al. (2016) assessed the impact of beef supply by using a spectral comparison of the pre-BSE and post-BSE shock periodograms of beef supply, and they found that beef supply showed 58% reduction in the peak of the beef supply cycle. Wigle et al. (2007) analyzed the economic impact due to BSE crisis using the simulation of three trade restriction scenarios in Canada. They found that Canadian economy lost almost $1 billion, but their losses decreased significantly after resuming beef trade with the US. McDonald and Roberts (1998) examined the economy-wide effect of BSE crisis in the UK using the Computable General Equilibrium (CGE) model. They attempted to compare government intervention (support buying and subsidies payments) scenario with no government intervention scenario. Their results showed that macroeconomic effect of the crisis might be small, but substitution and resource reallocation effects were considerable. Mainland and Ashworth (1992) analyzed the effect on the revenue from finished cattle due to BSE in the UK, using the regression model and ARIMA models. They estimated a loss of 8.6% and 11%, respectively during 1989–1990.

A few studies examined welfare effects on countries that are not major exporters (Nogueira et al. 2011, You and Diao 2007). For example, Nogueira et al. (2011) analyzed the economic effect of a hypothetical FMD outbreak on (domestic and international) trade in the Mexican cattle industry, using a discrete time dynamic optimization model. Their results showed that consumers would suffer the biggest lose while producers would either lose or gain as they could compensate possible loss from the increased price. You and Diao (2007) assessed the economic impacts of AI outbreak in Nigeria using a spatial equilibrium. They found that an AI outbreak could have potentially negative impacts on the poultry industry and farmers, estimating that Nigerian chicken production would fall by 21% and chicken farmers would lose US$ 250 million of revenue in the case of the worst-case scenario.

A limited number of empirical studies attempted to assess the impacts of animal diseases on domestic price or import demand (Kawasima and Sari 2010, Park et al. 2008, Lloyd et al. 2006). Kawasima and Sari (2010) analyzed the impact of a series of BSE outbreaks on Japanese beef imports. Their results revealed that the first BSE case in Japan, which occurred in September 2001, decreased the beef import index from 164.0 to 118.3 (equivalent to a 27.9% loss in beef imports). Park et al. (2008) assessed the impact of domestic and overseas disease outbreak on the prices of different meat types such as beef, pork, and chicken in the Korean meat market, using time series methods. They found that animal disease outbreaks generated a short-term price shock in the Korean market regardless of whether it occurred in Korea or abroad, and the type of disease. Lloyd et al. (2006) investigated the impact of food scares, particularly the BSE crisis in the UK. The results showed that the impact of the BSE crisis on farm prices was found to be more than double compared to retail prices, presenting a differential impact of food scares on retailers and producers.

Animal diseases can also impose significant economic losses not only in the agricultural industry but also in other relevant industries. William and Ferguson (2005) and Irvine and Anderson (2004) considered the impact of FMD outbreak on tourism industry in the UK. William and Ferguson (2005) found that the impact of FMD caused significant economic damages on the local leisure and tourism industries in the UK. Irvine and Anderson (2004) also found that small tourism firms in rural areas suffered severely, even in areas without diseases.

Regulation and policy measures

Effective regulation and policy measures play a key role in reducing potential threats to animal health and food safety. The effectiveness of prevention and control strategies against animal disease such as culling, vaccination, and other measures was analyzed by many scholars. A common finding of their studies is that the prevention strategies are generally ineffective and costly, but only effective when used complementally with other measures. For example, Pfeiffer et al. (2013) conducted an extensive literature review on the HPAI in the Greater Mekong sub-region and evaluated the cross-disciplinary efforts on initial disease control measures. They found that control measures such as culling and market closure were not effective since the measures didn't eradicate disease and were costly, leading to decrease in the incomes of already poor farmers. They claimed that other complementary measures such as controlled elimination, movement restrictions, improved hygiene and biosecurity, proper surveillance, and vaccination were also important to control any AI viruses. Hagerman et al. (2012), using a simulation modeling of FMD outbreak, examined the effect of the standard culling with and without emergency ring vaccination strategies on animal and economic welfare losses in California and Texas. They found that the standard culling with emergency vaccination would cause more animal and economic welfare losses. Thus, they argued that the selection and concentration of preventive and control strategy were important for eradicating FMD.

In particular, the vaccination policy received considerable scholarly attention with widespread critiques on its effectiveness (Jarvis and Valdes-Donoso 2015, Egbendewe-Mondzozo et al. 2013, Rasmusen 2010, Elbakidze et al. 2009, Rich and Winter-Nelson 2007, Mahul and Gohin 1999). For instance, Jarvis and Valdes-Donoso (2015) reviewed previous studies on the effectiveness of animal disease management strategies. They found that vaccination became politically unattractive over time and extensive vaccination were unlikely to eradicate the disease. Egbendewe-Mondzozo et al. (2013) assessed the economic effectiveness of vaccination in different poultry regions in Texas, using an integrated economic-epidemic partial equilibrium model. They found that flock density in the region and decision makers' risk aversion preference affected the effectiveness of vaccination. Rasmusen (2010), using game theory, highlighted the effects of free-riding as another problem of vaccination. They argued that farmers who knew that his neighbors vaccinated for reducing the disease hazard were likely to avoid vaccination in anticipation of free-riding on other farmers' vaccinations. Elbakidze et al. (2009) investigated several FMD strategies (early detection, enhanced vaccine availability, and enhanced surveillance) under the circumstance of FMD outbreak in Texas, using a linked epidemiologic-economic modeling framework. The results showed that early detection and surveillance were a cost-effective strategy, but vaccination was generally ineffective due to the rising costs and the declining value of vaccinated animals due to trade regulation. Rich and Winter-Nelson (2007) assessed vaccination and stamping out policies in the Southern Cone (i.e., Argentina, Uruguay, and Paraguay), using an integrated epidemiological-economic model. Their results indicated that a spatially differentiated policy that combined vaccination in Paraguay with stamping out elsewhere would generate the highest net revenue to the agricultural sector in the long term. Mahul and Gohin

(1999) evaluated the effectiveness of vaccination program in Brittany, France, using a theoretical framework based on the standard epidemiological model S-I-R. They found that vaccination program would cause unrecoverable additional losses.

On the other hand, with the rapid growth of global agricultural trade, a number of studies investigated the impact of trade restriction measures against animal diseases allowed by the Application of Sanitary and Phytosanitary Measures (the SPS agreement) of the World Trade Organization (WTO). These studies often viewed SPS measures as a type of food safety regulation and as a representative example of non-tariff barriers that impedes trade (Hensen and Caswell 1999). For example, Arita et al. (2015) identified the significant impact of SPS measures on the EU-US agricultural trade. They found that the EU SPS restrictions impeded the US beef, pork, and poultry exports, estimating that the ad valorem tariff equivalent (AVE) effects of the measures on beef, pork, and poultry were equivalent to a 23–24%, 81%, and 102%, respectively.

In particular, many scholars analyzed the economic impacts of trade bans after the BSE crisis, Panagiotou and Azzam (2010) studied welfare effects on the US beef industry following the BSE outbreak in the presence of overlapping trade restrictions (Canadian import ban and US export ban) and imperfect competition between Canada and the US. They found that (US) consumers were better off as a result of the partial ban on U.S. beef exports, but worse off under the total and partial ban on Canadian cattle imports. Wieck and Holland (2010) analyzed the short to medium-term effect of the import ban on the US economy using Computable General Equilibrium (CGE) model. They found that the US would lose $1.7 billion in GDP and 11,000 jobs due to Canadian import ban on US beef. Mutondo et al. (2009), using equilibrium displacement model, estimated that US beef producers lost $563.31 million due to Japanese and Korean import bans after BSE outbreak in the US. Marsh et al. (2008) studied the effects of the BSE outbreaks in Canada and the US in 2003 on US cattle prices. They showed that US producers received around 15% lower prices per cattle as a result of the ban of export to Japan and Korea. Tsigas et al. (2008) examined the economic impacts of the import bans on beef from Canada and the US in 2004 with a simulation framework using a partial equilibrium (PE) and a general equilibrium (GE) model. They found that the BSE bans would cause a long-term loss of $3.1 billion (equivalent to an 84% reduction in annual US beef export). Philippidis and Hubbard (2005), using the dynamic Global Trade Analysis Project (GTAP) model, investigated the impact of the export ban on UK beef with the three scenarios of the BSE ban (1996), the FMD ban (2001), and the long-run recovery (2020). They found that despite some impacts after removal of the export bans, exports and outputs would likely increase due to the remedial measures, but the economic-wide impacts both in the short- and long-term would be minimal.

Consumer valuation for food safety

Food scares have led to changes in consumers' consumption behaviors and the development of a global food policy, as fear of animal disease has led to continued discussion on food safety issues in the meat industry. Accordingly, there has been a growing body of research investigating consumer responses (changes in consumer awareness and consumption behavior) related to animal diseases and examining consumers' preferences and valuations for food safety.

Some studies explored consumers' valuation on food safety using consumer's willingness to pay (WTP) for safety labeling programs (Jin and Mu 2012, Ifft et al. 2011, Dickinson and Bailey 2005). Their results showed that consumers placed a high premium on safety. For example, Jin and Mu (2012) examined consumers' preferences and WTP for traceability labeling of poultry meat products in Beijing, China. They found that Chinese consumers were willing to pay approximately 10–11% points more for poultry products with traceability label. Ifft et al. (2011) evaluated consumers' WTP for safely produced chicken in Hanoi, Vietnam using a methodology of a field experiment. They found that Vietnamese consumers would be willing to pay for safety-labeled free-range chicken with an approximately 10–15% premium. In addition, they identified that the past choice of consumers on chicken breed did not influence the present choice of safety-labeled chicken. These results imply that safety branding can be a useful tool to address food safety issues, which encourages producers to adopt new practices and improve the safety of poultry production.

Dickinson and Bailey (2005) examined consumers' WTP for red meat traceability in the US, Canada, the UK, and Japan. The results showed that consumers were willing to pay a significant premium for traceability while showing even higher WTP for traceability-provided characteristics like additional meat safety and humane animal treatment guarantee.

Furthermore, there are a number of studies that examined consumers' WTP for BSE-tested beef and tried to identify whether BSE testing programs would be feasible (Lee et al. 2015, Lee et al. 2013, Lim et al. 2013, Lee et al. 2011, McCluskey et al. 2005, Latouche et al. 1998). These studies generally concluded that consumers preferred BSE-tested beef over beef with other attributes. For example, Lee et al. (2015) examined Korean consumers' WTP for a tax to support a mandatory BSE testing program by conducting a contingent valuation (CV) study. They found that Korean consumers would have a strong preference for a mandatorily tested domestic beef and even Korean consumers' WTP for the program would be higher than the estimated costs for implementing the program. Lee et al. (2013) examined possible heterogeneity in Korean consumers' WTP for beef products with BSE testing and country of origin labeling with respect to Korean consumers' level of risk perceptions and socio-demographic characteristics using a choice experiment. They found that Korean consumers would value BSE tested beef and prefer domestic beef to imported beef. Moreover, respondents with high-risk perception put more value on BSE-tested label rather than origin labeling, while respondents with low-risk perception showed the opposite tendency. Lim et al. (2013) examined the consumers' willingness to trade-off between US labeled steak, and Canadian and Australian beefsteaks. They found that US consumers strongly preferred domestic beef to imported beef. Moreover, consumers were significantly willing to pay for BSE-tested beef. Lee et al. (2011) investigated Korean consumers' willingness to pay for imported beef with traceability. They found that consumers were willing to pay a 39% premium for the beef with traceability over beef without traceability. McCluskey et al. (2005) examined the factors that might affect Japanese consumers' WTP premiums for BSE-tested beef. They found that attitudes towards food safety positively affected the WTP for BSE-tested beef. Latouche et al. (1998) investigated consumer behavior in Rennes, France after the BSE outbreak. They found that consumers preferred to have a better system of transparency or traceability on beef.

Information, risk perception and consumer behavioral change

Consumer's food choice is strongly influenced by psychological perception of product properties than the physical properties of products (Rozin et al. 1986). In particular, consumer's risk perception is an important factor to affect the attitudes and behaviors of consumers (Gstraunthaler and Day 2008). Information is another important factor to influence consumer perceptions since the accessibility of information has increased, but it is difficult to identify which information is correct (Swinnen et al. 2005). A large number of studies assessed the determinants of food consumption through examining how information and risk perception affect consumers' food choices (Gstraunthaler and Day 2008, Mazzocchi et al. 2008, Yang and Goddard 2011, Jin 2008, Schlenker and Villas-Boas 2009, Mazzocchi et al. 2004, Han et al. 1997, Mazzocchi 2006, Burton and Young 1996, Swinnen et al. 2005). For example, Gstraunthaler and Day (2008) assessed the drivers of poultry consumption change in the UK and evaluated the relationship between risk-based variables and behavioral change, based on psychological approach to risk perception and information. Their results showed that the change in consumption behavior was significantly correlated with three main determinants such as (1) knowledge, (2) assessment of the potential for crisis and (3) experience from past food-related incidents. Interestingly, the results revealed that the more knowledge consumers had, the less likely consumers were to change their behavior. However, this study found that the impact from the perceived level of government action or information provided by media or government had no influence on consumer behavioral change. Mazzocchi et al. (2008), using the SPARTA model with consumer survey data, investigated chicken consumption choices and found that the effects and interactions of consumer behavioral determinants varied across the European countries (France, Germany, Italy, the Netherlands, and the UK). Their results showed that risk perception and trust affected consumer choices due to food scares, whereas no relationship was found between trust in food safety and socio-demographic variables (e.g., age, education, income, etc.). Namely the impact of food safety information is dependent on the source and reliability rather than individual socio-demographic properties of the consumer. The results also indicated that consumer trust in food safety information did not always affect risk perception.

Furthermore, Yang and Goddard (2011) examined the impact of BSE on Canadian consumers' risk perceptions and attitudes toward beef. They found that both quantity and content of information on BSE significantly affected the demand for beef. Specifically, BSE information mentioning government negatively affected beef consumption, whereas consumer groups responded differently to a BSE information mentioning scientists. Jin (2008) found that South Korean consumers would avoid contracting potential health risks caused by BSE outbreaks, even though they did not suffer from BSE event in South Korea. It was mainly because Korean consumers have been aware of several mass media reports about Japanese BSE outbreaks. Schlenker and Villas-Boas (2009) used UPC scanner and futures data to examine the effects of media coverage of BSE on consumers. They found the negative effects of media coverage on beef consumption. Mazzocchi et al. (2004) assessed the impact of withholding the information about BSE crisis on consumer welfare in Italy. They found that the estimated welfare loss ranged from 12% to 54% of total expenditure on meat. Han et al. (1997) investigated Korean consumers' responses to information on the European BSE event and related CJD risks in Korea. They found that Korean consumers immediately reduced consumption of imported beef in response to information on BSE-related health risks. Burton and Young (1996) investigated the effects of BSE events on consumer's beef expenditures, using an index of media coverage. They found that media coverage and extensive publicity on BSE resulted in a 4.5% decline in market share for beef.

Swinnen et al. (2005) and Verbeke and Ward (2001) analyzed how the information market ultimately affects consumers' perceptions and purchasing behavior. Swinnen et al. (2005) found that consumers were likely to have imperfect information on most issues because of the costs of collecting information. Moreover, they found that negative news was more dominant than positive news. Verbeke and Ward (2001) found that the impact of TV publicity was negative on beef/veal consumption, but fresh meat advertising only had a small impact compared to negative media coverage in Belgium.

Policy Measure for Animal Disease

Mitigation measures

The economic costs and welfare implications of animal disease outbreaks indicates the need for the spread of animal diseases to be prevented and controlled within and across borders (Cairns et al. 2016). According to the World Organisation for Animal Health (OIE) guidelines, the OIE recommends mitigation measures against animal diseases such as early detection and accurate warning systems (for the initial measures), protection of disease-free countries, areas, or zones through strict import regulation, cross-border animal movement controls, and surveillance (for effective control), stringent biosecurity measures and good hygiene (by animal owners and producers), vaccination with a clear exit strategy[1] (for complementing culling for susceptible animals in endemic areas), and financial compensation systems (by government) for inducing early detection and transparent reporting of animal disease outbreaks.

Culling

Culling, as one of the control options for infectious diseases of livestock, is widely used to reduce the threat of disease virus transmission by killing potentially infected animals. If infected animal are detected, then the infected and contact animals are normally culled by the government. Over the last decade, culling strategies have been implemented in Europe, North America, and Asia to control highly contagious animal

---

[1] The vaccination policy should not be used for a long time. Thus, it is recommended to review the priority conditions at the local or national level in consideration of the possibility of generation of variant viruses and the possibility of infecting the human body and to use the necessary exit strategy to end the vaccination accordingly.

diseases such as BSE, AI, FMD, swine fever, etc. (Tildesley et al. 2009). However, the massive culling due to repeated outbreaks of animal disease has become socially unacceptable and the resulting economic impact discourages farmers to report infected animals and thus avoid culling. Jarvis and Valdes-Donoso (2015) highlight that many initial control measures such as culling programs have failed to eradicate diseases and have reduced the incomes of poor farmers due to the high cost of culling (Jarvis and Valdes-Donoso 2015). Thus, effective culling strategies must be selective to optimize epidemiological control and minimize economic and agricultural damages.

Vaccination

Vaccination is generally used to reduce the number of susceptible animals, prevent the spread of disease, and reduce the number of slaughtered animals (Egbendewe-Mondzozo et al. 2013). The OIE encourages the use of vaccination as a supplemental control measure, but also warns possible harm from vaccination since the viruses could become endemic in vaccinated animals and long-term circulation of the virus in vaccinated animals might result in both antigenic and genetic changes in the virus (OIE 2017). Therefore, the OIE strongly recommends that vaccination should be used for a limited duration when culling policy cannot be implemented because the disease could become endemic or it is difficult to detect infected animals. And many studies showed that vaccination was highly ineffective and costly, and only effective when it was used complementally with other measures.

Food safety measures

Recent food safety crises associated with animal diseases have generated an opportunity for consumers to rethink their attitudes and behaviors towards food consumption (Gellynck and Verbeke 2001). In response to consumers' concerns about new and continuing food safety problems, and consumers' demands for higher levels of food safety, governments and industries have strengthened regulatory programs that address more types of food safety-related attributes through the public and private food safety control systems such as SPS measures, certifications, traceability systems, testing polices, etc.

SPS measures

In response to the global needs to protect humans, plants, animals, and the environment, the WTO established the Agreement on the Application of Sanitary and Phytosanitary Measures (SPS Agreement) based on health and safety justifiable on scientific grounds and the Agreement on Technical Barriers to Trade (TBT Agreement) based on standards for quality specification (Wilson and Anton 2006). Particularly, in accordance with the purpose of ensuring food safety and human and animal health, the SPS Agreement sets out the international rules to discipline the regulatory measures that member countries adopt to ensure agricultural and food safety. For appropriate implementation, the SPS Agreement requires international standards, guidelines, and recommendations from the Codex Alimentarius Commission (Codex), the International Plant Protection Convention (IPPC), the International Office of Epizootics (OIE), and other international management standards (ISO 22000, HACCP, etc.). Based on these requirements, the SPS Agreement permits governments to maintain appropriate sanitary and phytosanitary protection measures.

The SPS measures that may fit into this category include cleaning, quarantines, inspections, tariffs, bans, etc. (Wilson and Anton 2006). A number of countries are implementing the SPS measures as an effective tool to ensure that foods are safe to consume, and to protect animals and plants from potential pests and diseases. Many SPS measures are used with scientific justification and clearly improve consumer and producer welfare. However, they are often perceived as discriminatory and protective measures in the guise of ensuring human, animal, or plant safety and can cause cross-border disputes.

Quarantine. Quarantine is a protective entry measure designed to protect humans, animals, plants, and the environment from importing exotic pests and diseases via possible carrier food products (James and Anderson 1999). The WTO allows member countries to use quarantine as an SPS import restriction measure. In recent years, quarantine is considered a risk management option that governments can apply to reduce potential risks from the importation of live animals and their products within the "national biosecurity" framework (Arthur 2004). However, it is also argued that stringent quarantine polices of developed countries can be a trade barrier for less developed countries due to the lack of specialized infrastructure, capability and expertise for quarantine.

Trade Ban. The SPS Agreement also allows countries to impose trade ban on livestock and livestock product imports as a science-based precautionary measure in the event of an outbreak in the importing country. In order to offer a scientific basis for such measures while restraining protectionism, the OIE provides a list of diseases that appear to have specific risks. When OIE-listed diseases[2] occur, countries can impose trade bans on specific countries infected with OIE-listed diseases without violating WTO rules. While trade restrictions aim to protect human health and ensure food safety from the risk of imported meat and poultry, countries infected with animal diseases suffer not only a loss in trade but also a decline in reputation of safety measures. Thus, SPS measures to restrict trade often cause disputes between countries. Nonetheless, the SPS measures such as trade restriction have important implications for food safety regulation at the national level, which allow food regulators to adopt only scientifically proven food control measures and to minimize the impact on trade (Hensen and Caswell 1999).

Certificate

HACCP and ISO 22000 are the most widely adopted food safety certification systems in the world. These systems are introduced as a complement to existing food safety systems.

---

[2] See OIE website for OIE-listed diseases in 2017. (http://www.oie.int/animal-health-in-the-world/oie-listed-diseases-2017/).

HACCP. Hazard Analysis Critical Control Point (HACCP) is a widely recognized system for reducing food hazards. HACCP is a quality management system to require companies to analyze food processing flow from raw materials to products. As HACCP is regarded to be more cost-effective than end-product testing (ICMSF 1988), many countries are adopting the HACCP system as a new approach to ensure the safety of food supply. For example, the EU has required food companies to implement an HACCP-based food safety control system since 1995. The US established mandatory regulation for meat and poultry in 1996. Recently developing countries have increasingly adopted the HACCP system. For example, the Chinese government has required a mandatory HACCP certificate for companies exporting meat and seafood products since 2002 (Wang et al. 2009). However, many studies investigating costs and benefits of voluntary or safety-related systems found that implementing safety-related systems was costly, especially for small companies.

ISO 22000. It is a food safety management system certified by International Organization for Standardization (ISO). ISO's food safety management standards help any organization in the food chain (producers, suppliers, manufacturers, distributors, retailers and food service organizations) identify and control food safety hazards. ISO 22000 is also used in the Food Safety Systems Certification (FSSC) Scheme FSSC 22000, which is a Global Food Safety Initiative (GFSI) approved scheme. Nowadays, many organizations and countries have adopted ISO 22000 certificate (FSSC 22000). As of 2017, over 15000 organizations in 140 countries achieved the FSSC 22000 certification. Some argue that food safety certificate systems such as the HACCP and the ISO 22000 often serves as non-trade barriers against the exports of developing countries that do not have financial and technical resources and thus fail to meet stringent requirements imposed by developed countries (Hammoudi et al. 2015), while others claim that such certificates can act as a catalyst to trade by facilitating developing countries' production innovation and improving production efficiency.

Traceability systems

Traceability systems generally include key elements such as identification, registration of animals, herds, meat processors, exporters, data capture, communication, and data management, and verification (Gellynck and Verbeke 2001). Traceability provides a lot of benefits for the food industry and consumers by enabling the industry to provide consumer assurance of food sources and safety and allow for disease control and monitoring, and infected source to be identified (Leat et al. 1998). In particular, recent food risks incurred from animal diseases have prompted consumers to push for further development of traceability systems by governments and industries.

Traceability systems have been mainly driven by the EU with traditionally strong consumer perception of food safety and quality than other regions. Thus the EU has implemented food labeling policies that is more focused on traceability, origin, and production process compared to other regions. In particular, European food safety scares have led to the development of "farm-to-fork" traceability systems and initiated a variety of food labeling policies such as protected designation of origin (PDO), protected geographical indication (PGI), and country-of-origin labeling (COOL) (Loureiro and Umberger 2007). These policies use geographical names in food labels, requiring a variety of traceability levels of origin and production processes. On the other hand, US food policy with regard to origin and traceability is far behind other countries such as the EU, Australia, Canada, and Japan. In recent years, however, there has been a growing demand for mandating the COOL system and implementing the mandatory US beef traceability system. In 2013, the mandatory COOL system was implemented in the US in a manner to accurately distinguish imported beef from US beef.

PDO[3] and PGI[4]. Recently labeling with the name of the origin region is becoming more noticeable due to a growing consumer demand for food information (Vecchio and Annunziata 2011). In response to consumers' desire to purchase safe and highly quality food with a clear geographical identity, the European Union Protected Food Names Schemes came into force in 1992 (Council Regulation (EEC) No. 2081/1991), offering labeling systems (e.g., PDO Orkney beef and PGI Welsh beef) (Heaton et al., 2008). In particular, the two quality signs, PDO and PGI, aiming at linking the quality characteristics of agricultural products to its geographical origin, are recognized as useful tools for producers to increase the value added and the market dominance of their products. However, in general, producers impose premium prices on PDO/PGI products because of strict conditions and more expensive raw materials (Belletti et al. 2007). Producers can benefit from the premium pricing with PDO/PGI labels while consumers can benefit from ensuring safe food purchases based on the information about the quality level of the product. It is evident that labels with geographical indications work as a quality assurance to consumers through greater transparency or traceability in the food chain (Latouche et al. 1998).

COOL. COOL regulation has been implemented to mitigate the public concerns on BSE outbreaks. The regulations are considered as an indirect way to protect the consumers from BSE scares, but an effective method to identify and track the cattle from the production to the marketing sectors. Several countries including the EU and the US have adopted the country of origin labeling system on beef. However, policy makers and producers in some countries still doubt the effectiveness and public demand for the labeling. There are some negative evaluations that labeling systems could lead to an increase in operating costs and reduction of global trade and lack of demand for labels, which might reduce consumer and producer welfare.

BSE-testing policies

Surveillance for BSE through testing is crucial not only to determine the incidence of BSE but also introduce guidelines for managing the disease. Although there is a considerable debate about the scope of BSE testing, BSE testing policies have been implemented in some countries. The Japanese government started BSE testing of all slaughtered cattle of all ages in 2001 but relaxed the requirement to cattle only 21 months or older in 2005. However, Japan's local authorities have continued to test

---

[3] PDO designated products are "produced, processed, and prepared within a given geographical area using recognized know-how" (EC Regulation 510/2006).

[4] PGI designated products "the geographical link must occur in at least one of the stages of production, processing or preparation. Furthermore, the product can benefit from a good reputation" (EC 2006).

beef for all domestic cattle regardless of age even since 2005. In 2001, the EU also implemented Regulation 999/2001 that requires BSE testing for cattle 30 months or older (Aldy and Viscusi 2013). However, the majority of countries with BSE-testing policies require testing only for BSE-suspected cattle due to the cost burden. There are still debates about the costs and the scope of BSE testing.

Current policy issues

The current policy issues in the multilateral system (e.g., the WTO SPS Committee) concern trade measures to ensure food safety from various risk aspects including animal diseases. In particular, non-compliance of international standards in importing countries, and long delays in completing risk assessments and resuming imports are pointed out as the most important issues. In bilateral or regional systems such as Free Trade Agreement (FTA), resolving regulatory conflicts on food safety is a major issue. For example, the Transatlantic Trade and Investment Partnership (TTIP) between the EU and the US well demonstrates an acute regulatory confrontation surrounding several key issues such as chlorine chicken wash, hormone-treated beef, geographical indications, and genetically modified foods (Josling and Tangermann 2016), which fundamentally come from the differences of risk perception (precautionary or after-care principle) and regulatory focus ("farm-to-fork" or risk-based approach).

How to improve better communication between consumers and producers

In general, most consumers don't know where their food has come from. Consumers know that farmers raise crops and livestock, someone processes and packs, and someone brings them to groceries and restaurants, but they do not know the details about what is involved in this process (Ikerd 2001). However, with recently growing consumer concerns about food safety related to animal diseases, various food safety systems have been developed in order to connect producers to consumers; however, consumers still remain uninformed about food safety for various reasons. Many of the consumer concerns stem from the fact that food suppliers are not providing proper and immediate information to consumers (Goddard 1999).[5] On the other hand, there has been the lack of public policies connecting producers to consumers and the absence of educational strategies valuing production process and quality (e.g., organic farming) as pointed out by Darolt and Constanty (2008). Lindgreen and Hingley (2003) observed Tesco[6] as a model of best practices to connect producers to consumers together. They conducted a case analysis on Tesco's approach to deal with consumers' concerns about food scares and animal welfare, using an interview survey of Tesco's meat suppliers. They concluded that Tesco's integrated and relationship-oriented, supply-chain approach to business-to-business dealing could offer substantial benefits to supply

---

[5] Cited in Lindgreen and Hingley (2003).

[6] Tesco is the single largest food retailer in the UK and the world's largest retailer with Walmart in the US and Carrefour in France. It has over 480,000 employees and operates 11 markets worldwide (as of 2017). Obtained from Tesco website (http://www.tesco-careers.com/Inside-Tesco.aspx#aboutus).

partners and enable them to deliver quality and performance improvements to end consumers. In addition, they highly valued the way that Tesco and supply chain jointly resolved specific consumer concerns on animal welfare and food safety. The Tesco case shows that a well-established communication network among all stakeholders is a key element to improve relationship between consumers and producers. Therefore, better communication efforts between producers and consumers are required. Better communication can be improved through enhancing interactive learning processes (Farnworth 2004), eliminating the grounds of negative media coverage, seeking more effective communication methods (Verbeke and Ward 2001), and actively utilizing traceability systems linking producers to consumers.

Unsolved Problems and Future Research

Analyzing regional disparities and reasons of animal diseases

Animal diseases have different geographic distribution patterns. For example, the outbreaks of AI are mainly seen in the Northern Hemisphere, and FMD outbreaks are mainly concentrated in Africa and Asia. Therefore, it is necessary to analyze the regional disparity in disease outbreaks and its reasons. In order to analyze the main causes of this regional disparity, it is necessary to closely examine the interrelationships of various factors such as natural environment (e.g., climate), culture (e.g., diet pattern, etc.) and policies.

Detailed design on market and consumer segmentation

Previous studies have shown that consumers respond differently to the BSE outbreaks depending on their segmented population groups (Yang and Goddard 2011). It indicates that it is becoming more difficult to predict consumers' behaviors in markets or there may be a significant change in meat consumption patterns in certain consumer groups. Therefore, future studies should be analyzed with more detailed market and consumer segmentation to better understand consumers' preferences in meat consumption. Furthermore, the responses of producers and consumers to unknown and uncontrolled risk also need to be investigated.

Proper program evaluation on policy interventions

It is necessary to conduct proper program evaluation on policy interventions to control animal diseases between treatment and control groups. For this, future research should evaluate internal and external validities of policy and consider endogeneity of government intervention since endogenous government intervention efforts would allow a better understanding of public management policies of animal disease (Wang and Hennessy 2015). Furthermore, in assessing policies related to animal diseases, the length and type of policies are often ignored. Therefore, as Hagerman et al. (2012) pointed out, future research should recognize the importance of assumption made on duration and types of related policies (e.g., trade ban and vaccination) since the economic impact of an epidemic may vary depending on which disease management policy is used for analysis (Elbakidze et al. 2009).

Enhancing international collaboration for interdisciplinary studies

An understanding of the impacts and interactions among economic systems and ecosystem such as disease spread among domesticated livestock and wildlife, and related environmental effects, is very important for future research (Forsyth 2005). From this perspective, an interdisciplinary approach requiring interaction between veterinarians, economists, and other social scientists can provide policy-makers with useful options given the complexity and the high political profile of potentially pandemic disease (Jarvis and Valdes-Donoso 2015). And better information and knowledge about the outbreak and disease epidemiology can help investigate the disease-related risk more realistically (You and Diao 2007). Furthermore, the integration of epidemiological and economic analysis can provide a rich ground for methodological development.

Expanding studies on both direct and indirect effects on related industries and regions

Many studies have mainly focused on direct effects such as economic damage and welfare impacts of the industry affected by animal diseases (e.g., poultry industry and livestock industry). However, animal disease outbreaks can indirectly influence other related industrial sectors including other agricultural commodities markets, tourism, restaurants, etc. and other regions. Therefore, future research needs to investigate the indirect effects of relevant industries as well as other animals and regions as a separate study (Egbendewe-Mondzozo et al. 2013, Tozer et al. 2015).

Improving economic valuations on animal diseases to reduce hypothetical bias

Future research needs to improve economic valuations on animal disease. For example, the willingness to pay (WTP) measurement should be conducted in a more realistic setting. Previous studies on consumers' WTP usually were based on survey methods to present a hypothetical scenario about an intervention and potential bias from hypothetical setting was inherent. Thus, future research needs to focus on measuring consumers' WTP while considering more realistic settings for a better understanding consumers' valuation on food safety by reducing hypothetical bias.

A larger system analysis using long-term and quantity data

As Ifft et al. (2011) pointed out, longer-term data is useful to assess production responses in the face of animal disease outbreaks even though short-term data allows to assess adjustment of small-scale producers. Thus, a long-term analysis is needed to better predict supply response and risk generation, and better design policy. Furthermore, quantity data analysis is also needed because much of the empirical research on the impacts of animal disease outbreaks focused on domestic prices in the meat supply chain due to lack of data, which neither explained the role of imported meat price in the domestic market nor investigated the relationship between domestic prices and import prices. Thus, as Park et al. (2008) explained, a larger system analysis including changes in the price and quantity of imported meat would provide a better understanding of the impacts of animal diseases and better policy implications.

Considering cultural factors and quality differentials

Consumer confidence in food safety can be reduced not only by specific scares or controversies, but also the erosion of cultural mechanisms that create and sustain confidence. So it is required to understand the role of cultural factors influencing consumer confidence in food safety (Beardsworth and Keil 1997). A better understanding of the interrelationship between cultural factors and consumer confidence will help to improve the effectiveness of food safety policy. Furthermore, future research needs to consider quality differentials such as the traceability systems, food safety standards, and promotional activities since quality differentials also form consumer bias (Kawashima and Sari 2010).

Conclusion

Animal diseases such as bovine spongiform encephalopathy (BSE), avian influenza (AI), and foot and mouth disease (FMD), known as highly contagious diseases, have had a huge impact on the relevant markets, industries, and also human health, causing considerable economic and social costs. Furthermore, recent rapid growth of global trade and travel increases the possibility of spreading animal disease and its subsequent damages. As approximately 60% of human infectious diseases result from animal diseases, human health and animal health are closely linked. Thus, growing public concerns and fears about animal diseases have required policy development and strategy for food safety as well as human and animal health. To respond to this, numerous policy measures have been implemented to prevent and control disease outbreak and food hazards such as culling, vaccination, trade ban, certificates, traceability systems, etc. Many scholars have studied the economic impacts of animal diseases, the effectiveness of regulation and policy measures related to animal disease and food safety, consumers' valuation on food safety, and consumers' risk perceptions and purchasing behaviors. Their findings show that animal diseases can cause significant economic welfare losses to the relevant markets and industries (e.g., livestock industry), and severe damage to other relevant industries (e.g., tourism industry). They also reveal that prevention and control measures such as culling and vaccination without other complementary measures are generally ineffective and costly. Furthermore, the findings reveal that consumers tend to put high valuation on food safety and consumers' risk perception influences their food choices. However, the impact of information largely depend on the source and reliability rather than individual socio-demographic properties.

There have been numerous efforts and policies to minimize the socio-economic impact through preventing animal diseases and food hazards. Nonetheless, policy failures and inefficiencies still exist and consumers still have a misperception of risk due to various reasons. Therefore, the major ways to minimize socio-economic impacts of animal diseases and to secure food safety are through proactive measures with more emphasis on animal welfare, elimination of unreliable media coverage, a better communication between producers and consumers, education on animal diseases and the value of food safety, and in-depth research based on a comprehensive approach, complementing the weaknesses of existing research techniques.

References

Aldy, J.E. and W.K. Viscusi. 2013. Risk regulation lessons from mad cows. Foundations and Trends® in Microeconomics 8: 231–313.

Arita, S., L. Mitchell and J. Beckman. 2015. Estimating the Effects of Selected Sanitary and Phytosanitary Measures and Technical Barriers to Trade on U.S.-EU Agricultural Trade. Economic Research Service Economic Research Report No. 199. United States Department of Agriculture (USDA).

Arthur, J.R. 2004. The role of quarantine in preventing the spread of serious pathogens of aquatic animals in Southeast Asia. pp. 25–33. *In*: Lavilla-Pitogo, C.R. and K. Nagasawa (eds.). Transboundary Fish Diseases in Southeast Asia: Occurrence, Surveillance, Research and Training. Proceedings of the Meeting on Current Status of Transboundary Fish Diseases in Southeast Asia: Occurrence, Surveillance, Research and Training, Manila, Philippines, 23–24 June 2004. Tigbauan, Iloilo, Philippines: SEAFDEC Aquaculture Department.

Beach, R.H., C. Poulos and S.K. Pattanayak. 2007. Farm economics of bird flu. Canadian Journal of Agricultural Economics/Revue canadienne d'agroeconomie 55: 471–483.

Beardsworth, A. and T. Keil. 1997. Sociology on the Menu: An Invitation to the Study of Food and Society. Routledge.

Belletti, G., T. Burgassi, A. Marescotti and S. Scaramuzzi. 2007. The effects of certification costs on the success of a PDO/PGI. pp. 107–121. *In*: Theuvsen, L., A. Spiller, M. Peupert and G. Jahn (eds.). Quality Management in Food Chains. Wageningen Academic Publishers, Wageningen.

Burton, M. and T. Young. 1996. The impact of BSE on the demand for beef and other meats in Great Britain. Appl. Econ. 28: 687–693.

Cairns, A., T. Tolhurst, K. Poon, A.P. Ker, S. Duff, D. Jacques and L. Yang. 2016. The economic impact of a foot-and-mouth disease outbreak for Ontario's beef sector. Can. J. Agr. Econ. (in press).

Darolt, M.R. and H. Constanty. 2008. 16th IFOAM Organic World Congress, Modena, Italy, June 16–20. 2008. Producers and Consumers Relationship Strategies in the Organic Market in Brazil. <http://orgprints.org/view/projects/conference.html> (accessed on 3 April 2017).

Davis, D.P. 2002. Emerging Animal Diseases: Global Markets, Global Safety. Summary of Workshop held on 15th January 2002. The National Academics' Board on Agriculture and Natural Resources of the Division on Earth and Life Studies. National Academy Press. Washington D.C.

Dickinson, D.L. and D. Bailey. 2005. Experimental evidence on willingness to pay for red meat traceability in the United States, Canada, the United Kingdom, and Japan. J. Agr. App. Econ. 37: 537–548.

Egbendewe-Mondzozo, A., L. Elbakidze, B.A. McCarl, M.P. Ward and J.B. Carey. 2013. Partial equilibrium analysis of vaccination as an avian influenza control tool in the US poultry sector. Agr. Econ. 44(1): 111–123.

Elbakidze, L., L. Highfield, M. Ward, B.A. McCarl and B. Norby. 2009. Economics analysis of mitigation strategies for FMD introduction in highly concentrated animal feeding regions. Rev. Agr. Econ. 931–950.

Farnworth, C.R. 2004. Quality relationships in the organic producer to consumer chain: From Madagascar to Germany. Forskningsnytt. 1: 12–14.

Forsyth, T. 2005. The political ecology of the ecosystem approach for forests. pp. 165–176. *In*: Sayer, J. and S. Maginnis (eds.). Forests in Landscapes: Ecosystem Approaches for Sustainability. Earthscan, London.

Gellynck, X. and W. Verbeke. 2001. Consumer perception of traceability in the meat chain. Ger. J. Agr. Econ. 50: 368–373.

Gstraunthaler, T. and R. Day. 2008. Avian influenza in the UK: Knowledge, risk perception and risk reduction strategies. Br. Food J. 110: 260–270.

Hagerman, A.D., B.A. McCarl, T.E. Carpenter, M.P. Ward and J. O'Brien. 2012. Emergency vaccination to control foot-and-mouth disease: Implications of its inclusion as a US policy option. App. Econ. Perspec. Policy. 34: 119–146.

Hammoudi, A., C. Grazia, Y. Surry and J.B. Traversac. 2015. Food Safety, Market Organization, Trade and Development. Springer.

Han, D.B., W.G. Hutchinson and M. Kim. 1997. Economic impact of health risk information: A case study of BSE in Korea. The Korea International Economic Association Winter Conference Seoul.

Heaton, K., S.D. Kelly, J. Hoogewerff and M. Woolfe. 2008. Verifying the geographical origin of beef: the application of multi-element isotope and trace element analysis. Food Chemistry 107: 506–515.

Hensen, S. and J. Caswell. 1999. Food safety regulation: An overview of contemporary issues. Food Policy 24: 589–603.

Ifft, J., D.R. Holst and D. Zilberman. 2011. Production and risk prevention response of free range chicken producers in Vietnam to highly pathogenic avian influenza outbreaks. Am. J. Agr. Econ. 93: 490–497.

Ikerd, J. 2001. Reconnecting consumers and farmers in the food system. University of Missouri. Presented at Conference on Reconnecting Consumers and Farmers, Sponsored by Citizens Policy Center, innovative Farmers of Ohio, and Ohio Citizens Action, Columbus, OH, March 24, 2001. <http://web.missouri.edu/ikerdj/papers/NAF4-Reconnecting.htm>.

Irvine, W. and A.R. Anderson. 2004. Small tourist firms in rural areas: agility, vulnerability and survival in the face of crisis. Int. J. Entrepreneurial Behave. Res. 10: 229–246.

James, S. and K. Anderson. 1999. Managing health risk in a market liberalizing environment: An economic approach. CIES Policy Discussion Paper 99/06.

Jarvis, L.S. and P. Valdes-Donoso. 2015. A selective review of the economic analysis of animal health management. Journal of Agricultural Economics 1–25.

Jin, H.J. 2008. Changes in South Korean consumers' preferences for meat. Food Policy 33: 74–84.

Jin, Y. and J. Mu. 2012. Avian influenza in China: Consumer perception of AI and their willingness to pay for traceability labeling. pp. 391–399. *In*: Zilberman et al. (eds.). Health and Animal Agriculture in Developing Countries. Food and Agriculture Organization of the United Nations and Springer Science+Business Media. LLC.

Josling, T. and S. Tangermann. 2016. TTIP and agriculture: Another transatlantic chicken war? Choice. AAEA. 2nd Quarter 31: 1–7.

Kawashima, S. and D.A. Puspito Sari. 2010. Time-varying Armington elasticity and country-of-origin bias: from the dynamic perspective of the Japanese demand for beef imports. Australian Journal of Agr. Resour. Econ. 54: 27–41.

Latouche, K., P. Rainelli and D. Vermersch. 1998. Food safety issues and the BSE scare: Some lessons from the French case. Food Policy 23: 347–356.

Leat, P., P. Marr and C. Ritchie. 1998. Quality assurance and traceability—The Scottish agri-food industry's quest for competitive advantage. Supply Chain Manage 3(3): 115–117.

Lee, S.H., J.Y. Lee, D.B. Han and R.M. Nayga. 2015. Are Korean consumers willing to pay a tax for a mandatory BSE testing programme? App. Econ. 47(13): 1286–1297.

Lee, S.H., J.Y. Lee, D.B. Han and R.M. Nayga. 2013. Assessing Korean consumers' valuation for BSE tested and country of origin labeled beef products. J. Rural Dev. 37: 185–205.

Lee, J.Y., D.B. Han, R.M. Nayga and S.S. Lim. 2011. Valuing traceability of imported beef in Korea: an experimental auction approach. Aust. J. Agr. Resour. Econ. 55: 360–373.

Lim, K.H., W. Hu, L.J. Maynard and E. Goddard. 2013. US consumers' preference and willingness to pay for country-of-origin-labeled beef steak and food safety enhancements. Can. J. Agr. Econ. 61: 93–118.

Lindgreen, A. and M. Hingley. 2003. The impact of food safety and animal welfare policies on supply chain management: The case of the Tesco meat supply chain. Br. Food J. 105: 328–349.

Lloyd, T.A., S. McCorriston, C.W. Morgan and A.J. Rayner. 2006. Food scares, market power and price transmission: the UK BSE crisis. Eur. Rev. Agr. Econ. 33: 119–147.

Loureiro, M. and W. Umberger. 2007. A choice experiment model for beef: What US consumer responses tell us about relative preferences for food safety, country-of-origin labeling and traceability. Food Policy 32: 496–514.

Mainland, D.D. and S.W. Ashworth. 1992. The effect of BSE on the revenue from beef fat stock. J. Agr. Econ. 43: 96–103.

Mahul, O. and A. Gohin. 1999. Irreversible decision making in contagious animal disease control under uncertainty: an illustration using FMD in Brittany. Eur. Rev. Agr. Econ. 26: 39–58.

Marsh, J.M., G.W. Brester and V.H. Smith. 2008. Effects of North American BSE events on US cattle prices. Rev. Agr. Econ. 136–150.

Mazzocchi, M., G. Stefani and S.J. Henson. 2004. Consumer welfare and the loss induced by withholding information: the case of BSE in Italy. J. Agr. Econ. 55: 41–58.

Mazzocchi, M. 2006. No news is good news: Stochastic parameters versus media coverage indices in demand models after food scares. Am. J. Agr. Econ. 88: 727–741.

Mazzocchi, M., A. Lobb, W. Bruce Traill and A. Cavicchi. 2008. Food scares and trust: a European study. J. Agr. Econ. 59: 2–24.

McCluskey, J.J., K.M. Grimsrud, H. Ouchi and T.I. Wahl. 2005. Bovine spongiform encephalopathy in Japan: consumers' food safety perceptions and willingness to pay for tested beef. Australian J. Agr. Resour. Econ. 49: 197–209.

McDonald, S. and D. Roberts. 1998. The economy-wide effects of the BSE crisis: A CGE analysis. J. Agr. Econ. 49: 458–471.

Mutondo, J.E., B.W. Brorsen and S.R. Henneberry. 2009. Welfare impacts of BSE-driven trade bans. Agr. Resour. Econ. Rev. 38: 324–329.

Nogueira, L., T.L. Marsh, P.R. Tozer and D. Peel. 2011. Foot-and-mouth disease and the Mexican cattle industry. Agr. Econ. 42: 33–44.

Panagiotou, D. and A.M. Azzam. 2010. Trade bans, imperfect competition, and welfare: BSE and the US beef industry. Can. J. Agr. Econ. 58: 109–129.

Park, M., Y.H. Jin and D.A. Bessler. 2008. The impacts of animal disease crises on the Korean meat market. Agr. Econ. 39: 183–195.

Pfeiffer, D.U., M.J. Otte, D. Roland-Holst and D. Zilberman. 2013. A one health perspective on HPAI H5N1 in the Greater Mekong sub-region. Comp. Immunol. Microbiol. Infect. Dis. 36: 309–319.

Philippidis, G. and L. Hubbard. 2005. A dynamic computable general equilibrium treatment of the ban on UK beef exports: A note. J. Agr. Econ. 56: 307–312.

Rasmusen, E. 2010. Games and Information. An Introduction to Game Theory. Malden, MA: Blackwell Publishing. 350 Main Street Commerce Place Malden, MA 02148 United States.

Rich, K.M. and A. Winter-Nelson. 2007. An integrated epidemiological-economic analysis of foot and mouth disease: Applications to the southern cone of South America. Am. J. Agr. Econ. 89: 682–697.

Rozin, P., M.L. Pelchat and A.E. Fallon. 1986. Psychological Factors Influencing Food Choice. John Wiley & Sons, Chichester and New York, NY.

Schlenker, W. and S.B. Villas-Boas. 2009. Consumer and market responses to mad cow disease. Am. J. Agr. Econ. 91: 1140–1152.

Swinnen, J.F., J. McCluskey and N. Francken. 2005. Food safety, the media, and the information market. Agr. Econ. 32: 175–188.

Tildesley, M.J., P.R. Bessell, M.J. Keeling and M.E.J. Woolhouse. 2009. The role of pre-emptive culling in the control of foot-and-mouth disease. Proc. R. Soc. B. Tranchard, S. 2016. Revision is ongoing for ISO 22000 on food safety management. <https://www.iso.org/news/2016/ 04/Ref2075.html> (accessed on 15 March 2017).

Tozer, P. and T.L. Marsh. 2012. Domestic and trade impacts of foot-and-mouth disease on the Australian beef industry. Aust. J. Agr. Resour. Econ. 56: 385–404.

Tozer, P.R., T. Marsh and E.V. Perevodchikov. 2015. Economic welfare impacts of foot-and-mouth disease in the Canadian beef cattle sector. Can. J. Agr. Econ. 63: 163–184.

Tsigas, M., J. Giamalva, N. Grossman and J. Kowalski. 2008. Commodity Trade Analysis in a General Equilibrium Framework BSE Restrictions on Beef Imports from the United States and Canada. Office of Economics and Office of Industries. U.S. International Trade Commission.

Twine, E.E., J. Rude and J. Unterschultz. 2016. Canadian cattle cycles and market shocks. Can. J. Agr. Econ. 64: 119–146.

Vecchio, R. and A. Annunziata. 2011. The role of PDO/PGI labelling in Italian consumers' food choices. Agr. Econ. Rev. 12: 80–98.

Verbeke, W. and R.W. Ward. 2001. A fresh meat almost ideal demand system incorporating negative TV press and advertising impact. Agr. Econ. 25: 359–374.

Wang, T. and D.A. Hennessy. 2015. Strategic interactions among private and public efforts when preventing and stamping out a highly infectious animal disease. Am. J. Agr. Econ. 97: 435–451.

Wang, Z., H. Yuan and F. Gale. 2009. Costs of adopting a hazard analysis critical control point system: case study of a Chinese poultry processing firm. Rev. Agr. Econ. 31: 574–588.

Wieck, C. and D.W. Holland. 2010. The economic effect of the Canadian BSE outbreak on the US economy. App. Econ. 42: 935–946.

Wigle, R., J. Weerahewa, M. Bredahl and S. Samarajeewa. 2007. Impacts of BSE on implications for the Canadian economy. Can. J. Agr. Econ. 55: 535–549.

Williams, C. and M. Ferguson. 2005. Recovering from crisis: Strategic alternatives for leisure and tourism providers based within a rural economy. Int. J. Pub. Sect. Manage. 18: 350–366.

Wilson, N.L.W. and J. Anton. 2006. Combining risk assessment and economics in managing a sanitary-phytosanitary risk. Amer. J. Agr. Econ. 88: 194–202.

World Organization for Animal Health (OIE). 2017. <http://www.oie.int/en/animal-health-in-the-world/web-portal-on-avian-influenza/early-detection-warning-diagnostic-confirmation> (Accessed on 9 March 2017).

World Trade Organization (WTO). 2005. Specific Trade Concerns G/SPS/GEN/204/Rev.5.

World Trade Organization (WTO). 2017. Current Issues in SPS. <https://www.wto.org/english/ tratop_e/sps_e/sps_issues_e.htm> (Accessed on 10 March 2017).

Yang, J. and E. Goddard. 2011. Canadian consumer responses to BSE with heterogeneous perceptions and risk attitudes. Can. J. Agr. Econ. 59: 493–518.

You, L. and X. Diao. 2007. Assessing the potential impact of avian influenza on poultry in West Africa: a spatial equilibrium analysis. J. Agr. Econ. 58: 348–367.

4

Consumer Perceptions of Genetically Modified Foods and GMO Labeling in the United States

William K. Hallman

Introduction

According to an analysis by the International Service for the Acquisition of Agri-biotech Applications (ISAAA) (2016), since they were introduced more than twenty years ago, Genetically Modified (GM) crops have been adopted worldwide faster than any other crop technology in the history of modern agriculture. In 2016, more than 185 million hectares (457 million acres) of GM crops were planted by approximately 18 million farmers and are harvested in 26 countries around the world (ISAAA 2016).

GM foods have been available for sale in the United States for more than two decades, beginning in 1994 with the introduction of the Flavr Savr tomato, the first commercially available GM food (Leary 1994). Since then, the United States has remained the largest producer of GM agricultural products. With nearly 73 million hectares (~180 million acres) planted in GM crop varieties, US farmers harvested more than 39% of the total global acreage of GM crops in 2016 (ISAAA 2016). Based on crop statistics collected by the Economic Research Service of the United States

Department of Human Ecology, School of Environmental and Biological Sciences, Rutgers, the State University of New Jersey; Cook Office Building, 55 Dudley Rd. New Brunswick, New Jersey, USA, 08901-8520.
E-mail: hallman@sebs.rutgers.edu

Department of Agriculture for the planting year 2016, 94% of all of the canola, 94% of the soy, 94% of all of the cotton, and 93% of the corn planted by US farmers was a GM variety (USDA-ERS 2016a). Moreover, in 2013 (the latest year for which statistics are available) over 99% of the sugarbeets harvested in the US were GM (Fernandez-Cornejo et al. 2016). Significantly, about 55% of the US domestic sugar supply is produced using sugarbeets (USDA-ERS 2016b).

Because these GM commodity crops are used to make oils, flours, starches, thickeners, sweeteners, and other key ingredients used in the food industry, the Grocery Manufacturer's Association has suggested that as much as 80% of the processed foods now consumed in the United States contains one or more ingredients derived from GM crops (GMA 2013). Furthermore, because American farmers were quick to adopt GM crop technology, the majority of processed foods in the US have likely contained ingredients derived from GM crops for more than 15 years. By the year 2000, more than 50% of the soybean acreage in the US was planted with GM varieties, and by 2005, more than half of the corn planted in the US was GM (USDA-ERS 2016b). As a result of the extensive use of these crops to make key food ingredients, the Genetically Engineered Organisms Public Issues Education (GEO-PIE) project at Cornell University had reported in 2001 that, "estimates suggest that 60 to 70% of foods in US markets contain at least a small quantity of some crop that has been genetically engineered" (GEO-PIE 2001, p.2).

Public Awareness and Knowledge of GM Foods

Despite the broad use of GM ingredients within the US food supply, surveys of public perceptions of GM foods have regularly shown that Americans have heard, read, and know very little about them. For example, the results of a nationally-representative online survey of 1,148 Americans found that 50% reported having heard or read "very little" or "nothing at all" about GM Foods, and one quarter reported that they had never heard of Genetically Modified Organisms (GMOs) prior to participating in the survey (Hallman et al. 2013). In addition, two-thirds (66%) of the participants in the same survey reported that they had *never* had a conversation about GMOs with anyone at any time in their lives (Hallman 2016). In fact, the percentage of Americans who say that they have had a conversation about GMOs or GM food with others has remained low for more than a decade. Nationally representative surveys of American consumers by Hallman et al. in 2001, 2003, 2004, and 2013, found that the number of Americans who reported having *ever* talked with someone else about GM foods was 31%, 38%, 37%, and 34% respectively.

Studies also suggest that Americans are not paying much attention to news stories related to GM food. Hallman et al. (2003) found in a nationally representative survey that only 19% of the participants could remember any event or news stories related to GM food when asked to do so in an open-ended question. Moreover, less than 1% were able to remember any details about any story they said that they could recall about GM foods. Similarly, in a nationally representative survey in 2004, participants were read a list of seven news stories that had appeared in the media, and then asked whether they recalled having heard each story. The responses to this recognition task suggested that none of the stories had much of an impact on American consumers.

The greatest level of recognition was associated with a story about demonstrations against GM foods in European countries; yet, only a third of the participants (36%) reported that they were familiar with this story (Hallman et al. 2004). This is consistent with data collected by the Gallup organization in 2003 and 2005 that found that the majority of Americans (59–60%) reported following the news about biotechnology "not at all", or "not too closely", while fewer than 10% reported following the news "very closely" (Runge et al. 2017). Similarly, the Pew Research Center (2016) found that only 6% of Americans follow news stories about GM foods "very closely".

Hallman et al. (2013) found that more than half of Americans (54%) also say they know very little or nothing at all about genetically modified foods. About one-third (32%) reported knowing "some", 11% reported knowing "a fair amount", and only 2% reported that they knew "a great deal" about GM foods. This is consistent with an online survey of 1004 participants conducted in 2015 by McFadden and Lusk (2016), who found that only 8% rated themselves as "very knowledgeable" about GM food, 32% rated themselves as "somewhat knowledgeable", while the remaining majority 60% rated themselves as "not knowledgeable" or "undecided".

Hallman et al. (2013) also found that fewer than half of Americans (43%) are aware that products containing GM ingredients are currently for sale in US supermarkets. Surprisingly, this percentage may have decreased in the past 15 years. Data collected in a nationally representative telephone survey of 1203 Americans in 2001 showed that only 41% were aware that GM foods were available in US supermarkets (Hallman et al. 2002). However, a similar survey of 1201 Americans conducted a decade earlier found that 52% thought that GM foods were on the market, and a nationally representative study of 600 Americans in 2004 found that 48% thought that GM foods were available for sale in the US (Hallman et al. 2003, 2004).

Moreover, while most Americans appear unaware that foods with GM ingredients are on supermarket shelves, Hallman et al. (2013) found that the majority of consumers who *do* believe that GM food products are available in the US are confused about which GM products are for sale (Fig. 1).

The 491 participants who reported that they believed that GM food products are available in US supermarkets were presented with a list of foods and asked to indicate which they thought were genetically modified and available to American consumers in the supermarket. In response, about three-quarters recognized that GM corn products are available, and almost six-in-ten endorsed the idea that products containing GM soybeans are available in US supermarkets. However, more than half (56%) indicated that they believed that GM tomatoes are available in US supermarkets. Yet, while the GM Flavr Savr tomato was introduced with significant fanfare in the US in 1994, GM tomatoes have not been sold in US supermarkets since 1997 (Martineau 2001). More than half also mistakenly endorsed the idea that GM wheat and chicken products were for sale in US supermarkets, though neither has been ever been approved for production in the US. Similarly, more than 40% endorsed the idea that GM apples and GM rice were on the market, and more than a third responded that GM salmon and GM oranges were available for sale. However, while apples genetically modified to resist browning were approved for sale in the US by the FDA in 2015, they only became available in a limited number of supermarkets in a few states in 2017, and so were not on the market at the time of the survey (FDA 2015a, Dewey 2017). Similarly,

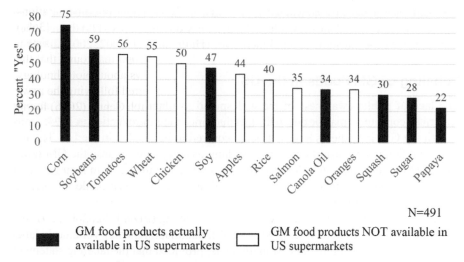

N=491

GM food products actually available in US supermarkets

GM food products NOT available in US supermarkets

Fig. 1. Perceived availability of GM foods in US supermarkets.

while genetically modified salmon were approved for consumption by the US FDA in 2015 (Pollack 2015), legislative barriers were imposed shortly thereafter, which blocked importation and sale of the fish in the US (Dennis 2016). As a result, no GM salmon were available for purchase in the US market at the time of the survey.

In contrast, only about one-third of the participants correctly recognized that GM canola oil was available in their supermarket, and less than one third correctly identified GM papaya, GM sugar, or GM squash as being available in the US marketplace. As already noted, by 2013, 94% of all of the canola and 99% of the sugarbeets planted were of GM varieties. In addition, limited quantities of papaya and squash genetically modified to be virus resistant have been available in the US since their approval for production in the mid-1990s, and a small number of acres were planted with these varieties at the time of the survey (James 2014, USDA-APHIS 2017). In sum, many of the GM products that are available in US supermarkets (i.e., corn, soy, sugar, and canola) are not recognized as such by even the minority of Americans who believe that GM products are for sale, while many non-GM products, including tomatoes, wheat, chicken, rice, salmon, and oranges, are mistakenly assumed by consumers to be GM.

Given the lack of awareness among US consumers that GM products are for sale, it isn't surprising that most Americans are not cognizant that they are eating GM foods. A nationally representative telephone survey of 2,002 Americans by the Pew Research Center (2016), found that about half of Americans (48%) say that little, or none of the food they eat contain GM ingredients, and only 11% of Americans believe that GM ingredients are present within most of the food they eat.

Consistent with these findings, Hallman et al. (2013) found that only about one-quarter of Americans (26%) believe that they have *ever* eaten *any* food containing GM ingredients. Surprisingly, despite the fact that GM ingredients have been ubiquitous in the US food supply for more than 15 years, this percentage has not significantly increased over the last decade. In their series of nationally representative survey of Americans, Hallman et al. found that in 2001 only about one-in-five (20%) thought they had ever personally consumed GM foods, in 2003, about one-in-four (26%) reported that they had done so, while in 2004, nearly one-in-three (31%) thought they had eaten GM food at some point in their lives (Hallman et al. 2002, 2003, 2004).

Labeling of GM Foods in the US

The lack of public awareness of the widespread availability of GM foods among American consumers may be due to the fact that since their introduction in 1994, foods containing GM ingredients have not been required to be labeled in the US. The US Food and Drug Administration (FDA) has long asserted that because the GM crops approved for consumption in the US are "substantially equivalent" to their non-GM counterparts, and that foods developed through genetic modification do not present any different or greater safety concerns than foods developed by traditional plant breeding techniques, they do not require special labeling (FDA 1992, 2001, 2015b).

Under FDA draft guidelines (2001, 2015b), manufacturers may voluntarily label their products as containing no GM ingredients, provided that the labeling statement does not express or imply that the non-GM food is superior. Yet, most Americans mistakenly believe that labeling GM food is mandatory in the United States (Hallman et al. 2013). Moreover, Lusk and Rozan (2008) found that consumers who believe a mandatory GM labeling policy exists were also more likely to believe GM food is unsafe to eat. In addition, those who consider GM foods to be unsafe say they check for GM food labels more often than those who believe they are safe to eat (Pew Research Center 2015).

In fact, the debate over mandatory labeling of genetically modified (GM) foods in the United States has had a long and contentious history. Those opposed to mandatory labeling point to the broad scientific consensus that GM crops and ingredients derived from them are as safe as those made from their conventionally bred counterparts, and as such, don't require special labeling. This consensus includes statements supporting the safety of GMOs made by the American Association for the Advancement of Science (2012), the American Medical Association (2012), the European Commission (2010), the Joint FAO/WHO Codex Alimentarius Commission (2009), the National Academies of Sciences, Engineering, and Medicine (2016), The World Health Organization (2014), and other national and international regulatory bodies and scientific organizations. Opponents of mandatory labeling also argue that foods derived from GM crops are nutritionally equivalent to those produced from non-GM crops (Snell et al. 2012) and that because of their "substantial equivalence", the US Food and Drug Administration (FDA) has not required the mandatory labeling of GM foods (FDA 1992, 2015). Moreover, FDA policies *do* require the labeling and renaming of GM foods if the GM product is no longer substantially equivalent to its non-GMO counterpart. However, while these regulations oblige manufacturers to alert consumers through labeling when the characteristics of a familiar food product have been substantially altered, the required labels do not need to specify that the change in the product was achieved through the process of GM (FDA 1992, 2015).

Opponents of required labeling strongly assert that a mandatory label has a 'signaling effect' (Sunstein 2016), which effectively serves as a "warning label" (Adler 2016). They argue that mandatory labels implicitly communicate the misleading message that GMOs are risky (Phillips and Grant 1998), falsely alarming consumers (AAAS 2012), and discouraging them from purchasing these products (Adler 2016, Charles 2016, Hallman and Aquino 2005, Phillips and Hallman 2013, Runge and Jackson 2000, Sunstein 2016). As a result, opponents of GM labeling argue that "Mandatory labels for GMO food are unscientific, unnecessary, and unconstitutional" (Adler 2016, p.26), amounting to compelled commercial speech (Vecchiarelli 2012) that unfairly stigmatizes products that contain GM ingredients (Adler 2016).

In fact, a Pew Research Center (2015) study shows that 57% of Americans believe that eating GM foods is "generally unsafe", and about half report that they always (25%) or sometimes (25%) look to see if products are genetically modified when they are food shopping. Less than one-third (31%) say they never look for such labels. The same study suggests that those who consider GM foods to be unsafe check for GM food labels more often. Similarly, Pew Research Center data (2016) shows that nearly 40% of Americans believe that "foods with genetically modified ingredients are generally worse than foods with no genetically modified ingredients", and that 89% of those deeply concerned about the issue of GM foods had purchased organic foods and 74% had purchased foods labeled GMO-free in the previous month.

Analyses of the potential economic impacts of mandatory labeling have also indicated that the costs of compliance would significantly increase the price of food products (McFadden 2017). These expenses include those related to maintaining the strict separation between GM, non-GM, and organic crops and the ingredients derived from them that is necessary to ensure the "identity preservation" (IP) required for accurate labeling (Carter and Gruère 2003, Hallman and Aquino 2005). The expense of non-GMO certifications and product testing necessary to meet the demands of consumers who wish to be assured that they are avoiding products with GM ingredients may also raise costs (Hemphill and Banerjee 2014). In addition, fears that consumers will reject foods with GM ingredients might cause firms to switch to non-GM ingredients, which would drive up the costs of these products (Alston and Sumner 2012, Carter et al. 2012).

In contrast, advocates of mandatory labeling argue that the costs associated with such labeling would be trivial relative to those routinely incurred by companies when making typical labeling changes (Costanigro and Lusk 2014). They also assert that there is no scientific consensus that GM products are safe to eat (Hilbeck et al. 2015, Krimsky 2015) and that consumers have a fundamental "right to know" whether the products they purchase contain ingredients derived from genetically modified organisms (GMOs) (Sand 2006).

Proponents of labeling also maintain that Americans overwhelmingly desire to retain "consumer sovereignty"; that is, the right to make food choices based on their own values (Thompson 1997). Wharton (2011) argues that these values have expanded beyond simple interest in the nutritional aspects of foods, and that consumer concerns now include a variety of ethical aspects of food production, including animal welfare, the use of GMOs, fair trade practices, climate change, and the sustainability of the natural resources required to produce particular food products. Similarly, Lazarides and Goula (2018) suggest that irrespective of the presumed equivalence between GM and non-GM foods in terms of quality, nutrition and safety, the production of GM crops raises other issues that may concern consumers. These include the potential impacts of GM crops on biodiversity, food security and the institution of agriculture, as well as issues of informed consumer choice, societal control over scientific research, and personal ethics.

For example, concerns about the perceived "naturalness" of GM crops has been shown to significantly affect the way that consumers think about their potential benefits (Phillips and Hallman 2013, Román et al. 2017, Siegrist et al. 2016). Frewer et al. (2013) have shown that "general" applications and plant-related applications are more acceptable than animal-related applications, while Scott et al. (2016) have shown evidence for absolute moral opposition to GM foods among some US consumers. They found that 71% of GM opponents (45% of the entire sample) were "absolutely" opposed to GM—that is, they agreed that GM technology should be prohibited regardless of its associated risks and benefits. They also found that these "moral absolutists" were significantly more disgusted by the proposition of consuming genetically modified food than were non-absolutist opponents or supporters of GM technology.

Overall, the Pew Research Center (2016) reports that "About one-in-six (16%) Americans care a great deal about the issue of GM foods. These more deeply concerned Americans predominantly believe GM foods pose health risks. A majority of this group also believe GM foods are very likely to bring problems for the environment along with health problems for the population as a whole" (p.43).

Advocates of mandatory labeling, such as Just Label It! (Justlabelit.org), a coalition of the organic foods industry, have also argued that in public opinion polls, Americans overwhelmingly favor the required labeling of GMOs. Surveys have indeed shown that the majority of Americans say they support mandatory labeling of GM foods, with some polls showing that as many as nine-in-ten Americans support it (Center for Food Safety 2017, Kopicki 2013). However, the results of these surveys likely misjudge the amount of public support for mandatory labeling due to biases inherent in the way the questions are asked. The problem is that people are likely to respond positively to questions asking whether they would support labeling that would provide more information about almost any food-related issue. For example, McFadden and Lusk (2017) found that when asked their participants the forced-choice question, "Do you support or oppose mandatory labeling for food containing GM ingredients", 84% indicated support, while 16% were opposed. However, in response to a similar question, 80% also indicated that they supported mandatory labels indicating whether foods contained DNA.

Hallman et al. (2013) have also shown that the number of people who express support for GM labeling depends on how you ask the question. They found that 82% of Americans say they sometimes, frequently, or always read food labels, suggesting that they are familiar with the contents of those labels. Yet, when asked to write down the information they would like to see on food labels that is not already there, only 7% of the participants wrote that they wanted information about GM ingredients. This was about the same percentage of participants (6%) who wrote that they wanted country of origin information on food labels. In a second section of the same survey, respondents were presented with a list of different kinds of information, and asked to rate each as to how important it was that the information appear on food labels. About six-in-ten (59%) said that it was "very" or "extremely" important to have information about

GM ingredients on a food label. This was a slightly smaller percentage of respondents who indicated that it was "very" or "extremely" important to have information on food labels concerning whether the product was produced using hormones (63%), pesticides (62%), or antibiotics (61%), whether it was grown or raised in the United States (60%), and the same percentage of those who indicated that it was important to indicate whether the product contains allergens (59%). Finally, in a third section of the same survey, participants were asked directly whether GM foods should be required to be labeled. In response, about three-quarters (74%) said "yes", 8% said no, and 18% were not sure. In effect, the same participants gave three different answers concerning GMO labeling in response to three different ways of asking the question (Hallman 2016).

Proponents of mandatory labeling also argue that because more than 60 other countries have GMO labeling laws, the lack of required labeling of GM foods in the US places the country out of the regulatory mainstream (O'Neil 2016, Pollan 2012). For example, the European Union (2003) requires mandatory labeling of products that contain GMO ingredients comprising more than 0.9% of the product.

Advocates also suggest that mandatory labeling would provide American consumers with greater transparency, offer increased choice, the ability to exercise dietary, religious, ideological, and/or philosophical preferences, and would enable more informed decision-making regarding the inclusion of GM ingredients in the products they choose to buy (Hallman 1995, 2000, Hallman and Aquino 2005, Hemphill and Banerjee 2014, Kopicki 2013, Phillips and Grant 1998).

However, opponents of mandatory labeling counter that because the USDA's National Organic Program Regulations do not permit the inclusion of GMOs in organic products (USDA-AMS 2013), consumers who wish to choose foods without GM ingredients can *already* do so by purchasing those that are labeled "USDA Organic". In addition, Third-Party Certification (TPC) of products as "organic" or as "non-GMO" can reassure consumers that what they are purchasing do not contain GM ingredients (Hatanaka et al. 2005, Roff 2009). As such, use of these voluntary "free from" labels, which assure consumers of the *absence* of GM ingredients in the food products that display them, obviates the need for mandatory labeling designed to require disclosure of the *presence* of GM ingredients in products which likely make up the majority of the US food supply.

Indeed, food companies have realized that using organic and non-GMO labels represents a potential marketing opportunity and an increasing number of companies have attempted to differentiate their products in the marketplace by claiming that they are organic or contain no GM ingredients. In addition, Bain and Dandachi (2014) suggest that food retailers and value-based food companies have come to recognize the reputational and economic value offered by new niché markets for non-GMO foods, resulting in an increasing number of private-sector efforts create non-GMO labels and brands. These include marketing efforts by retailers such as Whole Foods (Polis 2013), food manufacturers such as Ben and Jerry's (Hallenbeck 2014), and fast-food companies such as Chipotle (Alesci and Gillespie 2015) to go beyond simply labeling some of their products as non-GMO, but rather to eliminate GM ingredients from their entire product lines.

In fact, the number of products in the US market making claims that they are organic and/or non-GMO has rapidly increased since 2010 (Greene et al. 2016a). However, as Bain and Selfa (2017) point out, two non-GMO labels dominate the market "USDA Organic" and "Non-GMO Project Verified" (nongmoproject. org). Yet, the non-GMO status of a "USDA Organic" product is not obvious from the label, and many consumers are unware that these products are also "non-GMO". Additionally, consumers may not comprehend that non-GMO doesn't mean that the products are organic, which must meet additional requirements with respect to prohibited pesticides, herbicides, hormones, and antibiotics as well as complying with animal welfare standards, which make organic products more expensive to produce than those that are only non-GMO (Erbentraut 2016).

Organic producers, who have made significant investments in meeting USDA Organic standards and to go through its certification process are worried that non-GMO is seen by consumers as equivalent to organic. These concerns may be real, as McFadden and Lusk (2017) have shown in a willingness-to-pay study that consumers perceive Non-GMO Verified as a substitute for Organic. Further, data collected by the market-research firm Nielsen showed that food products labeled as non-GMO outsold those labeled as organic in 2015 (Bunge and Gasparo 2015). Moreover, the number of products bearing non-GMO labels appears to be increasing each year. In 2015, 15.7% of new food and beverage products introduced in the US market made non-GMO claims. By comparison the number of new products making such claims was 10.2% in 2014, 6.5% in 2013, 2.8% in 2012, 2.7% in 2011, and 1.6% in 2010 (Watson 2016). Not surprisingly, sales of "non-GMO" foods reached nearly $21.1 billion in 2016, an increase of more than $8 billion over 2012 sales figures (Keller 2017).

Consistent with this, the Non-GMO Project (2017) claims that its Non-GMO Verified label is the fastest growing label within the natural products market, accounting for $19.2 billion in annual sales and more than 43,000 verified products associated with more than 3,000 brands. However, Greene et al. (2016b) suggest that more than half of the Non-GMO Project Verified products are also certified organic, and as has been noted previously, the USDA Organic Program Regulations already prohibit organic products from containing GM ingredients, making Non-GMO Project verification redundant with USDA Organic certification.

In addition, critics contend that many Non-GMO Project Verified products do not contain ingredients that have GM counterparts, and so could never have been derived from GM crops. These include Non-GMO Project Verified oranges, tomatoes, grapes, and sea salt (which has no DNA to modify) (Senapathy 2017). Because *all* oranges, tomatoes, grapes, and sea salt in the US market are non-GMO, critics argue that placing the Non-GMO Project Verified seal on these products is misleading, unscientific, and is simply designed to create credence goods by tricking consumers into purchasing products that have imaginary value (Splitter 2017).

The labeling of such products as "non-GMO" points to the larger problem of defining this term. As Johnson (2015) notes, because GMO products lack a precisely specified common denominator, the idea of GMO does not "map neatly onto any clear category in the physical world", and is instead, "a cultural construct" that is used to talk about a set of concepts that do not have clear boundaries. For example, Senapathy (2017), suggests that in practice, GMO has become shorthand for any organism produced though "modern biotechnology". She argues that "according to this definition, the only GMO crops available in the U.S. are soybeans, corn (field and sweet), papaya, canola, cotton, alfalfa, sugarbeets and summer squash, with gene-silenced White Russet potatoes and Arctic apples available in some test markets". By this definition, the only products that should legitimately be considered as "non-GMO" in the US are those with ingredients derived from the conventionally bred varieties of these crops. That is, products should only be able to claim that they are "non-GMO" if they have a GMO counterpart.

Alternatively, Johnson (2015) suggests that many people equate GMOs with *transgenesis*; that is, moving genes between species. This was the technique used to genetically modify soy, corn, cotton, canola, alfalfa and sugarbeets to be herbicide resistant and to genetically modify corn and cotton to produce a protein designed to protect those crops from insect damage. These GM crops are currently the most widely planted, and have received the greatest attention from anti-GMO activists, and subsequently the greatest media attention. So, it makes sense that people view all GMOs as being produced through the introduction of foreign DNA into their genomes.

However, defining GMOs as the consequence of transgenesis ignores the modification of genetic traits through gene silencing; that is, techniques used to turn off a gene. For example, J.R. Simplot Co. used this technique to create its Innate™ potato, which is less susceptible to bruising, has less discoloration when sliced, and produces less acrylamide when it is cooked at high temperatures. This was achieved by adding fragments of genetic material from cultivated and wild potatoes, but no genetic material was added from unrelated organisms (Mellon 2013). A similar gene silencing technique was used to insert a gene sequence into Fuji, Granny, and Golden varieties of apples, which suppresses the production of an enzyme that causes plant cells to brown when exposed to oxygen. This allows the fruit to resist browning when cut or dropped (USDA-APHIS 2012, Vincent 2017).

Defining GMOs as the results of transgenesis also excludes products produced through the use of gene-editing techniques such as ZFN (Zinc-Finger Nuclease), TALEN (Transcription Activator-Like Effector Nuclease), and CRISPR (Clustered Regularly Interspaced Short Palindromic Repeats); systems that are used to revise or to regulate the existing genes of plants or animals without introducing foreign genetic material. For example, the gene-editing tool CRISPR-Cas9 was used to produce a non-browning version of the common white button mushroom (*Agricarus bisporus*), by deleting a small number of base pairs in the part of the genome that controls the production of polyphenol oxidase (PPO), an enzyme that causes browning (Waltz 2016a). Because the process doesn't introduce any foreign DNA, current USDA policies don't require that CRISPR-edited crops be subject to USDA regulatory oversight (Waltz 2016b).

Defining GMOs as the product of transgenesis also excludes important crops with genetic changes produced through the use of "mutation breeding", that is, through exposure to radiation or to chemical mutagens designed to generate mutations that may lead to desirable traits. Thus, it would exclude more than a thousand mutant varieties of staple crops grown on tens of millions of acres in more than 60 countries around the world (Joint FAO/IAEA Division of Nuclear Techniques in Food and Agriculture, Plant Breeding and Genetics Subprogramme 2017). These comprise a significant proportion of the world's commercial food crops, especially those grown in rural areas, including agriculturally important varieties of rice, corn, wheat, barley, sorghum, peanuts, peas, peppermint, and pears, grapefruit, bananas, cassava and sunflowers (Ahloowalia et al. 2004).

The National Bioengineered Food Disclosure Standard

After years of contentious debates at the state and federal levels about mandating the labeling of GM foods, the US Congress passed legislation to create a federal requirement to label food products containing GM ingredients (Charles 2016). Signed into law on 29 July, 2016, The National Bioengineered Food Disclosure Standard of 2016[1] (NBFDS) requires the labeling of all food products containing GM ingredients sold in the United States to disclose this fact. Although the law does not specify the date by which companies must disclose that their products contain GM ingredients, it instructs the US Department of Agriculture (USDA) to issue regulations necessary to implement the federal disclosure standard by 29 July, 2018.

The law was passed in part, to preempt Vermont's Act 120 (9 V.S.A. §§ 3041, 3048) which would have required manufacturers of foods containing GM ingredients to "label the package offered for retail sale, with clear and conspicuous words 'produced with genetic engineering" (General Assembly of the State of Vermont 2014) effective 1 July, 2016. The penalties for violating the law included daily fines of $1,000 for each item found on Vermont shelves that did not have the mandated label. Moreover, the law specified that the manufacturer would be fined even if it was not responsible for the presence of the product in the store. It was also enacted to a set of federal labeling standards that would preempt efforts by other states to pass their own individual labeling laws, which the food industry was concerned would lead to inconsistent and contradictory sets of regulations across States (Plumer 2016).

The NBFDS was passed as an amendment to the Agricultural Marketing Act of 1946 (7 U.S.C. 1621 et seq.). In doing so, responsibility for promulgating and enforcing GM food labeling regulations was placed within the USDA rather than the FDA, signaling that GM labeling is an issue of marketing and not of public health. It also gives the Secretary of Agriculture broad discretion in setting the standards for GM food labeling.

In particular, while the word "bioengineered" is used within the text of the NBFDS to refer to GM foods, ingredients and organisms, the law gives the Secretary of Agriculture the discretion to choose a different term to be used to label these products. This choice may be consequential, as research has shown that public opinion of GM foods is strongly influenced by what they are called. For example, using a word association task, Hallman et al. 2003 found that the term "biotechnology" evoked more positive first thoughts and images than either "genetic engineering" or "genetic modification", which were perceived more negatively. Similarly, using an experimental manipulation, Zahry and Besley (2017) found that the term 'agbiotech' evoked the most relative support for the technology. However, they also found that the effects

---

[1] Pub. L. 114-216, 130. Stat. 834 (29 July, 2016), codified 7 U.S.C. §§ 1639 et seq.

of framing the technology as genetic engineering may depend on who uses the term. They suggest that those seeking to promote the technology might benefit from using the term "genetic engineering", while those who oppose the technology might benefit from using framing it as "genetic modification".

In its guidance documents, the FDA (1992, 2015) asserts that because it encompasses the broad spectrum of genetic alterations that can be made in plants through a variety of techniques, including conventional crossbreeding, use of the term "genetic modification" is inaccurate. Instead, it prefers that manufacturers use the terms "modern biotechnology", "bioengineering", or "genetic engineering" (GE), to refer to the technology, and argues against using either "genetically modified" (GM) or "genetically modified organism" (GMO) to refer to foods derived from genetically engineered plants.

Consistent with this, Vermont's mandatory labeling law (Act 120) which the NBFDS superseded, required that manufacturers use the term "genetic engineering" when labeling their products. However, the GMA (2017) supports using the term "bioengineered" in the disclosure text that would be required by the NBFDS to identify food products containing GM ingredients. In contrast, the mandatory labeling requirements of the European Commission requires the use of the term "genetic modification" when labeling these same products (European Union 2003).

In addition to deciding how to refer to the technology, the Secretary of Agriculture has the authority to determine what products will be subject to mandatory labeling by defining what qualifies as a "bioengineered food". Within the text of the NBFDS, a "bioengineered" food is defined as a food "(A) that contains genetic material that has been modified through *in vitro* recombinant deoxyribonucleic acid (DNA) techniques; and (B) for which the modification could not otherwise be obtained through conventional breeding or found in nature" (§291(1) A-B). This definition is consistent with that used by the FDA in its 2015 guidance to industry concerning voluntary labeling of foods derived from genetically engineered plants. However, as has already been discussed, by restricting the definition to include only those organisms modified through *in vitro* recombinant DNA techniques, it excludes from labeling, foods derived from crops with genetic changes brought about through the use of gene silencing, gene editing, and mutation breeding.

In addition, because the definition in Section 291(1) specifies that a bioengineered food "contains" genetic material modified through *in vitro* recombinant DNA techniques, highly refined ingredients (HRIs) such as oils, sugars, syrups, and other products derived from genetically modified plants that are stripped of detectable DNA during processing may not qualify as bioengineered foods, and would therefore not be required to be labeled (Halloran 2016). However, the GMA (2017) has suggested that if HRIs are derived from GM plants, they should be identified as "Ingredients sourced from bioengineered crop(s)".

The NBFDS also specifically prohibits foods obtained from animals that consume genetically modified feed from being considered as bioengineered and it excludes foods served in restaurants from its requirements. It also exempts from labeling, any food product where meat is the predominant ingredient, even if the food contains other bioengineered ingredients (Halloran 2016).

The NBFDS also gives the Secretary of Agriculture authority to determine the amount of bioengineered ingredients in a food that would trigger mandatory labeling. However, determining this "threshold" is likely to prove difficult, as there is no established scientific basis for establishing what proportion of GM content should define a GM food (Weighardt 2006).

The NBFDS specifies that the requirements and procedures necessary to implement the federal disclosure standard should be in place within two years of the enactment of law (i.e., by 29 July, 2018). Controversially, it does not specify an effective date by which manufacturers must make such disclosures, nor does it provide for penalties that may be imposed on manufacturers for failing to disclose GM ingredients.

Policy Implications

This chapter reviews evidence that despite the widespread use of GM ingredients in the US food supply, most Americans have heard little and know little about GM food. Most don't follow news stories about GMOs, have never spoken with anyone about them, and don't realize that they have likely been eating foods with GM ingredients for more than two decades. Yet, many Americans fear that GM foods are unsafe to eat, are unsafe for human health, and are unsafe for the environment.

With the implementation of the NBFDS, there is an opportunity to change all of this. As a result, important decisions must be made by the US Secretary of Agriculture.

The decision concerning what to call foods produced through modern biotechnology may frame these technologies and the crops and foods produced through them in ways that may significantly affect public acceptance. Moreover, the definition of "bioengineered" given within the law may give the false impression that only *in vitro* recombinant DNA techniques should be of concern to consumers.

Decisions about what qualifies as "bioengineered" and must therefore be labeled, and what threshold quantities of bioengineered ingredients will trigger mandatory labeling are also likely to significantly influence consumer views of the extent of GM ingredients used in the US food supply. For example, excluding highly refined oils, syrups, and sugars derived from GM crops would greatly reduce the number of products that would be required to be labeled as containing "bioengineered" ingredients. Similarly, the exclusion of products with traits derived through the use of gene silencing, gene editing and mutation breeding may give consumers the misleading impression that far more food products are made from ingredients derived from "conventionally bred" crops than is actually the case. They might also assume that because a product does not disclose that it has "bioengineered ingredients", that it is, in fact, "Non-GMO" or even "GMO-Free".

Because of the likely consequences of these decisions, researchers from government, academia, and industry must fully evaluate their potential impacts before the regulations and procedures of the NBFDS are promulgated and enforced. To do otherwise risks additional decades of consumer confusion and fears about GM foods.

References

[AAAS] American Association for the Advancement of Science. 2012. Statement by the AAAS board of directors on labeling of genetically modified foods. [2015-12-09]. American Association for the Advancement of Science. <http://www. aaas.org/sites/default/files/AAAS GM statement.pdf> (accessed on 8 November 2017).

Adler, J.H. 2016. There is no consumer right to know. Regulation 39: 26–33.

Ahloowalia, B.S., M. Maluszynski and K. Nichterlein. 2004. Global impact of mutation-derived varieties. Euphytica 135: 187–204. <https://doi.org/10.1023/B:EUPH.0000014914.85465.4f> (accessed on 8 November 2017).

Alesci, C. and P. Gillespie. 2015. Chipotle is now GMO-free. CNN Money. <http://money.cnn. com/2015/04/26/investing/chipotle-gmo-free/> (accessed on 15 November 2017).

Alston, J. and D. Sumner. 2012. Proposition 37—California food labeling initiative: Economic implications for farmers and the food industry if the proposed initiative were adopted. <http://www.noprop37.com/ files/Alston-Sumner-Prop-37-review.pdf/> (accessed on 12 November 2017).

American Medical Association House of Delegates. 2012. Labeling of bioengineered foods. Council on Science and Public Health Report 2. <https://www.ama-assn.org/sites/default/files/media-browser/ public/hod/a12-csaph-reports_0.pdf> (accessed on 9 December 2017).

Bain, C. and T. Dandachi. 2014. Governing GMOs: The (counter) movement for mandatory and voluntary non-GMO labels. Sustainability 6: 9456–9476.

Bain, C. and T. Selfa. 2017. Non-GMO vs. organic labels: purity or process guarantees in a GMO contaminated landscape. Agr. Hum. Val. 34: 1–14.

Bunge, J. and A. Gasparro. 2015. Organic vs. Non-GMO labels. Who's winning? The Wall Street Journal. <https://www.wsj.com/articles/organic-vs-non-gmo-labels-whos-winning-49619118> (accessed on 21 July 2015).

Carter, C.A. and G. Gruère. 2003. Mandatory labeling of genetically modified foods: Does it really provide consumer choice? AgBioForum 6: 68–70.

Carter, C.A., G.P. Gruere, P. McLaughlin and M. MacLachlan. 2012. California's Proposition 37: Effects of mandatory labeling of GM Food. ARE Update 15: 3–8. University of California Giannini Foundation of Agricultural Economics. <https://giannini.ucop.edu/publications/are-update/issues/2012/15/6/ californias-proposition-3/> (accessed on 15 November 2017).

Center for Food Safety. 2017. U.S. Polls on GE Food Labeling. Center for Food Safety. <http://www. centerforfoodsafety.org/issues/976/ge-food-labeling/us-polls-on-ge-food-labeling> (accessed on 15 November 2017).

Charles, D. 2016. Congress just passed a GMO labeling bill. Nobody's super happy about it. National Public Radio (NPR). <http://www.npr.org/sections/thesalt/2016/07/14/486060866/ congress-just-passed-a-gmo-labeling-bill-nobodys-super-happy-about-it> (accessed on 1 October 2017).

Costanigro, M. and J.L. Lusk. 2014. The signaling effect of mandatory labels on genetically engineered food. Food Policy 49: 259–267.

Dennis, B. 2016. FDA bans imports of genetically engineered salmon—for now. Washington Post. January 29. <https://www.washingtonpost.com/news/to-your-health/wp/2016/01/29/fda-bans-imports-of-genetically-engineered-salmon-for-now/?utm_term=.43ef2522d587> (accessed on 12 November 2017).

Dewey, C. 2017. The apple that never browns wants to change your mind about genetically modified foods. *Washington Post*, January 23. <https://www.washingtonpost.com/news/wonk/ wp/2017/01/23/ the-apple-that-never-browns-wants-to-change-your-mind-about-genetically-modified-foods/?utm_ term=.8fbc869c4928> (accessed on 18 Feburary 2017).

Erbentraut, J. 2016. 'Non-GMO' does not mean organic, folks. Huffington Post. <https://www.huffingtonpost. com/entry/non-gmo-vs-organic_us_57a36208e4b0e1aac9150854> (accessed on 12 November 2017).

European Commission. 2010. A decade of EU funded GMO research (2001–2010). <http://ec.europa.eu/ research/biosociety/pdf/a_decade_of_eu-funded_gmo_research.pdf> (accessed on 9 November 2017).

European Union. Genetically modified organisms—traceability and labeling, 1830/2003 § (2003). Retrieved from <http://eur-lex.europa.eu/legal-content/EN/TXT/?uri=LEGISSUM:l21170> (accessed on 9 November 2017).

Fernandez-Cornejo, J., S. Wechsler and D. Milkove. 2016. The adoption of genetically engineered alfalfa, canola, and sugarbeets in the United States, EIB-163, U.S. Department of Agriculture, Economic Research Service. <https://www.ers.usda.gov/publications/pub-details/?pubid=81175> (accessed on 12 November 2017).

[FDA] Food and Drug Administration. 1992. Food for human consumption and animal drugs, feeds, and related products: Foods derived from new plant varieties; policy statement, 22984. Department of Health and Human Services, 57(104). <https://www.fda.gov/Food/ GuidanceRegulation/ GuidanceDocumentsRegulatoryInformation/Biotechnology/ucm096095.htm> (accessed on 3 November 2017).

[FDA] Food and Drug Administration. 2001. Guidance documents & regulatory information by topic— Draft guidance for industry: Voluntary labeling indicating whether food has or has not been derived from genetically engineered Atlantic salmon. <https://www.fda.gov/Food/GuidanceRegulation/ GuidanceDocumentsRegulatoryInformation/ucm469802.htm> (accessed on 3 November 2017).

[FDA] Food and Drug Administration. 2015a. FDA concludes Arctic apples and Innate potatoes are safe for consumption [Press release]. <https://www.fda.gov/NewsEvents/Newsroom/ PressAnnouncements/ ucm439121.htm> (accessed on 3 November 2017).

[FDA] Food and Drug Administration. 2015b. Guidance for industry: Voluntary labeling indicating whether foods have or have not been derived from genetically engineered plants. Center for Food Safety and Applied Nutrition. <https://www.fda.gov/RegulatoryInformation/ Guidances/ucm059098.htm> (accessed on 3 November 2017).

Frewer, L.J., I.A. van der Lans, A.R. Fischer, M.J. Reinders, D. Menozzi, X. Zhang, I. van den Berg and K.L. Zimmermann. 2013. Public perceptions of agri-food applications of genetic modification—a systematic review and meta-analysis. Trends in Food Sci. & Technol. 30: 142–152.

Genetically Engineered Organisms Public Issues Education (GEO-PIE). 2001. Genetically engineered foods: A consumer guide to what's in store. Cornell University. <http://blogs. cornell.edu/gmodialogue/ files/2013/06/GE_Foods-1s8zr15.pdf> (accessed on 9 December 2017).

[GMA] Grocery Manufacturers Association. 2013. Grocery Manufacturers Association launches www. FactsAboutGMOs.org. <http://www.gmaonline.org/news-events/newsroom/grocery-manufacturers-association-launches-wwwfactsaboutgmosorg/> (accessed on 15 July 2017).

[GMA] Grocery Manufacturers Association. 2017. Re: AMS Questions on Bioengineered Food Disclosure Law. Retrieved from <http://www.gmaonline.org/file-manager/GMA_Comments_ on_Bioengineered_ Food_Disclosure_Questions_from_AMS_-_FINAL.pdf> (accessed on 10 November 2017).

Greene, C., W. McBride, B. Cooke, G. Ferreira and H.F. Wells. 2016a. The outlook for organic agriculture. United States Department of Agriculture. <https://www.usda.gov/oce/forum/past_ speeches/2016_ Speeches/Greene.pdf> (accessed on 12 November 2017).

Greene, C., S.J. Wechsler, A. Adalja and J. Hanson. 2016b. Economic issues in the coexistence of organic, genetically engineered (GE), and Non-GE Crops (No. 232929). United States Department of Agriculture, Economic Research Service.

Hallanbeck, T. 2014. Ben and Jerry's says goodbye to GMOs. USA Today. <https://www.usatoday.com/ story/money/business/2014/06/15/ben-and-jerrys-says-goodbye-to-gmos/10542275/> (accessed on 20 November 2017).

Hallman, W.K. 1995. Public perceptions of Agri-biotechnology. Genetic Engineering News 15(13): 4–5. <http://dx.doi.org/doi:10.7282/T3VH5RBH> (accessed on 20 November 2017).

Hallman, W.K. 2016. Do American consumers want GM food labeling? It depends on how you ask the question. pp. 173–179. In: Thompson, G.A., S.E. Lipari and R.W.F. Hardy (eds.). NABC Report 27: Stewardship for the Sustainability of Genetically Engineered Crops: The Way Forward in Pest Management, Coexistence, and Trade. Proceedings of the 27th Annual Conference of the North American Agricultural Biotechnology Council, Hosted by the Pennsylvania State University, June 2 and 3, 2015. Ithaca, NY: North American Agricultural Biotechnology Council. LCCN: 2015948991. <http://nabc.cals.cornell.edu/Publications/Reports/nabc_27/NABC27Report.pdf> (accessed on 21 November 2017).

Hallman, W.K., A.O. Adelaja, B.J. Schilling and J. Lang. 2002. Public perceptions of genetically modified foods: Americans know not what they eat (Food Policy Institute Report No. RR-0302-001). New Brunswick, New Jersey: Rutgers University, Food Policy Institute. <http:// dx.doi.org/doi:10.7282/ T3VD71RX> (accessed on 21 November 2017).

Hallman, W.K. and H.L. Aquino. 2005. Consumers desire for GM labels: Is the devil in the details? Choices 20: 217–222.

Hallman, W.K., C.L. Cuite and X.K. Morin. 2013. Public perceptions of labeling genetically modified foods. Working Paper 2013-01. New Brunswick, New Jersey: Rutgers, The State University of New Jersey, New Jersey Agricultural Experiment Station. <http://dx.doi.org/ doi:10.7282/T33N255N> (accessed on 21 November 2017).

Hallman, W.K., W.C. Hebden, H.L. Aquino, C.L. Cuite and J.T. Lang. 2003. Public perceptions of genetically modified foods: A national study of American knowledge and opinion (Food Policy Institute Report No. RR-1003-004). New Brunswick, New Jersey: Rutgers University, Food Policy Institute. <http://dx.doi.org/doi:10.7282/T37M0B7R> (accessed on 20 November 2017).

Hallman, W.K., W.C. Hebden, C.L. Cuite, H.L. Aquino and J.T. Lang. 2004. Americans and GM food: Knowledge, opinion & interest in 2004 (Food Policy Institute Report No. RR-1104-007). New Brunswick, New Jersey: Rutgers University, Food Policy Institute. <http://dx.doi.org/ doi:10.7282/ T3KW5JFP> (accessed on 20 November 2017).

Halloran, J. 2016. Consumers Union letter to US Senate in opposition of S.764 to preempt state GMO labeling laws. <http://consumersunion.org/research/consumers-union-letter-to-us-senate-in-opposition-of-s-764-to-preempt-state-gmo-labeling-laws/> (accessed on 21 November 2017).

Hatanaka, M., C. Bain and L. Busch. 2005. Third-party certification in the global agrifood system. Food Policy 30: 354–369.

Hemphill, T.A. and S. Banerjee. 2014. Mandatory food labeling for GMOs. Regulation 37: 7–10.

Hilbeck, A., R. Binimelis, N. Defarge, R. Steinbrecher, A. Székács, F. Wickson, M. Antoniou, P.L. Bereano, E.A. Clark, M. Hansen and E. Novotny. 2015. No scientific consensus on GMO safety. Environ. Sci. Eur. 27: 4.

ISAAA. 2016. Global status of commercialized biotech/GM crops: 2016. ISAAA Brief No. 52. ISAAA: Ithaca, NY. <http://www.isaaa.org/resources/publications/pocketk/16/> (accessed on 19 November 2017).

James, C. 2014. Global status of commercialized biotech/GM crops: 2014. ISAAA Brief No. 49. ISAAA: Ithaca, NY. <http://www.isaaa.org/resources/publications/briefs/49/default.asp> (accessed on 19 November 2017).

Johnson, N. 2015. It's practically impossible to define "GMOs". Grist. <http://grist.org/food/ mind-bomb-its-practically-impossible-to-define-gmos/> (accessed on 22 November 2017).

Joint FAO/IAEA Division of Nuclear Techniques in Food and Agriculture, Plant Breeding and Genetics Subprogramme. 2017. Plant Breeding and Genetics. <http://www-naweb.iaea.org/ nafa/pbg/> (accessed on 22 November 2017).

Joint FAO/WHO Codex Alimentarius Commission. 2009. Codex Alimentarius: Foods derived from modern biotechnology. World Health Organization. <http://www.fao.org/docrep/011/ a1554e/a1554e00.htm> (accessed on 22 November 2017).

Keller, S. 2017. Insider insight to Non-GMO claim market trends. <http://foodsafety. merieuxnutrisciences. com/2017/03/07/insight-non-gmo-claim-market-trends/> (accessed on 5 May 2017).

Kopicki, A. 2013. Strong support for labeling modified foods. New York Times. <http://www. nytimes. com/2013/07/28/science/strong-support-for-labeling-modified-foods.html> (accessed on 5 May 2017).

Krimsky, S. 2015. An illusory consensus behind GMO health assessment. Sci. Technol. Hum. Values 40: 883–914.

Lazarides, H.N. and A.M. Goula. 2018. Sustainability and ethics along the food supply chain. pp. 41–61. *In*: Costa, P.P.R. (ed.). Food Ethics Education, Integrating Food Science and Engineering Knowledge into the Food Chain, Springer International Publishing.

Leary, W.E. 1994. FDA approves altered tomato that will remain fresh longer. New York Times. <http://www.nytimes.com/1994/05/19/us/fda-approves-altered-tomato-that-will-remain-fresh-longer.html> (accessed on 12 November 2017).

Lusk, J.L. and A. Rozan. 2008. Public policy and endogenous beliefs: the case of genetically modified food. J. Agr. Resour. Econ. 33: 270–289.

Martineau, B. 2001. First fruit: The creation of the Flavr Savr tomato and the birth of biotech foods. New York: McGraw-Hill.

McFadden, B.R. 2017. The unknowns and possible implications of mandatory labeling. Trends Biotechnol. 35: 1–3.

McFadden, B.R. and J.L. Lusk. 2016. What consumers don't know about genetically modified food, and how that affects beliefs. The FASEB J. 30: 3091–3096.

McFadden, B.R. and J.L. Lusk. 2017. Effects of the National Bioengineered Food Disclosure Standard: Willingness to pay for labels that communicate the presence or absence of genetic modification. Applied Economic Perspectives and Policy. <https://doi.org/10.1093/aepp/ ppx040> (accessed on 3 November 2017).

Mellon, M. 2013. Gene silencing: New products and new risks. Union of Concerned Scientists. <http://blog.ucsusa.org/margaret-mellon/gene-silencing-new-products-and-new-risks-147> (accessed on 9 November 2017).

National Academies of Sciences, Engineering, and Medicine. 2016. Genetically Engineered Crops: Experiences and Prospects. Washington, DC: The National Academies Press.

Non-GMO Project. 2017. Verification FAQs – The Non-GMO Project. <https://www. nongmoproject.org/product-verification/verification-faqs/> (accessed on 19 November 2017).

O'Neil, C. 2016. Top 10 reasons to oppose the Senate DARK act. AgMag. <http://www.ewg.org/agmag/2016/02/top-10-reasons-oppose-senate-dark-act> (accessed on 20 July 2017).

Pew Research Center. 2015. Americans, politics and science issues. <http://assets.pewresearch. org/wp-content/uploads/sites/14/2015/07/2015-07-01_science-and-politics_FINAL-1.pdf> (accessed on 19 November 2017).

Pew Research Center. 2016a. The new food fights: U.S. public divides over food science. <http://assets. pewresearch.org/wp-content/uploads/sites/14/2016/12/19170147/PS_ 2016.12.01_Food-Science_ FINAL.pdf> (accessed on 19 November 2017).

Phillips, P.W.B. and I. Grant. 1998. GMO labeling: Threat or opportunity? AgBioForum 1: 25–30.

Phillips, D.M. and W.K. Hallman. 2013. Consumer risk perceptions and marketing strategy: the case of genetically modified food. Psychol. Market 30: 739–748.

Plumer, B. 2016. The controversial GMO labeling bill that just passed Congress, explained. Vox. <http://www.vox.com/2016/7/7/12111346/gmo-labeling-bill-congress> (accessed on 2 November 2017).

Polis, C. 2013. Whole Foods GMO labeling to be mandatory by 2018. The Huffington Post. <http://www.huffingtonpost.com/2013/03/08/whole-foods-gmo-labeling2018_n_2837754. html> (accessed on 12 November 2017).

Pollack, A. 2015. Genetically engineered salmon approved for consumption. New York Times. <https://www.nytimes.com/2015/11/20/business/genetically-engineered-salmon-approved-for-consumption. html> (accessed on 15 November 2017).

Pollan, M. 2012. Vote for the Dinner Party. New York Times Magazine. <http://www. nytimes. com/2012/10/14/magazine/why-californias-proposition-37-should-matter-to-anyone-who-cares-about-food.html> (accessed on 22 November 2017).

Roff, R.J. 2009. No alternative? The politics and history of non-GMO certification. Agr. Hum. Val. 26: 351–363.

Román, S., L.M. Sánchez-Siles and M. Siegrist. 2017. The importance of food naturalness for consumers: Results of a systematic review. Trends in Food Sci. Technol. 67: 44–57.

Runge, C.F. and L.A. Jackson. 2000. Labelling, trade, and genetically modified organisms—A proposed solution. J. World Trade. 34: 111–123.

Runge, K.K., D. Brossard, D.A. Scheufele, K.M. Rose and B.J. Larson. 2017. Attitudes about food and food-related biotechnology. Public Opin. Quart. 81: 577–596.

Sand, P.H. 2006. Labelling genetically modified food: The right to know. Rev. Eur. Comp. Int. Environ. Law. 15: 185–192.

Scott, S.E., Y. Inbar and P. Rozin. 2016. Evidence for absolute moral opposition to genetically modified food in the United States. Perspect. Psychol. Sci. 11: 315–324.

Senapathy, K. 2017. The Non-GMO Project is ruining my shopping experience. <https://www. forbes.com/sites/kavinsenapathy/2017/05/31/the-non-gmo-project-is-ruining-my-shopping-experience/#515469881a60> (accessed on 14 November 2017).

Siegrist, M., C. Hartmann and B. Sütterlin. 2016. Biased perception about gene technology: How perceived naturalness and affect distort benefit perception. Appetite 96: 509–516.

Snell, C., A. Bernheim, J.-B. Berge, M. Kuntz, G. Pascal, A. Paris and A.E. Ricroch. 2012. Assessment of the health impact of GM plant diets in long-term and multigenerational animal feeding trials: a literature review. Food Chem. Toxicol. 50: 1134–1148.

Splitter, J. 2017. Whole Foods would look a lot different if it were science-based. The Cut. <https://www. thecut.com/article/a-science-based-whole-foods-would-look-a-lot-different. html> (accessed on 8 November 2017).

Sustein, C.R. 2016. On mandatory labeling, with special reference to genetically modified foods. Univ. Penn. Law Rev. 165: 1043–1095.

Thompson, P.B. 1997. Food biotechnology's challenge to cultural integrity and individual consent. Hastings Center Report 27: 34–38.

[USDA-AMS] U.S. Department of Agriculture, Agricultural Marketing Services. 2013. Can GMOs be used in organic products? <https://www.ams.usda.gov/publications/content/can-gmos-be-used-organic-products> (accessed on 17 October 2017).

[USDA-AMS] U.S. Department of Agriculture, Agricultural Marketing Service. 2015. Process Verified Programs, Official Listing of Approved USDA Process Verified Programs. <http://www.ams.usda.gov/services/auditing/process-verified-programs> (accessed on 17 October 2017).

[USDA-APHIS] U.S. Department of Agriculture, Animal and Plant Health Inspection Service. 2012. Questions and answers: Okanagan Specialty Fruits' Non-browning Apple (Events GD743 and GS784). <http://www.aphis.usda.gov/publications/biotechnology/2012/faq_okanagan_ apple.pdf> (accessed on 19 November 2017).

[USDA-APHIS] U.S. Department of Agriculture, Animal and Plant Health Inspection Service. 2017. Petitions for Determination of Nonregulated Status. <https://www.aphis.usda.gov/aphis/ ourfocus/ biotechnology/permits-notifications-petitions/petitions/petition-status> (accessed on 26 November 2017).

[USDA-ERS] U.S. Department of Agriculture (USDA), Economic Research Service. 2016a. Recent trends in GE adoption. <https://www.ers.usda.gov/data-products/adoption-of-genetically-engineered-crops-in-the us/recent-trends-in-ge-adoption.aspx> (accessed on 26 November 2017).

[USDA-ERS] U.S. Department of Agriculture (USDA), Economic Research Service. 2016b. Sugars & sweeteners, background. <https://www.ers.usda.gov/topics/crops/sugar-sweeteners/ background/> (accessed on 26 November 2017).

Vecchiarelli, S.N. 2012. Mandatory labeling of genetically engineered food: constitutionally, you do not have a right to know. San Joaquin Agr. Law Rev. 22: 215–239.

Vincent, J. 2017. The first GMO non-browning apples will go on sale in the US next month. The Verge. <https://www.theverge.com/2017/1/19/14321944/gmo-non-browning-apples-on-sale-us-arctic> (accessed on 25 November 2017).

Waltz, E. 2016a. Gene-edited CRISPR mushroom escapes US regulation. Nature 532: 293.

Waltz, E. 2016b. CRISPR-edited crops free to enter market, skip regulation. Nature Biotechnol. 34: 582.

Watson, E. 2016. Mintel GNPD data: 15.7% of new US food/bev products made non-GMO claims in 2015 vs. 2.8% in 2012. <http://www.foodnavigator-usa.com/Markets/Mintel-GNPD-label-claims-trends-Non-GMO-vegan-all-natural> (accessed on 11 November 2017).

Weighardt, F. 2006. European GMO labeling thresholds impractical and unscientific. Nature Biotechnol. 24: 23–25.

Wharton, C. 2011. Food beyond nutrition: bringing politics and ethics into nutrition curriculum. Teaching Ethics 11(2): 15–24.

World Health Organization. 2014. Frequently asked questions on genetically modified foods. <http://www.who.int/foodsafety/areas_work/food-technology/faq-genetically-modified-food/en/> (accessed on 25 November 2017).

Zahry, N.R. and J.C. Besley. 2017. Genetic engineering, genetic modification, or agricultural biotechnology: does the term matter? Journal of Risk Research. <http://www.tandfonline.com/ doi/full/10.1080/13 669877.2017.1351470> (accessed on 25 November 2017).

5

Consumer Concerns about Radioactive Contamination

Empirical Analysis of the Vegetable Wholesale Market in Kanto Region

Shigeru Matsumoto[1,*] and *Masashi Yamamoto*[2]

Introduction

Social science research on food safety in the past few decades has revealed that consumers are strongly concerned about food safety. In addition, previous research has also confirmed that typical consumers have very poor knowledge about food safety management.

When hearing about a food contamination incident, consumers who do not know the details are often gripped by panic. Their confusion gets compounded when scary rumors are spread. Even if government agencies inspect the foods concerned and publicize safety messages, consumers are unlikely to go back to their previous purchasing patterns. In terms of the complexity of health effects to consumers and the magnitude of damages to agricultural production, the radioactive contamination caused by the nuclear accident of Fukushima, Japan, can be considered one of the most serious food contamination incidents in history.

Scientists agree that consuming food contaminated with radioactive material increases the amount of radioactivity a person is exposed to and increases his/her cancer risk. However, there is great diversity in the opinions regarding the magnitude

---

[1] Department of Economics, Aoyama Gakuin University, Room 828, Building 8, 4-4-25 Shibuya, Shibuya, Tokyo, Japan, 150-8366.

[2] Center for Far Eastern Studies, University of Toyama, Gofuku 3190, Toyama, Toyama, Japan, 930-8555.
  E-mail: myam@eco.u-toyama.ac.jp

* Corresponding author: shmatsumoto@aoyamagakuin.jp

of the increased health risk. Some scientists argue that the impact of Cesium-137, which is the main radioactive contaminant due to its relatively long half-life (over 30 years), is relatively minor when taking into account the fact that people ingest Potassium-40, which is a radioactive isotope of potassium, in their daily life (Hayashi and Midorikawa 2013). Other scientists claim that such an argument oversimplifies the health-risk analysis by ignoring the possibility that Cesium-137 bio-accumulates in muscle tissues and other organs (Heley 2014).[1] A consensus regarding the health risk from internal exposure is less likely to be reached among scientists in the near future.

The nuclear accident has also caused devastating damage to the agricultural production in Fukushima and neighboring prefectures. Japanese authorities implemented a 20 km exclusion zone and a 20–30 km indoor standby zone around the nuclear power plant immediately after the nuclear accident. Agricultural production inside these zones has become impossible since then. According to the calculations by the Ministry of Agriculture, Forestry, and Fisheries (MAFF 2011), more than 20,000 hectares of agricultural land were abandoned and 13,745 cows, 44,340 pigs, and 1.9 million chickens were slaughtered.

After the detection of food contamination, shipments of agricultural products produced in the affected prefectures were suspended. According to the estimation by MAFF (2011), the total loss of vegetable production exceeded 40 billion yen in 2011. The Japanese government ordered the Tokyo Electric Power Company (TEPCO) to pay compensation for damages. As of 31 May 2015, the amount of compensation demanded from TEPCO reached 190.6 billion yen of which TEPCO had already paid 77% of the requested amount.[2]

Scholars have investigated the effect of radioactive contamination on consumer valuations of food. Some conducted surveys soon after the Fukushima nuclear accident to determine consumer reactions toward the radioactive contamination. They estimated the consumers' willingness to pay (WTP) in order to avoid food contaminated by radioactive material or food produced in the area affected by the nuclear accident. Their analyses showed that typical consumers pay non-negligible amounts of money to avoid risky food.

The previous survey studies predicted that consumer would avoid agricultural products produced in the area affected by the nuclear accident. In a survey by MAFF (2012a), about 30.9% of food industry personnel answered that they would switch from purchasing agriculture, forestry, and fisheries products from areas with radioactive contamination to purchasing them from the non-affected areas. Although it was expected that the nuclear accident had influenced consumer purchasing behavior, scholars have not yet examined how the market share of the production regions had changed after the nuclear accident. In this study, we focus on vegetables and examine how the nuclear accident changed consumer valuations across production regions.

The structure of this chapter is as follows. Section 2 provides an overview of the Fukushima nuclear accident and examines its impact on the Japanese agricultural market. Section 3 reviews the findings of previous studies that examined the effects

---

[1] Fuse (2012), head of the Kusugi Medical Clinic, stated at a New York City press conference of physicians that the risk from internal exposure is 200–600 times greater than the risk from external exposure.

[2] The sales loss caused by harmful rumors was also compensated for.

of the nuclear accident on consumer food selection. We use the data from the Kanto wholesale market for our empirical study, which covers the agricultural sales in Tokyo and surrounding large cities. Section 4 summarizes the data characteristics. We employ the Almost Ideal Demand System (AIDS) model in order to evaluate the impact of the nuclear accident on consumer food selection. Section 5 reports the empirical results. Our results suggest that (1) consumers shifted buying vegetables from the affected areas to the non-affected areas when the vegetables in the non-affected areas were affordable, and (2) vegetable demand became more price sensitive after the nuclear accident. Section 6 draws several policy implications and concludes the chapter.

Background Information

Radioactive contamination by the Fukushima nuclear plant accident

On 11 March 2011, an earthquake of magnitude 9.0 occurred in the Pacific Ocean off the coast of Japan's Tohoku region. The damage from this earthquake was tremendous; the National Police Agency (2016) reported that 15,893 persons had died and 2,556 persons were missing, while the Cabinet Office of Japan (2011) estimated that the economic damage was about 16.9 trillion yen, which corresponded to approximately 3.4% of Japan's GDP.

The massive tsunami followed by the earthquake disabled the power supply and cooling system of the Fukushima Daiichi nuclear plant. Three of the six reactors at the plant were severely damaged and they spread hydrogen and radioactive materials mainly in the Tohoku and Northern Kanto regions.[3] The geographical locations of these two regions and the Fukushima Daiichi plant are shown in Fig. 1.

Since the nuclear accident, the Japanese government has used airplanes to measure air radiation dose rates throughout the country. According to the 2014 monitoring results released by the Japan Atomic Energy Agency (2014), only the air dose rates of Fukushima, Tochigi, and Gunma prefectures exceeded 0.1 μ Sv per hour, which represents the air dose rates typically observed in the major cities in North America and Europe. Hence, external exposure is expected to be relatively minor outside the heavily contaminated areas in Fukushima Prefecture, so much more attention has been devoted to the internal exposure through food intake.[4]

In March 2011, immediately after the nuclear accident, the Ministry of Health, Labor, and Welfare (MHLW), Japan set provisional regulation values for radioactive

---

[3] As of May 2012, the Tokyo Electric Power Company (TEPCO) (2012) estimated that the amount of radioactive materials released into the atmosphere was as follows: Noble gases, 5×10^{17} Bq; Iodine-131, 5×10^{17} Bq; Cesium-134, 1×10^{16} Bq; and Cesium-137, 1×10^{16} Bq. The Nuclear and Industrial Safety Agency estimates that the release of radioactive material to the atmosphere represented approximately 10% of the Chernobyl accident, which was the only other accident to reach a Level 7 rating under the International Nuclear and Radiological Event Scale (International Atomic Energy Agency 2012). A massive amount of radioactive material was also washed-out to the ocean.

[4] Soil contamination remains a very serious problem, especially in Fukushima prefecture. It is expected that the cost of decontamination efforts will exceed $42.75 billion (Hotaka and Naito 2013). We assume that the conversion rate is 1 USD to 120 yen throughout this chapter.

Fig. 1. Geographical locations.

substances in food to ensure food safety. Although the provisional regulation was stricter than the ones in Codex,[5] the EU, and the US, the Japanese people still expressed concerns about radiation safety standards. Consequently, the Ministry decided to strengthen the regulation values in April 2012 (see Table 1).

Soon after the accident, incidents of radioactive contamination were reported in Fukushima and its neighboring prefectures. Following the detection of radioactive substances at levels above the provisional limits, MHLW ordered Fukushima, Ibaraki,

---

[5] It is the global reference point for consumers, food producers and processors, national food control agencies and the international food trade.

Table 1. Regulation levels of radioactive cesium (Bq/kg).

	Drinking water	Milk	General food	Baby food
Japan				
From April 2012	10	50	100	50
Provisional	200	200	500	200
US (Compliance Policy Guide Sec. 560.750)	1200	1200	1200	1200
EU (Euratom: No. 3954/87)	1000	1000	1000	400
CODEX (STAN 193-1995)	1000	1000	1000	1000

Source: MHLW (2012).

Tochigi, and Gunma prefectures to suspend shipments of spinach and kakina[6] on 21 March 2011.[7] MHLW also ordered Fukushima Prefecture to suspend shipping fresh raw milk for a similar reason. In summer of 2011, beef contamination caused by feeding polluted rice straws to cattle was discovered in Fukushima, Miyagi, Iwate, and Tochigi prefectures and MHLW banned the shipment of beef originating from those areas (Hatsuzawa and Takano 2016). Following the detection of radioactive contamination that exceeded the provisional limit, the government banned shipments of rice harvested in the Onami district of the city of Fukushima on 17 November 2011.

To address consumer safety concerns, the Japanese government established an extensive monitoring system. Prefectures have been inspecting the food items that are likely to contain radioactive cesium at a predetermined frequency and the central government has been publicizing the test results. The test results revealed that the radioactive levels of most food dropped very quickly. Although the proportion of foods produced in Fukushima Prefecture that exceeded the regulation limit was 3.3% in 2011, it dropped to 0.6% in 2014 (Gibney 2015). Outside the heavily contaminated areas in Fukushima Prefecture, only edible wild plants, game meat, fish, and wild mushrooms exceeded the regulation limits after 2014.

Impact of the nuclear accident on agricultural production

The Tohoku and Northern Kanto regions, which were severely damaged by the nuclear accident, are major agricultural production areas in Japan. In 2010, immediately before the nuclear accident, these two regions accounted for 27.2% of the total cropland in Japan (MAFF 2010a). Approximately 25.0% of Japanese farm households lived in these two regions (MAFF 2010b) and they represented 26.2% of the country's total shipments (MAFF 2010c). Figure 2 presents the change in agricultural production value of four prefectures, where a series of radioactive contamination incidents were

---

[6] Kakina is a plant of the Brassicaceae family.
[7] MHLW later expanded the shipping suspension to other vegetables grown throughout the entire Fukushima Prefecture, including non-head-type leafy vegetables, head-type leafy vegetables (such as cabbage, iceberg lettuce, etc.), flower-head brassicas (such as broccoli, cauliflower, etc.), and turnips.

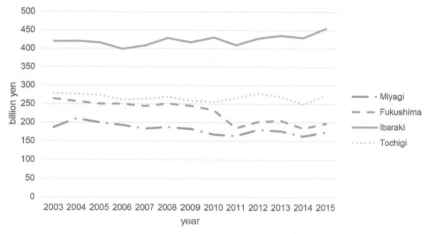

Fig. 2. Change in agricultural production values.

reported. It shows that the agricultural production value dropped in three prefectures in 2011. The largest drop was observed in Fukushima Prefecture, where the production value decreased by 20.6%.

Several years after the accident, the agricultural production values of Miyagi, Ibaraki, and Tochigi prefectures returned to the levels prior to the accident. However, the agricultural production value of Fukushima Prefecture remained relatively low. The acreages of major agricultural products decreased at the time of the nuclear accident. For instance, the area of cucumbers planted in Fukushima Prefecture decreased from 887 ha in 2010 to 761 ha in 2012, while that for tomatoes decreased from 473 ha in 2010 to 398 ha in 2012 (MAFF 2012b). The prices of major agricultural products also decreased during this period. At the Tokyo Metropolitan Central Wholesale Market (TMCWM), the average price of cucumbers produced in Fukushima Prefecture decreased from 266 yen/kg in 2010 to 188 yen/kg, while the price of tomatoes decreased from 368 yen/kg to 294 yen/kg (TMCWM 2012). Both acreage and price reductions led to the decrease in agricultural production values.

Related Studies

Studies outside Japan

The worst nuclear accident in history occurred in 1986 at the Chernobyl Power Plant in Ukraine. Few studies have examined the impact of the Chernobyl accident on agricultural supply and demand.

Hanemann et al. (1992) examined the impact of the nuclear accident on the behavior of moose hunters in Sweden. They conducted two contingent valuation surveys in 1986 and 1987. The 1986 survey focused on moose hunting in the 1985 season (i.e., before the nuclear accident), while the 1987 survey focused on moose hunting in the 1986 season (i.e., after the nuclear accident). They compared the WTP for moose hunting before and after the nuclear accident, and found that the average WTP dropped by more than $200 due to the Chernobyl accident. Based on the estimation results, they estimated that Swedish moose hunters suffered a loss of almost $5 million in the 1986 hunting season.

Bostedt (2001) examined the effect of the nuclear accident on reindeer husbandry in the Sami, Sweden. Considering the fact that reindeer husbandry by the Sami people is operated by family members and most herders only get part of their incomes from reindeer husbandry, he constructed a reindeer supply function based on the time allocation model. By analyzing the data from 1973–74 to 1995–96, he showed that the reindeer supply function has a backward-sloping shape. He also showed that reindeer demand sharply dropped after the Chernobyl accident.

Stated preference analysis

Monma (2013) reviewed the studies of agricultural and food consumption that evaluated the impacts of the Fukushima nuclear accident. He reported that the previous studies had focused on two fields: (1) the development of a radiation monitoring system and decontamination, and (2) the study of the radioactive contamination of agricultural products, consumer behavior, and harmful rumors.

Several scholars have used the stated preference methods to identify consumer concerns about the radioactive contamination of food. Some scholars conducted a series of online surveys soon after the nuclear accident to examine how consumer concerns about radioactive contamination have changed over time.

Hangui (2013) conducted an online survey of 3030 married women in Fukushima Prefecture, the Tokyo metropolitan area, and the Kansai (western) area in Japan. He focused on cucumbers, which are the most popular vegetable produced in Fukushima Prefecture. First, he investigated the acceptance of cucumbers from Fukushima using a five-point Likert scale and found that the acceptance declined as the distance from Fukushima Prefecture increases. The highest score was recorded among Fukushima residents and the lowest score was recorded among Kansai residents. Second, he divided the subjects into two groups to examine whether consumer acceptance could be altered through the provision of scientific information. The experimental results demonstrated that consumer acceptance improved after people viewed a slide show about the radioactive material in food. Finally, he conducted a choice experiment on cucumbers from Fukushima, Gunma, and Miyagi prefectures for the same subjects. He confirmed that the subjects chose cucumbers in the following order—Miyagi Prefecture, Gunma Prefecture, and Fukushima Prefecture.

Yamane (2013) conducted a mail survey of 12,000 households living in eight major Japanese cities from 9 December 2011 to 26 December 2011.[8] He conducted a beef choice experiment that included such food attributes as cesium concentration, inspection method (self-inspection at the grocery store and professional inspections by importers or local slaughterhouses), production area (four prefectures whose shipments had been restricted, including the US and Australia) and price. He found that consumers avoided beef produced in the prefectures affected by the nuclear accident, even when the cesium concentration levels of the beef were controlled for. This result is important. Individual farmers in Fukushima Prefecture can tackle the radiation

---

[8] The response rate was 7.36%.

problem by strengthening the measure for radiation contamination. In contrast, they cannot change the consumer reputation of their production area by themselves. Yamane also found that households with a pregnant woman or purchasing cancer insurance tend to avoid such beef more strictly.

Ujiie (2013) conducted a series of eight online surveys of 1000 married women living in the Tokyo and Osaka metropolitan areas from March 2011 to February 2013. He used the contingent valuation method (CVM) to estimate the willingness to accept (WTA) for spinach from Fukushima and Ibaraki prefectures.[9] He found that consumer concerns remained very high, even in the last survey conducted in February 2013; 42.3% of subjects in the Tokyo metropolitan area and 51.4% of subjects in the Osaka metropolitan area answered that they would not purchase Fukushima spinach, regardless of the price. In the empirical analysis, he divided the WTA into two parts: (1) health risk aversion, and (2) origin evaluation. He found that the health risk aversion was similar in both the Tokyo and Osaka areas, but the origin evaluation varied between the two areas.

Ito and Kuriyama (2017) conducted a series of three choice experimental surveys from June 2011 to February 2012. The number of subjects in each online survey was about 500 and they were recruited from the Kanto (east) and Kansai (west) regions in Japan. The authors focused on rice, spinach, beef, and Pacific saury, and presented their subjects with four hypothetical foods that varied in several aspects, including price, radiation exposure, and production area. They then asked the subjects to choose their most preferred among four alternatives in order to estimate the conditional means of the marginal willingness to pay (MWTP) for reducing radiation exposure. They found that the average consumer spent money to reduce radiation exposure, but there was significant heterogeneity among consumers.[10] Unlike other studies, Ito and Kuriyama (2017) did not observe a systematic difference in the MWTP between survey areas. Lastly, they found that the characteristics of changes in averting behavior during the survey period varied, depending on the type of food.

Aruga (2016) conducted an online survey from 30 January 2014 to 4 February 2014. A total of 6945 sample responses were collected from all 47 prefectures in Japan. He used a CVM to estimate the WTA for the agricultural products (rice, apples, cucumbers, beef, pork, eggs, and shiitake mushrooms) from the regions near the nuclear plant. He observed a larger WTA among the subjects who lived at a greater distance from the Fukushima Daiichi nuclear plant and had children under the age of fifteen. In contrast, he observed a smaller WTA among the older subjects who were knowledgeable about radiation and radioactive materials.

The empirical findings from the above-mentioned survey studies can be summarized as follows:

1. Although a typical consumer spends money to lower the radiation risk, there is significant heterogeneity in averting behavior across consumers.

---

[9] Shipments of spinach were banned after the nuclear accident. In his preliminary study, Ujiie (2012) also studied raw milk.

[10] In a preliminary study, Kuriyama (2012) estimated that the WTP for the radiation risk reduction associated with rice consumption was 35 yen per μSv.

2. A consumer expressing concern about one specific food also expresses concern about other foods.
3. Consumers living farther from Fukushima Prefecture have a more negative perception of food from the affected areas.
4. Even after controlling for radiation dosages, consumers pay attention to the production region.
5. Consumer acceptance of the food produced in the affected area has been gradually improving.

Although these findings are important for understanding consumer valuations about radioactive contamination, there are at least two limitations of the previous survey studies. First, the survey questions do not directly address the actual regulations of radioactive contamination. Although the scholars asked the subjects to compare foods with different contamination levels, the government cannot directly control the contamination levels of food. Rather, the government determines the regulation value and bans the entire shipment once it detects that the radiation level of the food samples exceeds that value. Second, and more importantly, all surveys were conducted after the nuclear accident. Hence, we cannot compare consumer valuations before and after the nuclear accident, and we cannot measure the impact of the nuclear accident on consumer food selections.

Market Data Analysis

Tajima et al. (2016) is the only study that has formally analyzed the market data to quantify the impact of the nuclear accident. They conducted a hedonic regression analysis to assess the impact of the nuclear accident on fresh vegetable prices. They analyzed the monthly price data of asparagus, bean sprouts, broccoli, cucumbers, green beans, and tomatoes obtained from the Tokyo Metropolitan Central Wholesale Market. The authors found that the prices of vegetables grown in Fukushima Prefecture decreased by 10–36% after the nuclear accident and such price reductions persisted for three years after the accident.

One of the strongest assumptions in their hedonic study is that price differences arise entirely from the change in consumer preferences. However, there are many other factors that had changed before and after the nuclear accident. Some of those factors are likely to affect the food price. In this study, we compare the market shares of agricultural products before and after the nuclear accident, and then evaluate both the impact of the price change and the impact of the change in consumer preferences.

Data

The objective of this paper is to test whether the Fukushima nuclear power plant accident affected consumer preferences based on a market data analysis. We use the wholesale market data of vegetables, which are widely available to the public.

Monthly data of Japan's major wholesale markets are available from MAFF's website and include production area information.[11] We focus on the Kanto Wholesale Market, which covers almost the entire greater Tokyo region, and collect the wholesale vegetable price data in the Kanto region from 2003 to 2015. The Tokyo Metropolitan Central Wholesale Market is the main share of the Kanto Wholesale Market. The Kanto Wholesale Market covers 26% of Japan's wholesale transaction volume and 29% of the country's transaction value.

Among 16 kinds of publicly available vegetable data, we choose eggplant and tomatoes for our analysis. We chose these two vegetables because Fukushima Prefecture maintains a particular level of market share in the Kanto Wholesale Market.[12]

We used Deaton and Muellbauer's famous AIDS model for our empirical analysis. Although the AIDS model analyze the market share of substitutes, the number of parameters needed to be estimated increases exponentially as the number of substitutes increases. To reduce the number of parameters, we categorized all of the 47 prefectures into four groups and examined the change in market share among these four groups. The four groups are composed of two treatment (affected) prefectures, one control (non-affected) prefecture, and the remaining 44 prefectures. The combinations of the treatment/control prefectures for the two vegetables are as follows.

- Eggplant: Fukushima and Ibaraki (treatment), Kochi (control)
- Tomato: Fukushima and Ibaraki (treatment), Aichi (control)

Table 2 presents the descriptive statistics of the data. The number of samples for tomatoes is 624 (156 months × 4 prefectures), while that for eggplant is smaller and unbalanced. When the transaction quantity is less than four tons, the price is not recorded, in consideration of data reliability. This data recording condition makes the eggplant data unbalanced.

Table 2. Descriptive statistics.

	Unit	Egg plant			Tomato		
		Number	Mean	Standard Deviation	Number	Mean	Standard Deviation
Volume	Ton	610	2,041.96	2,436.36	624	3,994.95	5,324.50
Price	Yen/kg	553	343.71	97.35	624	361.96	95.25
Expenditure	Mil. yen	553	720.21	750.12	624	1,280.53	1,628.79
Share		553	0.28	0.27	624	0.25	0.31

Source: MAFF (2017): http://www.maff.go.jp/j/tokei/kouhyou/seika_orosi/.

Empirical Models

For the empirical analysis, we use the AIDS model originally developed by Deaton and Muellbauer (1980). This model has been broadly applied in demand analyses of a variety of agricultural products, including meat, timber, and other goods. One of the

---

[11] See http://www.maff.go.jp/j/tokei/kouhyou/seika_orosi/index.html#r.

[12] Data is not complete for all vegetables. Data preparation is one of the reasons for choosing only two vegetables.

reasons for the popularity of using the AIDS model in a demand analysis is that the model was developed to be consistent with consumer demand theory.

In an AIDS model analysis, the researcher estimates the following share equation:

$$w_i = \alpha_i + \sum_j \gamma_{ij} \log p_j + \beta_i \log(\tfrac{x}{P}) \tag{1}$$

where w_i is the market share of group i's vegetable, p_j is the price of group j's vegetable, and x denotes the total expenditure for the vegetable. Finally, P is a price index given by

$$\log P = \alpha_0 + \sum_k \alpha_k \log(p_k) + \frac{1}{2} \sum_k \sum_j \gamma_{kj} \log p_k \log p_j \tag{2}$$

We estimate the parameters α, β, and γ in the empirical analysis.

We examine how the market share changes with the changes in price or expenditure by estimating Eqn. (1). However, Eqn. (1) is a nonlinear equation and is not easily estimated, although Deaton and Muellbauer (1980) propose to replace Eqn. (2) with a linear price index function. Other authors have criticized that the proposed function is no longer consistent with consumer demand theory. See Holt and Goodwin (2009) for details.

Considering the assentation, we follow Blundell and Robin (1999) and use an iterated linear least squares estimator (ILLSE) to estimate Eqs. (1) and (2). It is likely that the shape of the vegetable demand function for eggplant and tomatoes before (January 2003 to February 2011) and after (March 2011 to December 2015) the nuclear accident are not the same, and examine how the parameters for the demand function of those vegetables have been changed.

The AIDS estimation was conducted using statistical software R and its micEconAIDS package developed by Henningson (2014). The package allows us to impose the necessary restrictions when estimating the parameters through the ILLSE method.

Estimation Results

Data examination

Before estimating Eqn. (1), we need to examine whether the long time-series data that we will use in this analysis are stationary. If the data are not stationary, then we could have spurious correlations which may obscure the true relationship between those vegetables. The augmented Dickey–Fuller (ADF) test is one of the common methods to examine this problem. The null hypothesis is that there is a unit root. We conducted the ADF test on the price and volume data for each vegetable. Note that the data tested is monthly and that the data is aggregated in terms of prefectures (national level). The test successfully rejected the null hypothesis at the 1% level for the two vegetables. Based on the test result, we substantiate that our estimation method does not have a unit-root problem.

Impact of nuclear contamination on expenditure sensitivities

In economic analyses, it is common to calculate the expenditure elasticity of demand in order to examine the responsiveness of the quantity demanded of a good to a change in the consumer expenditure of the commodity in question. One of the advantages of the AIDS model is the easy ability to construct price elasticities from the estimated parameters. The expenditure elasticity is

$$\eta_{ix} = \frac{\beta_i}{w_i} + 1$$

It is the value of the percentage change in quantity divided by the percentage change in total expenditure. High elasticity means that the demand is more sensitive to total expenditure.

Table 3 gives the list of our estimated expenditure elasticities. Our analysis is mainly devoted to Fukushima and Ibaraki prefectures, which were severely affected by the nuclear accident. These prefectures have also been known as agricultural prefectures for a long time. The estimated expenditure elasticities presented in Table 3 are all positive and those for Fukushima and Ibaraki are greater than unity. These results mean that the market shares of Fukushima and Ibaraki will increase more than proportionately as total expenditures increase. When comparing the expenditure elasticities of Fukushima and Ibaraki before and after the nuclear accident, we find that there is only a significant decline for eggplant. This is probably because the eggplant control (Kochi) has a dominant market share in the Kanto Wholesale Market. The quantity of Kochi-grown eggplant is two to ten times larger than Fukushima and Ibaraki combined. This means that there would not be a problem without Fukushima- and Ibaraki-grown eggplant in the Kanto Wholesale Market, but there would be a huge problem for tomatoes. Perhaps our results suggest that consumers buy Fukushima- and/or Ibaraki-grown vegetables only when they cannot find affordable alternatives.

Table 3. Expenditure elasticities.

Groups	Eggplant		Tomato	
	Before	After	Before	After
Fukushima	2.713**	1.175***	3.104***	3.731***
	(0.907)	(0.250)	(0.366)	(0.694)
Ibaraki	2.809***	1.172***	1.618***	1.345**
	(0.320)	(0.074)	(0.215)	(0.457)
Control	−1.199***	1.194***	−0.517***	−0.382
	(0.245)	(0.023)	(0.126)	(0.235)
Remaining	1.696***	1.160***	0.995***	1.002***
	(0.083)	(0.026)	(0.025)	(0.054)

Note: The control prefecture is Kochi for eggplant and Aichi for tomatoes.
*, **, and *** indicate statistical significance at the 10%, 5%, and 1% levels, respectively. The numbers in parentheses are standard errors of the coefficient estimates.

Impact of nuclear contamination on own-price sensitivities

When prices increase, consumers will demand less. In economic analysis, it is common to calculate the price elasticity of demand to show the responsiveness of the quantity demanded of a good to a change in its price. Assuming that consumer income before and after the price change is the same, the uncompensated (Marshallian) price elasticity becomes

$$\eta_{ij} = -\delta_{ij} + \frac{\gamma_{ij} - \beta_i(\alpha_j + \sum_k \gamma_{jk} \log p_k)}{w_i}$$

where δ_{ij} is the Kronecker delta term, in which $\delta_{ij} = 1$ if $i = j$, and $\delta_{ij} = 0$ otherwise. It is the price elasticity of good i for a change in the price of good j. Hence, it is the value of the percentage change in quantity divided by the percentage change in price. High elasticity means that the demand is more price sensitive.

Table 4 presents own-price elasticities. All elasticities before the nuclear accident are negative. This result is consistent with the consumer theory. In contrast, the elasticities of Kochi-grown eggplant are estimated to be positive. The positive sign means that the demand will go up when the price goes up, which is counterintuitive. This must be the consequence of some distortion outside the market, although we have not yet identified the reason.

Except for the positively estimated case, the magnitude of the own-price elasticities has increased after the nuclear accident. The result for Miyazaki-grown tomato is interesting, in the sense that the demand function changes from non-elastic to elastic. It was a renowned brand before the nuclear accident, as the demand decreased only slightly, even when the price increased. As consumers began spending more time on food shopping, due to serious concerns about their own safety after the nuclear accident, they also began making more deliberate decisions about the price of agricultural products. This change in consumer behavior made the demand for vegetables more price elastic.

Table 4. Own-price elasticities (Marshallian).

Groups	Eggplant		Tomato	
	Before	After	Before	After
Fukushima	−1.017	−1.294***	−1.238	−4.520*
	(1.225)	(0.061)	(0.849)	(1.949)
Ibaraki	−1.456***	−1.973***	−2.463***	−3.264***
	(0.208)	(0.118)	(0.364)	(0.702)
Control	−1.135***	3.931***	−0.302	−1.463**
	(0.330)	(0.121)	(0.444)	(0.455)
Remaining	−1.400***	−4.046***	−1.073***	−1.207***
	(0.329)	(0.084)	(0.046)	(0.053)

Note: The control prefecture is Kochi for eggplant and Aichi for tomatoes.
*, **, and *** indicate statistical significance at the 10%, 5%, and 1% levels, respectively. The numbers in parentheses are standard errors of the coefficient estimates.

Impact of nuclear contamination on cross-price sensitivities

The upper part of Table 5 summarizes the cross-price elasticities for the case of tomatoes; that is, the quantity change of tomatoes grown in prefecture i with respect to the price change of tomatoes in prefecture j. Note that we only present the cases that had significant results both before and after the nuclear accident. There are obvious

Table 5. Cross-elasticities (Hicksian).

	Coefficient	Standard Error	p-value
Tomatoes			
Fukushima Quantity/Remaining Price			
Before	1.489	0.675	0.028
After	2.530	1.400	0.073
Ibaraki Quantity/Remaining Price			
Before	2.000	0.376	< 0.001
After	2.771	0.646	< 0.001
Remaining Quantity/Fukushima Price			
Before	0.096	0.044	0.028
After	0.136	0.076	0.073
Remaining Quantity/Ibaraki Price			
Before	0.213	0.040	< 0.001
After	0.216	0.050	< 0.001
Eggplant			
Ibaraki Quantity/Kochi Price			
Before	1.550	0.640	0.017
After	−1.986	0.320	< 0.001
Kochi Quantity/Ibaraki			
Before	0.408	0.170	0.017
After	0.486	0.079	< 0.001
Kochi Quantity/Remaining Price			
Before	0.861	0.462	0.064
After	1.977	0.210	< 0.001
Remaining Quantity/Kochi Price			
Before	0.370	0.1977	0.069
After	−2.271	0.241	< 0.001

differences in the cross-price elasticities in the upper two rows in the table. After the nuclear accident, the demand for tomatoes from Fukushima and Ibaraki became more sensitive to the price changes in the remaining prefectures. However, the table shows that the converse is not true; the price sensitivities of the remaining prefectures did not change very much before and after the nuclear accident.

The bottom part of Table 5 summarizes the cross-elasticities for the case of eggplant. The notable changes are the two instances where the sign of the cross-price elasticity changed from positive to negative. This means that, although each pair of products had a substitution relationship before the nuclear accident, they began to have a complementary relationship after the accident.

One possible reason for this change is that eggplant from Ibaraki Prefecture lost its reputation as a result of the nuclear accident. The eggplant grown in the dominant production area (Kochi Prefecture) further strengthened market power. Indeed, even in the peak period (June to October), the market share of Ibaraki Prefecture after the nuclear accident was much lower than that before the accident. Consumers no longer care about the harvest season of eggplant, but they do care about where the vegetable comes from.

Policy Implications

How do consumers react to food contamination when the risk is barely publicized? In this chapter, we reviewed consumer responses to the Fukushima nuclear accident. Previous survey studies predicted that a typical consumer is willing to pay money to avoid agricultural products produced in the areas affected by the nuclear accident. In this study, we analyzed wholesale market data and examined whether the nuclear accident altered consumer valuations of selected vegetables. The results of our analysis revealed that the nuclear accident had altered the demand for vegetables. Specifically, we found that the demand for vegetables has become more price elastic after the nuclear accident. Therefore, even a small price increase will lead to a large decrease in vegetable demand. In addition, it is expected that supply-side shocks, such as weather changes, cause larger price volatilities than before.

Previous survey studies also predicted that consumers would greatly shift their products' origins in order to avoid agricultural products grown in contaminated areas. However, consumers may not be able to do so, given the constraints of agricultural production. Some agricultural products are essential commodities. Hence, even if people have concerns about radiation risks, they will continue to purchase and consume agricultural products from the affected areas when food from the non-affected areas is not available. Our estimation results suggest that consumers only changed their products' origins when agricultural products from the non-affected areas were affordable. We need to consider the supply-side constraint when also estimating the consumers' willingness to accept the risk of radioactively contaminated food in survey studies.

Appendix

Let us assume that the expenditure function ($= \log c$) is based upon the Price Independent Generalized Logarithmic (PIGLOG) class of preference, so that we can aggregate over consumers:

$$\log c(u,p) = (1 - u) \log[a(p)] + u \log [b(p)] \tag{A1}$$

where u is the utility level and p is a price vector. Further, let us specify $\log [a(p)]$ and $\log [b(p)]$ as follows:

$$\log a(p) = a_0 + \sum_k a_k \log(p_k) + \frac{1}{2}\sum_k\sum_j \gamma_{kj}^* \log p_k \log p_j \tag{A2}$$

$$\log b(p) = \log a(p) + \beta_0 \prod_k p_k^{\beta_k} \tag{A3}$$

where α, β, γ_{kj}^* are parameters to be estimated. By applying Shephard's lemma through the differentiation of Eqn. (1) with respect to a logarithmic price, Deaton and Muellbauer (1980) obtained the following share equation:

$$w_i = a_i + \sum_j \gamma_{ij} \log p_j + \beta_i \log \left(\frac{x}{P}\right) \tag{A4}$$

where x denotes the total expenditure, $\gamma_{ij} = \frac{1}{2} (\gamma_{ij}^* + \gamma_{ji}^*)$

To be consistent with consumer demand theory, the parameters of the AIDS model must satisfy the following conditions:

- Additivity: $\sum_i \alpha_i = 1$, $\sum_i \beta_i = 0$, $\sum_i w_i = 1$
- Homogeneity: $\sum_i \gamma_{ij} = 0$
- Symmetry: $\gamma_{ij} = \gamma_{ji}$

We imposed these restrictions when estimating our parameters.

References

Aruga, K. 2016. Consumer responses to food produced near the Fukushima nuclear plant. Env. Econ. Policy St. 19: 677–690.

Blundell, R. and J.M. Robin. 1999. Estimation in large and disaggregated demand systems: An estimator for conditionally linear systems. J. App. Econometrics 14: 209–232.

Bostedt, G. 2001. Reindeer husbandry, the Swedish market for reindeer meat, and the Chernobyl effects. Agr. Econ. 26: 217–226.

Cabinet Office of Japan. 2011. Estimated damage of the Pacific Ocean off the coast of Japan's Tohoku region (24 June 2011).

Deaton, A. and J. Muellbaur. 1980. An almost ideal demand system. Am. Econ. Rev. 70: 312–326.

Fuse, J. 2012. Head of Tokyo-area Medical Clinic: Risk from internal exposure is 200–600 times greater than risk from external exposure. New York City Press Conference. <http://enenews.com/head-of-tokyo-area-medical-clinic-risk-from-internal-exposure-is-200-600-times-greater-than-risk-from-external-exposure-video> (accessed on 4 May 2012).

Gibney, E. 2015. Fukushima data show rise and fall in food radioactivity, Giant database captures fluctuating radioactivity levels in vegetables, fruit, meat and tea. Nature News (27 February 2015).

Hanemann, W., P.O. Johansson, B. Krisröm and L. Mattsson. 1992. Natural resources damages from Chernobyl. Environ. Resour. Econ. 2: 523–525.

Hangui, S. 2013. Consumers' response to the information provided about contamination of the food by radioactive materials. J. of Rural Econ. 85: 173–180.

Hatsuzawa, T. and T. Takano. 2016. Characteristics of the evacuation area and the spatial distribution of radioactive pollution in Fukushima prefecture. Chapter 11. *In:* Karan, P.P. and U. Suganuma (eds.). Japan After 3/11: Global Perspectives of the Earthquake, Tsunami, and Fukushima Meltdown. The University Press of Kentucky, USA.

Hayashi, T. and M. Midorikawa. 2013. Status of radioactive contamination of food. J. for the Integrated Stud. of Dietary Habits. 24: 7–10 (in Japanese).

Heley, M. 2014. Fukushima what you need to know. North Atlantic Books, Berkeley California.

Henningsen, Arne: micEconAids - Demand analysis with the Almost Ideal Demand System (AIDS) suggested by Deaton and Muellbauer (1980). Links to an external site. "R" package available at CRAN.

Holt, M.T. and B.K. Goodwin. 2009. The almost ideal and translog demand systems. Chapter 2. pp. 37–59. *In:* D. Slottje (ed.). Quantifying Consumer Preferences. Contribution to Economic Analysis. Emerald Group Publishing Limited, UK.

Hotaka, T. and W. Naito. 2013. Cost analysis of decontamination efforts in the decontaminated area in Fukushima prefecture. <https://www.aist-riss.jp/wp-content/uploads/2014/12/5d22c054334dc5ff4f6d1e14e6d636c1.pdf> (accessed on 21 June 2017).

International Atomic Energy Agency. 2012. Fukushima Nuclear Accident Update Log (24 May 2012).

Ito, N. and K. Kuriyama. 2017. Averting behaviors of very small radiation exposure via food consumption after the Fukushima Nuclear Power Station accident. Am. J. Agr. Econ. 99: 55–72.

Japan Atomic Energy Agency. 2014. Airborne Monitoring. <http://emdb.jaea.go.jp/emdb/en/selects/b143/> (accessed on 20 June 2017).

Kuriyama, K. 2012. Radioactive substances and food purchasing behavior: a choice experiment analysis. (Hoshaseibusshitu to Shkuhin Kobai Kodo: Sentaku Jikken Ni Yoru Bunseki Kara. Special edition. Nogyo to Keizai (Agriculture and the Economy): 30–38 (in Japanese).

Ministry of Agriculture, Forestry, and Fisheries. 2010a. Cultivated and Planted Land Statistics.

Ministry of Agriculture, Forestry, and Fisheries. 2010b. Census of Agriculture and Forestry.

Ministry of Agriculture, Forestry, and Fisheries. 2010c. Agriculture Production Income Statistics.

Ministry of Agriculture, Forestry, and Fisheries. 2011. Impact of the Fukushima nuclear accident on agricultural and fishery production (15 April 2011).

Ministry of Agriculture, Forestry, and Fisheries. 2012a. FY2011 Annual Report on Food, Agriculture and Rural Areas in Japan.

Ministry of Agriculture, Forestry, and Fisheries. 2012b. Statistics on Production and Shipment of Vegetable.

Ministry of Agriculture, Forestry, and Fisheries. 2017. Fruit and vegetable wholesale market survey. <http://www.maff.go.jp/j/tokei/kouhyou/seika_orosi/> (accessed on 20 June 2017).

Ministry of Health, Labour, and Welfare. 2016. Radioactive materials in foods -current situation and protective measures. <http://www.mhlw.go.jp/shinsai_jouhou/dl/20131025-1.pdf> (accessed on 25 June 2017).

Ministry of Health, Labour, and Welfare. 2016. Regulation levels of radioactive substances in food in foreign countries. <http://www.mhlw.go.jp/stf/shingi/2r9852000001ip01.html> (accessed on 20 June 2017).

Monma, T. 2013. Research trends of radio contamination influences on agriculture and food consumption behavior. J. of Rural Econ. 85: 16–27 (in Japanese).

National Police Agency. 2016. Damage of the Pacific Ocean off the coast of Japan's Tohoku region and police measures.

Tajima, K., M. Yamamoto and D. Ichinose. 2016. How do agricultural markets respond to radiation risk? Evidence from the 2011 disaster in Japan. Reg. Sci. Urban Econ. 60: 20–30.

Tokyo Electric Power Company. 2012. The estimated amount of radioactive materials released into the air and the ocean caused by Fukushima Daiichi nuclear power station accident due to the Tohoku-Chihou-Taiheiyou-Oki earthquake, as of May 2012 (24 May 2012).

Tokyo Metropolitan Central Wholesale Market. 2012. Market transaction record. <http://www.shijou.metro.tokyo.jp/torihiki/geppo/> (accessed on 20 June 2017).

Ujiie, K. 2012. Consumer's evaluation on radioactive contamination of agricultural products in Japan–Decomposition of WTA into a part due to radioactive contamination and a part due to area of origin. Food Syst. Res. 19: 142–155.

Ujiie, K. 2013. Transition of consumer's evaluation on radioactive contamination of agricultural products in Japan. J. of Rural Econ. 85: 164–172.

Yamane, F. 2013. Consumers' purchasing behavior and choice of information source under the radioactive contamination of beef. Proceeding of the Annual Meeting of Behaviormetric Society of Japan, pp. 320–323.

PART II
International Trade

6

International Trade and Credence Goods

John C. Beghin

Introduction

This chapter presents an economist's perspective on credence goods and international trade. It focuses on the economic impact of policies regulating credence goods and attributes in an open economy context. We discuss the impact of such policies on trade and welfare. Such policies are sometime protectionist. They can also be legitimate ways to address market imperfections arising with credence goods. The chapter discusses several simple tests to detect protectionism.

It also shows that when a credence attribute leads to market imperfections, it is difficult to come up with robust policy prescriptions. Readers are reminded that trade and welfare may move in opposite directions and that changes in trade flows are often not informative for deriving welfare effects of policies affecting credence goods. The chapter further discusses issues related to trade frictions between trade partners trying to facilitate trade flows when they have heterogeneous policies to address the credence good problem. Several arrangements are discussed.

Policies addressing credence attributes usually fall into three categories under the "MAST" classification of UNCTAD, all three under the broad class of technical measures affecting trade. They are sanitary and phytosanitary measures, technical barriers to trade, and pre-shipment inspection and other formalities. Despite their multitude and heterogeneity, and fortunately for the economist, many of these measures can be analyzed using a common simple framework which we present in the second section. Although the framework is simple and tractable, the key implications emerging

Department of Agricultural and Resource Economics, North Carolina State University, 3350 Nelson Hall.
  E-mail: jcbeghin@ncsu.edu

from it is that effects of policies addressing credence attribute are ambiguous for both trade and welfare. They may inhibit or stimulate trade, and they may increase or decrease welfare. In addition, trade and welfare effects may go in opposite directions and trade effects have little predictive power for welfare effects.

Another issue often present in the analysis of these policies affecting credence attributes is their potential protectionism. The policy interventions may not be motivated entirely by the desired to protect the public or the environment (Maskus et al. 2000). They may seek to keep foreign production out of the country in the guise of addressing market imperfections. It is hard to sort these cases out.

The chapter first discusses their potential protectionism of policy interventions addressing credence attributes. The issue is to separate legitimate reasons to intervene in a market for a credence good from protectionist motives. Even when there is market imperfection linked to a credence good, some market interventions could be ill-designed and could constitute protection. Then in the following section, we present the simple framework to analyze trade and welfare effects of these policies. Some empirical evidence and well-known policy cases are discussed throughout.

This discussion chapter sets the stage for the following chapter in this international trade part of the book and *per se* does not break new grounds. It draws on and distills several recent scholarly efforts by the author and his co-authors [Beghin et al. (2015), Beghin and Xiong (2017), and Beghin and Orden (2013), among others].

The Potential Protectionism of Policies Addressing Credence Attributes

The strong emergence of these policies in the last 20 years, often in the form of required standards and informational requirements brings the allegation that the policies are protectionist. Protectionism means that a domestic industry benefits at the expense of foreign competitors thanks to the policy put in place. How can we assess this protectionist claim?

The first step is to determine if the policy addressing the credence attribute in question is legitimate in the following sense. The credence attribute itself must be legitimate, and must address some external effect or another market imperfection in production or consumption such as an environmental or health hazard unknown to some parties or some other asymmetric information situation. The established way to gauge this aspect is to use a criteria defined by the World Trade Organization (WTO). Hooker and Caswell (1998) call this the "science test". The defined credence attribute and associated policy should be based on science and some risk assessment to establish the risk of negative outcome. If the credence attribute and the policy do not address a market imperfection, that is, there is no scientific evidence that a credence attribute constitutes or leads to a market imperfection, then the policy is protectionist. There is an exception to this first principle. If the science is being actively researched but not yet established, then a precautionary policy could be put in place temporarily. A precautionary policy is also deemed protectionist by the WTO if it is maintained over a long period of time without actively undertaking the research to establish the science (e.g., the EU GMO policies).

Assume that a policy passes this first step of the "science test". Next, one should look at a "discrimination test". The latter is referred to as national treatment in the WTO jargon. Policies or technical barriers addressing a credence attribute should not make any difference between imported and domestic products which are "like" products. Like products means close substitutes. A different policy for imported products (say a more stringent pesticide residue than for a domestic substitute) would signal protectionism (Baldwin 2000).

The U.S. regulation on country of origin label (COOL) is a related example of imposing a policy requirement which discriminates imported live animal products compared to domestic live animals. COOL combined issues at the intermediate and final consumer demand levels. Meat processors and retailers would have been required to indicate on the retailed meat where the animals were born, raised and slaughtered. Consumers mostly worry about sanitary conditions at the processing time. Many U.S. producers raise cattle without segregating them by place of birth. Given the NAFTA-based integration of the meat industry, animals often cross borders between birth and slaughter. The COOL regulations was deemed to impose excessive cost to producers and processors to keep a system of identity preservation along the supply chain and would have handicap foreign partners in the US supply chain and create protection for U.S. livestock interest. COOL's required record keeping and verification imposed a disproportionate burden on livestock producers and processors of livestock compared to the information received by consumers. Some important exemptions from labeling for restaurants and in processed foods suggested that consumers' information was not the key motivation for the regulation. So the COOL regulation imposed unequal treatment on Canadian and Mexican livestock producers relative to U.S. livestock producers (animals born and raised in the US). Note that this case was pretty convoluted and took years to be resolved. A far cry from a "quick test".

Following the science-based test and the domestic treatment requirement, a third useful practical test to establish protectionism is the "transparency test". If a credence attribute has to be addressed by a policy, this policy should be transparent to trade partners in the following sense. Trade partners should be notified of the policy in a timely and predictable manner. They should also be consulted about the policy. The WTO also allows trade partners to have consultations on these policies and to express concerns when they find some element of the policy to be objectionable. There are many dimensions to transparency, and the devil is in the details. Policies and procedures could be unnecessarily arcane and complex, for example.

Still in the case of market imperfections such as hazardous substances used in food supply chains (say pesticides used in agriculture), policies addressing the potential health hazard could be compared to recommended policies by international agencies such as Codex Alimentarius. The WTO recommends such practice of relying on international standards. Li and Beghin (2014) have proposed such systematic comparison using indices of policy stringency for pesticide residues. The indices serve as aggregator of various policies and one can compare systematic deviations from international policy recommendations. Some countries could be at higher risk than other countries and more draconian policies may be explained by that higher risk. However, if stringency is systematically much higher than international norms over many products and hazardous substances, one can reasonably conclude that policies are overly stringent. If compliance cost is much easier for incumbent domestic firms then the policies are likely to be protectionist. Such deviations were used by Hooker and Caswell (1999) to detect protectionism.

A fourth and last test is more controversial and not widely accepted by economists. It is the "least trade restrictive test". This criterion is (mis)used by the WTO in part because the WTO tends to focus on production and profit and not on consumer welfare. Maximizing social welfare including the cost of market imperfections and the well-being of consumers may or may not be achieved by choosing least-trade restrictive policies (Disdier and Marette 2010, Marette 2016). So, comparing policies and attempting to rank them by order of increasing trade restrictiveness (in the sense of this test) will not lead to sound policy design. One should rather apply the least-welfare restrictive policy as in Baldwin (1970) and (2000), Fisher and Serra (2000), Marette and Beghin (2010), and Marette (2016). The latter criteria developed by these authors is conceptually clear, but difficult to implement empirically. Results are likely to be dictated by modeling assumptions and to be sensitive to the initial setup used to analyze a situation.

These authors start from a setup in which the social planner wants to correct a market imperfection (say some hazard from unsafe consumption) in an open economy in which home and foreign firms supply the home market. The optimum policy to address the market imperfection associated with the credence attribute will maximize global welfare. Policies which do not maximize global welfare, are suboptimal and may be protectionist by providing higher profits to home firms. This can take place in absence of rent-seeking behavior.

There is no explicit demand for protection. If there is rent seeking, lobbying of domestic firms and consumers may or may not lead to policies being protectionist depending on the relative cost of compliance between home and foreign firms (Swinnen and Vandemoortele 2008, 2009, 2011). So even in simple conceptual models, establishing robust conditions to identify protectionism is difficult even though the concept itself is limpid. Then the empirical implementation of such protectionism criteria would compound the ambiguity of results of the conceptual model (Marette 2016).

To conclude this section, the first three simple tests (science, discrimination, and transparency) offer the most potential to empirically gauge protectionism of policies addressing credence goods. More elaborate approaches are fraught with ambiguity and implementation challenges.

A Simple Framework to Analyze Policies Addressing Credence Attributes in Trade

The economics of policies addressing credence attributes of a good in an open economy context are described graphically. We use a simple supply and demand partial equilibrium framework. We explain the potential effects of a policy addressing credence goods or attributes leading to imperfection on supply and demand and at the border for a single market, in a small country with a parametric world price. This approach follows the literature (Josling et al. 2004, Fugazza 2013, Van Tongeren et al. 2009, Maskus et al. 2000, Beghin and Xiong 2017).

Supply effects

Policies addressing credence attributes typically increase cost of production by requiring additional steps and procedures (e.g., HACCP procedures to minimize health risk). Foreign and home firms may not be identical in technology and cost structure but we know that their marginal costs will shift upward. This is shown in Fig. 1 for an importable good in a small open country.

We follow Beghin and Xiong's notation for supply, demand and imports. Domestic supply is denoted by y, domestic demand by x, imports by m. *Pol* indicates the new policy put in place, and world price is wp. The figure shows the shift upward of the parametric world import supply induced by the added cost of the policy addressing the credence good, $t(Pol)$, at the border of the small country. The horizontal import supply curve moves from wp to $wp+t(Pol)$. This assumed that the cost of the policy is added at the border, which is a convenient assumption for our discussion.

The domestic supply also shifts from y to y' (to the left) for the added cost of the additional procedures. The two vertical shifts may have different effects on unit cost. It is easy to see that the policy could either benefit domestic producers if the shift of domestic supply is smaller than the shift in the import supply or vice versa. The figure shows the policy favoring domestic producers.

As a special case, a policy could reduce a detrimental external effect born by domestic suppliers prior to the policy being implemented. That could be a pest or invasive species for which imports are a vector. This special case would then have an upward shift in the foreign supply by $t(Pol)$ as in Fig. 1, and an outward shift of the domestic supply from the reduced externality. In this case, the impacts on domestic and foreign supply will have opposite directions as shown in Fig. 2.

Imports would then decrease unambiguously although welfare would likely increase because of the reduction in the externality affecting production. Consumers would be worse off because of the measure affecting the unit cost of imports. Welfare increases and trade decreases. The trade impact is not informative on the direction of the welfare change.

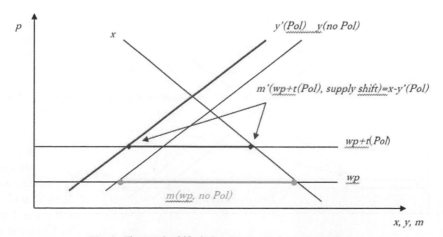

Fig. 1. The supply shifts induced by a cost-increasing policy.

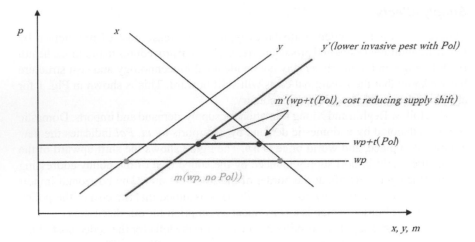

Fig. 2. Supply shifts with a policy targeting invasive pest brought by imports.

Demand effects

Assume now that the credence attribute affects consumers (e.g., a long-term health effect from food safety attributes or valuation of sustainability attributes). The policy addressing this attribute influences consumption. Consumers may become aware of the higher quality with the policy in place. Consumption will increase all things being equal (an outward shift of demand). Conversely, if the policy forces disclosure of some negative credence attribute ("this food is bad for you"), then demand may shift inward. Figure 3 shows the first case of demand enhancing policy. Figure 4 shows the demand reducing effect of a health warning.

Domestic supply is kept stationary to simplify the discussion. In reality the policy will induce some additional cost for domestic supplier. In Fig. 3, imports are larger under the demand enhancement. Imports are smaller in Fig. 4 with the health warning.

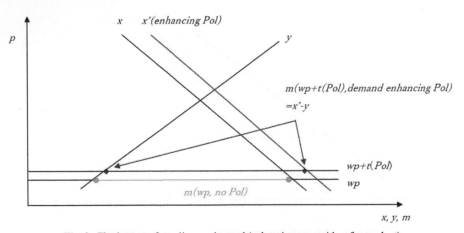

Fig. 3. The impact of a policy on demand (enhancing case with safer product).

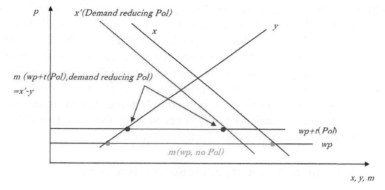

Fig. 4. The impact of a policy on demand (decreasing case with health warning).

Again no logical link exists between the trade effect of the policy and its welfare effect, assuming that the policy is optimum—it reduces the market imperfection optimally to maximize social welfare inclusive of external effects on health.

Combined supply and demand effects

When put together these effects on demand and supply compound the ambiguity of results and the complex relationship between welfare and trade. Unit cost tends to increase as shown in Cadot and Gourdon (2016), since added cost occurs at the border. The ambiguity and complex policy-welfare-trade link call for empirical assessment of the policies motivated by credence goods. Figure 5 illustrates the resulting ambiguity. Demand is enhanced by a safer product regulation (say reduced pathogen contamination with HACCP regulation) with the contemporaneous increase in unit cost from implementing the regulation.

Foreign suppliers have to follow similar procedures. Depending on the relative shifts of home and foreign supplies, trade could expand or be reduced. The unit cost increases in all cases with the regulation addressing the credence attribute. This is the only prediction that is robust.

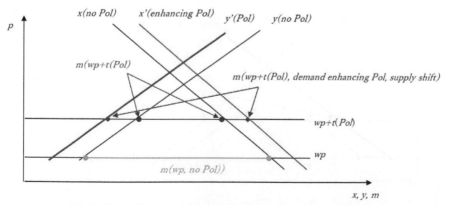

Fig. 5. The ambiguous net impact of credence good policy on imports.

Welfare effects depend on the decrease in profits in production and change in consumer welfare inclusive of the externality. In some cases, the external effect occurs for consumers or producers in other markets, but not directly in the market which is regulated. For example, a regulation on waste water treatment in animal production (a process standard) may not affect demand for meat but will benefit some economic agents living in the proximity of animal production. The absence of demand shift in the targeted market is not a sufficient condition to conclude that protection is taking place. Partial equilibrium analysis can still account for these welfare effects with proper accounting for the market imperfections at work.

Partial equilibrium analysis of protectionist regulations

Credence attributes may not lead to any market imperfection even though there is asymmetric information in the market or when they are left unregulated. For example, many aspects of food production processes are unknown to consumers but they do not lead to market imperfection if left unknown or unregulated. Let's call this type of credence attribute innocuous. Policies that regulate and constraint innocuous credence attributes are protectionist because there isn't any market imperfection associated with them. The justification for corrective action on resource allocation does not exist. When risks to the environment or health are negligible or when asymmetric information does not lead to quality deterioration, we are in this situation. From causation we know that demand will not shift following the implementation of such policy regulating an innocuous credence attribute. The motivation of such policy will be to transfer rents to either domestic producers or/and to the regulating body who will benefit from influence by winners and losers created by the policy.

A simple case is shown in Fig. 6. The policy increases the cost of imports without inducing further cost on domestic producers. An example would be requirements to over-mitigate the risk of invasive pest like in the WTO apple dispute between Japan and the US. In the latter, requirements on buffer zone in U.S. orchards were prohibitively expensive and made U.S. apple exports noncompetitive in Japan (Calvin et al. 2008).

Figure 6 shows the increase in unit cost of imports, and the lack of effect of the policy on consumers or producers (there was no risk change for either). In this case,

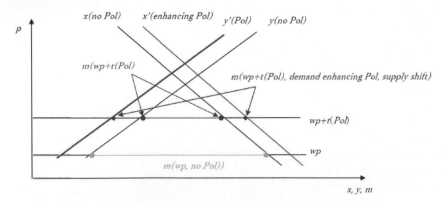

Fig. 6. The impact of a protectionist regulation of an innocuous credence attribute.

we have trade and welfare moving in the same direction. Producer surplus is higher under the policy by increasing the price by a tax equivalent of the policy. Consumer welfare falls because of the tax equivalent. Society welfare falls by the trapezoid defined by imports under the policy and imports without regulations. Rent seeking activities may capture or waste the rectangle *t(pol)*m[wp+t(Pol)]*.

The Regulatory Heterogeneity across Countries

Many countries have developed their own framework for regulating credence attributes. Some countries have aligned themselves on international standards like food safety standards established by Codex Alimentarius. By and large, these regulatory frameworks are heterogeneous across countries and create frictions, impeding trade and reducing exchange opportunities. How can this issue be addressed?

To complicate matters, countries may differ in their valuation of external effects like the effects on health and the environment or may not value animal welfare or may have different understanding of what fair trade attributes should be. So it is not clear that these countries share common valuations and preferences over credence goods. The latter would simplify the choice of a single policy that would be optimum across borders. This is rarely the case in practice.

However, a single or harmonized policy may not be necessary, thanks to a principle of equivalence advocated by the WTO. Under the SPS Agreement of the WTO, importing members should accept another country's policy or regulation as equivalent to their own if an exporter of that country can establish that his/her own country's regulation provides a similar level of desired protection of consumers or the environment. Establishing the equivalence can take work and is another friction cost.

Beyond establishing equivalence, if two countries trade and regulate differently, what can they do to minimize frictions? By frictions, we mean how a country could reduce trading costs and the cost of adapting products to meet the partner's regulations. There are several ways to address this heterogeneity of regulations on credence attributes. Many countries have harmonized their regulatory frameworks and policies regarding credence attributes, through custom unions like EU member states, bilateral or regional trade agreements (e.g., in NAFTA), and Mutual Regional Agreements (MRAs) on these measures (e.g., NZ-Australia on food standards). Many trade agreements include transparency requirements on regulatory framework to minimize unnecessary frictions but without setting common regulations. The transparency mitigation was noted in the previous discussion of protectionist regulation.

The evidence on the effects of these harmonization attempts on trade is mixed. Intuitively, harmonized standards and MRAs facilitate trade by creating new trade among participating countries and sometime with excluded countries, albeit weakly in some cases. In harmonization, the level of the harmonized regulation is paramount. If the harmonized policy is set too high, harmonization could hinder trade opportunities. These mixed effects have been shown for NAFTA countries, for example. The US and Mexico have many regulations harmonized but which are less stringent than Canadian policies. The level of harmonization (up to the Canadian level or down to the U.S.-Mexico level) has different implications for trade flows between the three countries.

Some early investigations suggest that harmonizing to international standard levels tends to create new trade. MRAs once established are probably the most cost effective but they could affect consumers in the importing countries if the exporter's regulation is judged overly lax or culturally different. For that reason, they may be difficult to negotiate, as for example in the case of regulation of food credence attributes between the EU and the US. There are major disagreements on adequate protection of consumers' interest on a host of attributes (GMO material in food, use of chlorine in meat processing, pasteurized milk in cheese, and geographical and regional labels and product names, among others).

A corollary issue is how to streamline the regulatory framework for credence attributes in global value and supply chains. Products during their transformation cross or crisscross borders in various forms. They are subject to redundant regulations and to extensive private standards large food processors and retailers have imposed on these supply chains. The issue how to insure that credence attributes such as food safety are maintained while removing the redundancy of regulations and while respecting the coexistence of public and private standards. This is an ambitious task requiring some public-private partnerships along the supply chains using such as "value/supply-chain councils" suggested by Hoekman (2013) and (2014).

Concluding Remarks

This chapter presented an economist's perspective on credence goods and international trade. Credence goods emerge because of market imperfections which call for regulation and policy interventions in markets. The chapter focused on the economic impact of policies regulating credence goods and attributes in an open economy context.

The chapter discussed the potential protectionism of such policies. Such policies are sometime protectionist. They can also be legitimate ways to address market imperfections arising with credence goods. The chapter discusses several simple tests to detect protectionism. It also shows that when a credence attribute leads to market imperfections, it is difficult to come up with robust policy prescriptions to avoid protectionism.

The chapter also looked at the impact of such policies on trade and welfare. Readers were reminded that trade and welfare may move in opposite directions and that changes in trade flows are often not informative for deriving welfare effects of policies affecting credence goods. The chapter further discussed issues related to trade frictions between trade partners trying to facilitate trade flows when they have heterogeneous policies to address the credence good problem. Several arrangements were discussed.

Acknowledgement

The author would like to thanks the OECD for its financial support, Tsunehiro Otsuki and Shigeru Matsumoto for their hospitality and their invitation to the OECD-CRP sponsored symposium: "Food Credence Attributes: How can we design policies to meet consumer demand?" Tokyo, Japan, 18–20 May 2017. The author also thanks Bo Xiong, Jason Grant, David Orden, Jo Swinnen, Stephan Marette, Anne-Celia Disdier, and Jean Christophe Bureau for discussions related to this chapter.

References

Baldwin, R.E. 1970. Nontariff distortions of international trade. Brookings Institution.

Baldwin, R.E. 2000. Regulatory protectionism, developing nations, and a two-tier world trade system. Brookings Trade Forum 2000: 237–293.

Beghin, J. and D. Orden (eds.). 2012. NTMs, agricultural & food trade, & competitiveness. A special issue. World Econ. 35(8).

Beghin, J.C., M. Maertens and J.F.M. Swinnen. 2015. Nontariff measures & standards in trade & global value chains. Annu. Rev. Resour. Econ. 7: 425–450.

Beghin, J.C. and B. Xiong. 2017. Quantifying standard-like non-tariff measures and assessing their trade and welfare effects. Mimeo.

Beghin, J. (ed.). 2017. Nontariff Measures and International Trade, in the World Scientific Studies in International Economics Series of World Scientific and Imperial College Press, 2017.

Cadot, O. and J. Gourdon. 2016. Non-tariff measures, preferential trade agreements, and prices: new evidence. Rev. World Econ. 152: 227–249.

Calvin, L., B. Krissoff and W. Foster. 2008. Measuring the costs and trade effects of phytosanitary protocols: A U.S.–Japanese apple example. Rev. Agr. Econ. 30: 120–135.

Fisher, R. and P. Serra. 2000. Standards and protection. J. Int. Econ. 52: 377–400.

Fugazza, M. 2013. The economics behind non-tariff measures: Theoretical insights and empirical evidence. No. 57. United Nations Conference on Trade and Development.

Hoekman, B. 2013. Adding value. Financ. Dev. 50(4) December: 22–24.

Hoekman, B. 2014. Supply chains, mega-regionals and multilateralism: a roadmap for the WTO. Res. Pap. 27, Robert Schuman Cent. Adv. Stud.

Hooker, N.H. and J.A. Caswell. 1999. A framework for evaluating non-tariff barriers to trade related to sanitary and phytosanitary regulation. J. Agr. Econ. 50: 234–246.

Josling, T., D. Roberts and D. Orden. 2004. Food regulation and trade: Towards a safe and open global system. Institute for International Economics, Washington DC.

Li, Y. and J.C. Beghin. 2014. A protectionism index for non-tariff measures: An application to maximum residue levels. Food Policy 45: 57–68.

Marette, S. and J. Beghin. 2010. Are standards always protectionist? Rev. Int. Econ. 18: 179–192.

Marette, S. 2016. Non-tariff measures when alternative regulatory tools can be chosen. J. Agr. Food Ind. Org. 14: 1–17.

Maskus, K.E., J.S. Wilson and T. Otsuki. 2000. Quantifying the impact of technical barriers to trade. World Bank Policy Research Working Paper 2512.

Swinnen, J.F.M. and T. Vandemoortele. 2008. The political economy of nutrition and health standards in food markets. Appl. Econ. Perspect. Policy 30: 460–468.

Swinnen, J.F.M. and T. Vandemoortele. 2009. Are food safety standards different from other food standards? A political economy perspective. Eur. Rev. Agr. Econ. 36: 507–523.

Swinnen, J.F.M. and T. Vandemoortele. 2011. Trade and the political economy of food standards. J. Agr. Econ. 62: 259–280.

Van Tongeren, F., J.C. Beghin and S. Marette. 2009. A cost-benefit framework for the assessment of non-tariff measures in agro-food trade. No. 21. OECD Publishing.

7

Food Safety Standards and Trade Patterns

Tsunehiro Otsuki,[1,*] *Keiichiro Honda*[2] and *Bin Ni*[3]

Introduction

Consumption safety is one of the most important attributes consumers consider when purchasing food products. Food safety standards are not only enforced for a specific country's domestic food products, but also for imported ones to guarantee domestic consumers' acceptable level of dietary intake of hazardous substances such as pesticide residues, food additives, toxic substances, and microbial contaminants. Despite their primary purpose of protecting consumer safety, food safety standards can detract the food trade. As such, suppliers in countries with less stringent standards bear greater costs when exporting to countries with more stringent standards. Therefore, stringent mandatory standards can raise concerns of constituting non-tariff barriers, particularly from developing countries whose technological capacity of standard compliance is limited (Wilson (2002)). The World Trade Organization's (WTO) Sanitary and Phytosanitary (SPS) and Technical Barriers to Trade (TBT) Agreements state that trading countries should avoid standard regulations becoming non-tariff trade barriers by deterring trade unfairly.

---

[1] Osaka School of International Public Policy, Osaka University, 1-31 Machikaneyama, Toyonaka, Osaka, Japan, 560-0043.

[2] Department of Administrative Studies, Prefectural University of Kumamoto, 3-1-100, Tsukide, Higashi-ku, Kumamoto, Japan, 862-8502.
E-mail: khonda@pu-kumamoto.ac.jp

[3] Department of Accounting and Finance, Faculty of Business Administration, Toyo University, 5-28-20 Hakusan, Bunkyo-ku, Tokyo, Japan, 112-8606.
E-mail: jiadaniel@hotmail.com

* Corresponding author: otsuki@osipp.osaka-u.ac.jp

The process to building international consensus on food safety standards should thus be based on reliable information on how food safety standards affect the economy, because the mechanism through which standards affect consumers, suppliers, and other stakeholders is extremely complex. The SPS and TBT Agreements are based on the idea that food safety regulations should balance the protection of human, animal, and plant health with the benefits generated through trade, such as the profits received by suppliers in exporting countries. However, international consensus cannot be reached without defining an appropriate level of standards. Furthermore, knowledge about the mechanism through which food safety standards affect each country and the respective stakeholders, as well as the size of the benefits and costs, are all indispensable.

Although significant attention has been paid to the benefits and costs of food safety standards in the economic literature, the detailed mechanism of their effects remains to be identified. The challenge posed by considering the diversity of the impacts of standards across countries, products, and stakeholders characteristics adds to the debate. Most studies have hitherto focused on the sign of the standards' effects on trade flows without paying attention to the possibility of multiple channels through which the standards affect the economy. As a result, such studies may misguide policy makers as the benefits and costs may likely vary across countries, products, types of standards, and availability of substitutes. Additionally, the heterogeneity of consumer preferences and producer ability to comply with the standards add to the complexity of the issue. As a result, policies should be based on a set of empirical analyses incorporating all these aspects.

This chapter reviews empirical studies on the effect of food safety standards and also synthesizes the empirical results of related market- and consumer-level studies to provide integrated policy recommendations. Further, it empirically analyzes the impact of the maximum residue limits of veterinary drugs on meat product trade flows, based on Xiong and Beghin's (2014) generalized gravity model.

The remainder of the chapter is organized as follows. The second section provides a survey of the relevant studies. The third section further presents the results of three selected empirical studies. The fourth section discusses the implications of these analyses. The final section concludes the paper.

Literature Review

Macro-level studies using gravity models

An increasing number of empirical studies that estimate the trade impact of food safety standards have been published owing to the recent development of datasets and analytical methodologies. Their majority has used the gravity model or its extended variants to analyze the impacts of standards on bilateral trade due to the relative availability of datasets and the methodological tractability of the models.

The gravity model has also been used widely in empirical studies to assess the countrywide aggregate impact of tariffs and non-tariff measures, as well as other trade policies on bilateral trade, as it can effectively isolate the variation in bilateral trade flows due to regulations and other importer- or exporter-specific factors, based on panel-data econometric estimation methods. Under its standard specification, bilateral trade flows are regressed on importer and exporter economies' sizes, and the distance between importers and exporters as follows:

$$\ln y_{ij} = \beta_0 + \beta_1 \ln GDP_i + \beta_2 \ln GDP_j - \beta_3 \ln Dist_{ij} + \mathbf{X\beta} + \varepsilon_{ij}$$

where y_{ij} is the value of bilateral trade flow between countries i and j, GDP proxies the size of the economy in each country, $Dist_{ij}$ is the distance between the two countries, and ε is the error term. Additional regressors are often included in the model. In our case, let them be the elements of matrix $\mathbf{X}$, including variables for colonial relationships, common language usage, common borders, and regional trade agreements. Regulation variables such as those on food safety standards are also incorporated into $\mathbf{X}$ to estimate the sign and size of their impact.

Earlier studies that applied the gravity model to assess the impact of food safety standards employed the above standard specification of the model. Maximum residue limits (MRLs) for hazardous substances in foods have often been used as a measure of stringency of food safety standards. Most studies have found that they exhibit negative impacts on trade flows, particularly from developing countries. Otsuki et al. (2001a) analyzed the effect of European Union's (EU) MRLs for aflatoxin on the imports of groundnuts from African countries, and found a negative effect resulting from the tightened aflatoxin standard. Otsuki et al. (2001b) expanded product coverage to include cereals, dried fruits, and nuts for the same set of importing and exporting countries, and generally found a similar negative effect of the standard.

Additionally, other studies that deal with MRL standards and food products generally found negative standard effects. For instance, Wilson et al. (2003) observed a negative effect of the tightened MRLs for tetracycline, an antibiotic veterinary drug, on beef exports from major beef exporting countries. Chen et al. (2008) found a negative impact of tightened pesticide MRLs for chlorpyrifos on China's food and agricultural products, while Wei et al. (2012) identified a negative effect of pesticide residue standards on China's tea exports. There are also studies that focus on food safety standards other than MRLs, for example, Melo et al. (2014), Fontagné et al. (2015), Grant et al. (2015), and Crivelli and Groeschl (2016) all found a negative impact of food safety standards on trade.

Some recent gravity model studies point to the critical caveats of standard gravity models [Bergstrand (1985), Anderson and van Wincoop (2003), Santos Silva and Tenreyro (2006)]. Among other factors, the lack of theoretical orientation and failure to consider samples with zero trade values have frequently been subject to criticism. To solve the former issue, Anderson and van Wincoop (2003) introduced multilateral resistance terms to avoid potential bias: they measure the average barrier between an importer and exporter. To overcome the latter caveat, the widely used approaches include a sample selection model using two-step estimation, as proposed by Heckman (1979), and models that allow zero values of the dependent variable, such as the Poisson pseudo-maximum likelihood (PPML) method, proposed by Santos Silva and Tenreyro (2006), and its variants.[1]

---

[1] Counterparts of the PPML include, for example, the negative binomial pseudo-maximum-likelihood model, zero-inflated Poisson pseudo-maximum-likelihood model, zero-inflated negative binomial pseudo-maximum-likelihood model, and generalized negative binomial model.

Some recent studies employed the improved approaches of the gravity model. For instance, Ferro et al. (2013) examined the impact of importers' food safety standards on exports to 61 importing countries, finding a negative effect of the tightening standards on plant product trade. Xiong and Beghin (2012) re-examined the gravity model analysis of Otsuki et al. (2001a), incorporating multilateral resistance and zero trade values. They found no significant negative effect of the EU MRLs on aflatoxins in 2002 on African groundnut exports. However, they did find that domestic supply factors could have constrained African groundnut exports. Disdier and Marette (2010) applied the same method to estimate the impact of MRLs on antibiotics for the exports of crustaceans to the U.S., the EU, Canada, and Japan. Overall, the coefficient of the MRL variable is smaller than that of the standard gravity specification, implying the possibility of upward bias in the gravity model, which ignores multilateral resistance and samples with zero trade values.

Concerning other or more general sets of food or agricultural products, Disdier et al. (2008) applied the theory-oriented gravity model to examine the effects of SPS standards on the food and agricultural imports from Organisation for Economic Co-operation and Development (OECD) countries. They found that tightening SPS decreases trade. Drogué and DeMaria (2012) examined the impact of the pesticide residue limits of importing countries on apple and pear exports worldwide, finding that the more similar the regulations of importing and exporting countries are, the greater the trade flows. Winchester et al. (2012) analyzed the impact of pesticide residue standards on all plant products, and determined that the heterogeneity of regulations can deter the trade in both developing and developed countries.

While the gravity model, including its variants, only allows us to estimate the impact of regulations on trade flows, some studies conveniently extended the methodology to allow the distinction between the demand and supply impacts of standards or conduct welfare analysis.

For example, Xiong and Beghin (2014) applied a general equilibrium model to characterize the separate impacts of MRLs on the import demand and the trading partners' supply of plant products. From the demand side, a utility function with constant elasticity of substitution is used to derive the import demand of consumers from importing countries. On the other hand, the exporting country's supply is derived using product revenue maximization under a constant elasticity of transformation technology. They also assumed that the difference in MRLs between trading partners is a source of trade costs as follows. Exporters from countries with lax or no MRL regulations have to bear additional costs to reach the stringent limits required by importing countries. However, exporters who have been subject to more stringent MRLs on their domestic markets are more likely to comply with regulations at a negligible cost. The equilibrium trade value, as well as the equilibrium price, are then obtained for each product sector and any pair of countries by solving the above problems simultaneously. This solution gives the basic structure of a generalized gravity model, in which the demand-enhancing and trade-cost effects of MRLs can be separately identified by measuring the differential impacts of MRLs on exporters. By using bilateral trade flow data from 61 exporting countries to 20 high-income OECD countries, they found that MRLs enhance import demand and reduce export supply. Additionally, the demand-enhancing effect was greater for shipments from developed countries compared to those from developing ones. More importantly, MRLs were found to hinder the export flows from developing countries more than those from developed ones. This implies that developing countries comparatively face greater disadvantages in meeting food safety standards.

Several other studies based on the gravity models estimate the effect of standards on welfare. For instance, Disdier and Marette (2010) conducted an *ex-post* simulation analysis of the impact of standards on each country's welfare, finding that tightening regulations reduced the welfare of foreign exporters, but increased the total welfare of all countries. Wieck et al. (2012) developed a calibrated spatial simulation model (combined with a gravity model) to examine the effect of the demand and supply curve on welfare. They observed that a ban has a nearly prohibitive trade impact, whereas permissions to export uncooked meat to the non-infected regions of the infected country are trade enhancing. Cororaton and Peterson (2012) assessed the effect of the import ban imposed by the U.S. against Argentinian lemons under different scenarios. Overall, the losses in consumer welfare because of the import ban were larger than the gains from producer surplus.

Market-Level Studies of Demand Response to Food Safety Standards

The gravity models, despite their advantage of handling data for a wide range of countries, have limitations in analyzing the responses of markets to food safety standards, as well as those of consumers and suppliers, for an individual country. The demand response of a market can be better analyzed by allowing interrelations between products, such as substitution under the budget constraint, along with complementarity. A demand system analysis is thus suitable for determining the response of demand for differentiated products. Fourmouzi et al. (2012) used a demand system for both organic and non-organic fruit and vegetables. The estimation provides empirical evidence of interrelationships between organic and non-organic products as relevant cross-price elasticities. Own-price elasticities indicate that organic fruit and vegetables are more price elastic than their non-organic counterparts, and lower social class households with children have the most own-price elastic demand. Moreover, cross-price elasticities indicate relatively strong loyalty to organic products.

Source-differentiated demand data allow us to analyze the consumer response to food safety regulations. The interrelationships between products are particularly of interest when focusing on the same product that differs among the countries of origin with different food safety standards. Honda et al. (2013) used the source-differentiated almost ideal demand system (AIDS) model to analyze the response of Japanese demand for poultry from different countries of origin to its MRLs on veterinary drug residues. The results indicated an asymmetric demand response across countries of origin to the MRL regulations. The detail of these results will be subsequently discussed along with the discussion of the result of our gravity model analysis.

Micro-level studies on consumer's willingness-to-pay for food safety standards

Analyses based on micro-level data have the advantage of addressing the typical limitations of the macro- and market-level studies. Non-market valuation methods, such as contingent valuation and conjoint analysis, estimate the value of particular of product or activity characteristics through their changes. McCluskey et al. (2005) used contingent valuation to determine consumers' willingness to pay (WTP) for BSE-tested beef relative to non-tested beef, finding that consumers are willing to pay an average premium of at least 50% for the BSE-tested product. However, the unbounded valuation of beef attributes may have caused an upward bias in WTP when measured by the contingent valuation method. Therefore, the fundamental drawback of contingent valuation is its inability to allow for substitution among multiple attributes, as tastes may be instead substituted by safety attributes to some extent.

Conjoint analysis is also a non-market approach that assesses the impact of environmental changes using consumer survey data based on stated preferences, and can be extended to analyze WTP for favorable food safety standards. While contingent valuation requires individuals to state the values of particular characteristics, conjoint analysis allows them to choose from different choice sets identified through a set of attributes.

Several studies examined the WTP for food safety as one of the food attributes used in conjoint analysis. Bernard et al. (2007) employed conjoint analysis to estimate the US consumers' probability of purchasing chicken with different sets of attributes, including genetic modification, irradiation, antibiotics, and price. Chung et al. (2009) also used conjoint analysis to estimate Korean consumers' WTP for several attributes of imported beef, including country of origin, marbling, freshness, genetically modified organism (GMO) free, and antibiotic free. Gwin et al. (2012) estimated the WTP for the various attributes of beef in the U.S. considering grass-fed versus conventional grain-fed animals. All these studies included price in the regression analysis to estimate WTP.

Otsuki (2007) estimated the WTP of Japanese consumers regarding compliance with the food safety standards set by national regulatory bodies. The study employs conjoint analysis based on a choice experiment dataset of the demand for poultry of Japanese consumers. Several choice alternatives regarding product safety and general attributes, whose levels are systematically set, elicit consumers' preference toward these attributes. The safety attributes include suppliers' compliance with the Japanese standards for antibiotic residues and radioactive contamination, and past avian influenza incidence. The results of the mixed logit model indicated positive and large WTPs for antibiotic residues and radioactive safety standards. Moreover, the WTPs for antibiotic residues and radioactive safety standards are found to be higher for richer consumers, women, and respondents with children. The details will be subsequently discussed along with the result of our gravity model analysis.

Estimating the Effects of Food Safety Standards

Data

We have combined several datasets for our econometric analysis. The datasets used for estimation cover the period 2012–2016, and we have included as many countries as possible as per the datasets. The data on trade values have been obtained from UN COMTRADE. The products of interest include beef, pork, and chicken, categorized by the HS six-digit level. The data on the MRLs of veterinary drug residues have been obtained from the Global MRL database. The data on the distances between importing and exporting countries, commonness of official language, presence of colonial relationship, and common borders have been constructed based on CEPII. The data on the presence of regional trade agreements (RTA) have been constructed based on the RTA list of WTO.

The number of regulated veterinary drugs in Japan is considerably large and, thus, we aggregated their MRLs into an MRL index according to the formula proposed in Li and Beghin (2014):[2]

$$MRL_{jk} = \frac{1}{N_{(k)}} \sum_{n_{(k)}=1}^{N_{(k)}} \exp\left(\frac{MRL_{codex,kn_{(k)}} - MRL_{jkn_{(k)}}}{MRL_{codex,kn_{(k)}}} \right)$$

where $MRL_{jkn_{(k)}}$ is the MRL adopted by country j for product k, and targeting veterinary drug $n_{(k)}$; $MRL_{codex,kn_{(k)}}$ is the MRL recommended by the Codex Alimentarius' guideline for international food safety standards for the same product and veterinary drug combination; and $N_{(k)}$ is the total number of veterinary drugs applicable to product k for all the variables used for the econometric analysis. The more stringent the MRL is, the higher the value of MRL_{jk}. The descriptive statistics are presented in Table 1.

Table 1. Descriptive statistics.

	# of obs.	Mean	Std. dev.	Min	Max
Trade Value$_{sijt}$ (in thousands USD)	64,472	6.606	41.025	0.000001	1648.678
ln (Trade Value$_{sijt}$ + 1)	413,488	0.102	0.490	0	7.408
Total Export$_{sit}$ (in thousands USD)	413,488	70.444	275.651	0.000001	4450.733
ln (Total Export$_{sit}$)	413,488	0.113	3.860	−13.816	8.401
Distance$_{ij}$ (in km)	413,488	6779.529	4740.932	59.617	19772.340
ln (Distance$_{ij}$)	413,488	8.429	1.043	4.088	9.892
Language Dummy$_{ij}$	413,488	0.117	0.322	0	1
Colony Dummy$_{ij}$	413,488	0.0651	0.247	0	1
Contingency Dummy$_{ij}$	413,488	0.0353	0.185	0	1
RTA Dummy$_{ijt}$	413,488	0.533	0.499	0	1
MRL$_{sjt}$	413,488	1.123	0.212	0.379	1.586

Source: Authors' calculations.

---

[2] We would like to thank John Beghin for sharing the data on the index.

Estimation strategy

We apply a generalized gravity model to investigate how the variations of MRLs by country affect meat product trade. Following the methodology proposed in Xiong and Beghin (2014), we separately identify the demand-enhancing and trade-cost effects of the veterinary MRL standards, using the generalized gravity model. The estimable version of their generalized gravity model is derived by solving a general equilibrium model for consumers in importing countries and producers in exporting countries. The main econometric equation takes the form of a standard gravity model. The MRL variable enters two different terms as additional regressors. Therefore, the specification becomes:

$$\ln(T_{sijt}) = \mathbf{X}_{ijt}\boldsymbol{\beta} + \ln(E_{sit}) + \gamma \max\{MRL_{sjt} - MRL_{sit}, 0\} + \delta MRL_{sjt}$$
$$+fe_{it} + fe_{jt} + fe_{hit} + \varepsilon_{sijt}$$

where T_{sijt} is the actually observed bilateral trade flow from exporting country i to importing country j for product s in year t. $\mathbf{X}_{ijt}$ is a vector of control variables, which includes log distance between countries i and j, a common language dummy variable that takes the value 1 if the two countries use the same official language, a common border dummy variable that takes the value 1 if the two countries are geographically adjacent, a colonial dummy variable that takes the value 1 if the two countries had a colonial relationship in history, and an RTA dummy variable that takes the value 1 if the two countries have at least one signed RTA between them. The log of the total value of export E_{sit} of product s is included to control the exporting country's production capacity. The two terms containing the MRL index variable *MRL* capture the trade-cost and demand-enhancing effects introduced by Xiong and Beghin (2014), which are subsequently detailed. The term fe_{jt} represents the fixed effects for importing country j in year t, and fe_{hit} represents the fixed effects for product category under HS chapter h in exporting country i in year t.

Variable MRL_{sjt} is the MRL index for product s and country j in year t as previously explained. The term $\max\{MRL_{sjt} - MRL_{sit}, 0\}$ captures the trade-cost effect. Here, $MRL_{sjt} - MRL_{sit}$ is the deviation of importing country j's MRL from exporting country i's MRL. When the importing country's MRL is more stringent than that of the exporting country, $MRL_{sjt} - MRL_{sit}$ is positive, and is likely to impose an additional trade cost on the exporting country. On the other hand, the exporting country is unlikely to face additional costs if the importing country's MRL is less stringent than that of the exporting country. This implies that the term should be 0. Therefore, the term $\max\{MRL_{sjt} - MRL_{sit}, 0\}$ conveniently deals with both cases. The term MRL_{sjt} captures the demand-enhancing effect. Because a higher MRL index implies greater food safety, it can increase import demand.

We adopt the PPML method to estimate the generalized gravity model as it can also deal with a null trade flow in the dataset (349,016 out of 413,488 observations take the value 0). According to Xiong and Beghin's (2014) estimation procedure, the export regime equation should also be estimated. However, we do not include this equation in our estimation because we choose the PPML method to avoid discarding zero trade samples from the main equation. Additionally, for the robustness check, we use ordinary least squares (OLS) with dependent variable log $(T_{sijt} + 1)$ instead to avoid a log of zero.

Estimation results

Columns (1) and (2) of Table 2 show the results of OLS and PPML respectively, when all importing countries are included. The coefficients for the core variables in the gravity model, such as total export, distance, and the RTA dummy, are found to generally have the expected signs in both OLS and PPML. The parameter corresponding to the trade-cost effect (γ) is negative and significant under both estimation methods, which is consistent with the results of Xiong and Beghin (2014). The parameter corresponding to the demand-enhancing effect (δ) retains positive signs under both OLS and PPML, but it is significant in only OLS. In terms of the relative magnitude of the trade-cost and demand-enhancing effects, our result shows that the former effect is comparatively greater. This contrasts with the results of Xiong and Beghin (2014),[3] who found a greater demand-enhancing. Given that our magnitude of the trade-cost effect is similar to that of Xiong and Beghin (2014), the contrasting result is the smaller demand-enhancing effect for meat products compared to plant products. This may be because consumers do not appreciate the lower residues of veterinary drugs in meat products as much as they do the lower residues of pesticides in plant products. This may have to do with consumers' perception of risk: they might care more about pesticide residues as a health risk.

Columns (3) and (4) respectively show the result of the OLS and the PPML using only a sample of high-income importers, because sensitivity to food safety levels might vary depending on the income levels of importing countries.[4] The trade-cost effect is negative and significant, as is for the full sample. The demand-enhancing effect is also found to be positive and significant under OLS and positive but not significant under PPML. The magnitude of both the trade-cost and demand-enhancing effect parameters decrease for the subsample analysis. This is possibly because developed countries (high-income) have already greater MRL index value, which lowers the observed values of these parameters. For example, developed countries are thought to have a slightly higher average MRL index level (1.20 for high- and 1.08 for low-to-middle-income countries). Therefore, they consume meat products with a high average level of safety, which is thought to reduce the marginal impact of safety improvements.[5]

The dominance of the trade-cost effect of standards over the demand-enhancing effect implies that the coefficient of the MRL index, which is the net effect of the above two effects, is likely to be negative if estimated using the standard gravity model without separating these effects. This is consistent with the standard gravity model analysis of Wilson et al. (2003): a more stringent MRL standard on tetracycline as a

---

[3] According to Xiong and Beghin (2014, p.1198, Table 3), the coefficient of the trade cost effect is –0.250 (statistically significant at the 1% level), whereas that of the demand-enhancing effect is 0.707 (statistically significant at the 1% level).

[4] See the Appendix for the list of countries in our study. High-income countries are in bold.

[5] The possibility of the effect of MRLs varying with income levels motivates the use of an interaction term between the MRL terms and per capita GDP, which we will explore in our future research.

Table 2. Estimation results.

	(1)	(2)	(3)	(4)
	All countries		High-income importer	
	OLS	PPML	OLS	PPML
ln(Total Export$_{sit}$)	0.0297***	0.903***	0.0404***	0.939***
	(0.000281)	(0.0136)	(0.000548)	(0.0195)
ln(Distance$_{ij}$)	−0.0532***	−0.571***	−0.0805***	−0.447***
	(0.00150)	(0.0466)	(0.00311)	(0.0596)
Language Dummy$_{ij}$	0.0299***	0.304***	0.0212***	0.411***
	(0.00322)	(0.0743)	(0.00677)	(0.108)
Colony Dummy$_{ij}$	−0.104***	−0.171***	−0.129***	−0.160*
	(0.00419)	(0.0612)	(0.00756)	(0.0873)
Contingency Dummy$_{ij}$	0.526***	0.826***	0.580***	0.951***
	(0.0103)	(0.0629)	(0.0159)	(0.0862)
RTA Dummy$_{ijt}$	0.0677***	0.280***	0.127***	0.522***
	(0.00262)	(0.0882)	(0.00721)	(0.143)
Trade cost shifter (γ)	−0.273***	−2.260***	−0.237***	−2.752***
	(0.00791)	(0.315)	(0.0141)	(0.492)
Demand shifter (δ)	0.187***	0.290	0.109***	0.0589
	(0.00794)	(0.210)	(0.0159)	(0.388)
Constant	0.162***	3.701***	0.880***	1.431*
	(0.0259)	(0.577)	(0.0412)	(0.772)
Observations	413,488	64,472	138,154	30,349
R-squared	0.199	0.485	0.242	0.523

Note: The numbers between parentheses are standard errors. ***, **, and * imply statistical significance at the 1, 5, and 10% levels, respectively.

veterinary antibiotic in importing countries tends to reduce beef trade. However, it is possible that the demand-enhancing effect is not large enough to offset the trade-cost effect in Wilson et al.'s (2003) analysis, resulting in the net negative effect.

Synthesis of market- and consumer-level studies

The above analysis shows that the theory-oriented generalized gravity model can successfully identify the distinct effects of food safety standards on import demand and export supply. The results largely support the expected signs of the standards' effects. However, further empirical analysis from market- or micro-level studies on consumers would complement these results. Such analyses can estimate parameters as to capture consumer preferences on imported food products across different countries of origin, or over different combinations of food attributes. For example, the market-level demand system analysis can imbed a substitution effect that occurs between source countries with different MRL levels.

In a market-level study on meat products, Honda et al. (2013) used source-differentiated AIDS to estimate the response of Japanese demand for poultry from various countries of origin to the importing country's MRLs for veterinary drugs. The study demonstrated a highly heterogeneous response over various source countries.

Table 3 shows their simulation on the change in the demand shares of domestically produced poultry and from four major exporting countries and the rest of the world, based on their estimated parameters from the AIDS model. When the safety regulations of the importing country become more stringent, consumers feel safer to consume products from countries with less stringent food safety standards, which have been previously thought to be unsafe. This applies to poultry from Thailand and Brazil. Then, the reverse applies to that from the U.S. and China, or domestically produced.

This type of demand system analysis is not suitable to investigate the demand-enhancing effect at an aggregate level, but is suitable to investigate the heterogeneous demand response for source-differentiated products. If the demand-enhancing effect is present in the gravity model and one focuses on the demand-side effect by setting the supply-side effect aside, we can conclude that this is a force to increase demand in the importing country, but that the signs of the change in demand for poultry from a particular country may be affected by the distribution of consumer preferences.

In its consumer-level study dealing with meat products, Otsuki (2017) estimated the Japanese consumers' WTP, or price premium, for the guaranteed food safety of poultry in terms of compliance with Japanese veterinary drug MRLs and other attributes, using conjoint analysis. Although the implication is not exactly the same, WTP can be said to be closely related to the demand-enhancing effect. The positive WTP for food safety attributes implies a potential increase in prices when food safety is warranted by border regulations. Table 4 presents the WTP results for each of the food attributes, as well as their variations according to respondent income, gender, and having children. The first column shows the WTP estimates, and the WTP for compliance with Japanese antibiotic veterinary drug MRLs is 86.6% for the average unit price (JPY 200). The study also estimated the WTPs for compliance with Japanese radioactive standards, no history of avian influenza infection, domestic production, and freshness. Among them, the WTP for compliance with radioactive standards is close to that for antibiotic MRL. The remaining WTPs are relatively low.

Moreover, the third to the last columns show the changes in WTPs according to income and other respondent characteristics. The change associated with income is calculated based on a one-standard-deviation higher income (JPY 3.4 million) compared to average (JPY 5.5 million), and the increase in the WTP is found to be rather low (1.8% of the average unit price). Female respondents and those who have children are found to increase their WTPs by 5.5% and 16.8% by the unit price,

Table 3. Simulated demand shares of imported and domestically produced poultry under alternative MRLs.

Source countries	Base prediction	MRL 10% less stringent	MRL 10% more stringent
China	0.0985	0.109 (+0.0108)	0.0866 (−0.0227)
Thailand	0.0639	0.0579 (−0.0060)	0.0706 (+0.0127)
U.S.	0.0478	0.0537 (+0.0059)	0.0413 (−0.0124)
Brazil	0.0940	0.0686 (−0.0254)	0.122 (+0.0535)
Japan	0.695	0.711 (+0.0154)	0.678 (−0.0325)
Rest of world	0.000422	−0.000247 (−0.0007)	0.00116 (+0.0014)

Source: Adopted from Table 4 in Honda et al. (2013).

Table 4. Estimated WTP for food safety attributes (in JPY per 100 g).

	WTP (JPY)	% in unit price	Effect of an increase in income by one standard deviation* (JPY)	% in unit price	Effect of gender (female) (JPY)	% in unit price	Effect of having children (JPY)	% in unit price
Compliance with antibiotic MRL	173.2	86.6	+3.7	1.8	+11.1	5.5	+33.5	16.8
Compliance with radioactive standards	153.2	76.6	+3.6	1.8	+31.8	15.9	+21.4	10.7
No avian influenza	89.4	44.7	+1.8	0.9	+16.1	8.0	−1.0	−3.5
Domestically produced	31.9	15.9	+7.7	3.9	+35.7	17.9	+10.5	5.3
Freshness	3.6	1.8	+3.8	1.9	+9.8	4.9	+3.4	1.7

Source: Adopted from Table 7 in Otsuki (2017).
Note: Percentage in price is calculated based on JPY 200 per 100 g.
*One standard deviation is JPY 341 million in the used sample.

respectively, implying that women and those who have children tend to appreciate food safety more. Therefore, the study also demonstrates the heterogeneity of WTPs for food safety over respondents' income levels and gender, and whether they have children.

This micro-level study complements the gravity model one by demonstrating that the demand-enhancing effect is generally heterogeneous across individual consumers and, thus, the aggregate demand-enhancing effect in a particular country can also depend on the distributions of income levels, gender, and having children. Although the implication of income levels is rather limited, the demand-enhancing effect is expected to be higher for the high-income countries.

Conclusions

Despite their primary purpose to protect consumer safety, food safety standards can possibly detract food trade by imposing additional costs on exporting suppliers. The process of building international consensus on food safety standards should be based on reliable information on how food safety standards affect the economy, because the mechanism through which standards affect consumers, suppliers, and other stakeholders is extremely complex. This chapter provides a survey of the empirical studies concerning the effect of food safety standards.

Most empirical studies support the negative effects of food safety standards on food trade, but the analyses lacking theoretical foundation are misleading: most importantly, they fail to identify demand- and supply-side effects.

A theory-oriented generalized gravity model based on Xiong and Beghin (2014) is used to estimate the impact of food safety standards on meat product trade, and suggests statistically significant negative trade-cost and positive demand-enhancing effects although the latter are not significant depending on the estimation scheme. Unlike Xiong and Beghin's (2014) results for plant product trade, our estimated trade-cost effect for meat product trade is found to be greater than the demand-enhancing effect. Therefore, extra attention should be paid to avoid impeding meat product trade by raising the stringency of veterinary drug standards, unlike in the case of plant products. Referring to the results of market- and consumer-level studies on meat product imports, a potentially greater demand-enhancing effect in high-income countries may encourage imports with greater safety, thereby possibly rewarding suppliers' investments in compliance with the more stringent standards.

The empirical analyses conducted in this chapter and other relevant existing studies suggest further research in the following areas. Most importantly, gravity model studies should improve their flexibility as to incorporate the heterogeneity of countries in terms of stages of development and the greater range of product in one system, while allowing consumption substitution. Our findings demonstrate that the trade-cost and demand-enhancing effects can vary across countries and types of products. As market-level and micro-level studies are useful in complementing the macro-level ones, there is a limitation in their capacity to extend to comprehensive range of countries and products due to the limited availability of datasets.

Appendix

List of country

Exporter and importer (high-income importers are marked in bold):

Algeria, **Argentina**, **Australia**, **Austria**, **the Bahamas**, **Belgium**, Brazil, Bulgaria, **Canada**, **Chile**, China, Colombia, Costa Rica, **Croatia**, **Cyprus**, **Czech Republic**, **Denmark**, Dominican Republic, Ecuador, Egypt, El Salvador, **Estonia**, **Finland**, **France**, **Germany**, **Greece**, Guatemala, Honduras, **Hong Kong**, **Hungary**, India, **Ireland**, **Israel**, Italy, Jamaica, **Japan**, Jordan, Kenya, **Korea, Rep.**, **Latvia**, Lebanon, **Lithuania**, **Luxembourg**, Malaysia, **Malta**, Mexico, Morocco, **Netherlands**, **New Zealand**, Nicaragua, Pakistan, Panama, Peru, **Poland**, **Portugal**, Romania, **Russian Federation**, **Singapore**, **Slovak Republic**, **Slovenia**, South Africa, **Spain**, Sri Lanka, **Sweden**, Thailand, **Trinidad and Tobago**, Tunisia, **United Arab Emirates**, **United Kingdom**, **United States**, Vietnam
(71 countries, including 44 high-income countries)

Only importer:

Barbados, **Bermuda**, Indonesia, the Philippines, Turkey, **Venezuela**
(6 countries, including 3 high-income countries)

References

Anderson, E.J. and E. van Wincoop. 2003. Gravity with gravitas: A solution to the border puzzle. Am. Econ. Rev. 93: 170–192.

Bergstrand, H.J. 1985. The generalized gravity equation, monopolistic competition, and the factor-proportions theory in international trade. Rev. Econ. Stat. 71: 143–153.

Bernard, J.C., J.D. Pesek Jr. and X. Pan 2007. Consumer likelihood to purchase chickens with novel production attributes. J. Agr. Appl. Econ. 39: 581–596.

Chen, C., J. Yang and C. Findlay. 2008. Measuring the effect of food safety standards on China's agricultural exports. Rev. World Econ. 144: 83–106.

Chung, C., T. Boyer and S. Han. 2009. Valuing quality attributes and country of origin in the Korean beef market. J. Agr. Econ. 60: 682–698.

Cororaton, C.B. and E.B. Peterson. 2012. Potential of regional and seasonal requirements in US regulation of fresh lemon imports. World Econ. 35: 1022–1036.

Crivelli, P. and J. Groeschl. 2016. The impact of sanitary and phytosanitary measures on market entry and trade flows. World. Econ. 39: 444–473.

Disdier, A.C., L. Fontagné and M. Mimouni. 2008. The impact of regulations on agricultural trade: Evidence from the SPS and TBT. Am. J. Agr. Econ. 90: 336–350.

Disdier, A.C. and A. Marette. 2010. The combination of gravity and welfare approaches for evaluating nontariff measures. Am. J. Agr. Econ. 92: 713–726.

Drogué, S. and F. DeMaria. 2012. Pesticides residues and trade: the apple of discord? Food Policy 37: 641–649.

Ferro, E., T. Otsuki and J. Wilson. 2013. The Effect of Product Standards on Agricultural Exports from Developing Countries. World Bank Policy Research Working Paper No. 6518, the World Bank, Washington, D.C.

Fontagné, L., G. Orefice, R. Piermartini and N. Rocha. 2015. Product standards and margins of trade: Firm level evidence. J. Int. Econ. 97: 29–44.

Fourmouzi, V., M. Genius and P. Midmore. 2012. The demand for organic and conventional produce in London, UK: A system approach. J. Agrc. Econ. 63: 677–693.

Grant, J.H., E. Peterson and R. Ramniceanu. 2015. Assessing the impact of SPS regulations on U.S. fresh fruit and vegetable exports. J. Agr. Resour. Econ. 40: 144–163.

Gwin, L., C.A. Dhuram, J.D. Miller and A. Colonna. 2012. Understanding markets for grass-fed beef: Taste, price and purchase preferences. J. Food Distribution Res. 42: 91–111.

Heckman, J. 1979. Sample selection bias as a specification error. Econometrica 47: 153–161.

Honda, K., F. Kimura and T. Otsuki. 2013. Demand response for imported and domestic poultry meat products to food safety regulations in Japan: An application of the almost ideal demand system model. Osaka University, mimeo.

Li, Y. and J.C. Beghin. 2014. Protectionism indices for non-tariff measures: An application to maximum residue levels. Food Policy 45: 57–68.

McCluskey, J.J., K.M. Grimsrud, H. Ouchi and T.I. Wahl. 2005. Bovine spongiform encephalopathy in Japan: Consumers' food safety perceptions and willingness to pay for tested beef. Aust. J. Agr. and Resour. Ec. 49: 197–209.

Melo, O., A. Engler, L. Nahuehual, G. Cofre and J. Barrena. 2014. Do sanitary, phytosanitary, and quality-related standards affect international trade? Evidence from Chilean fruit exports. World. Dev. 54: 350–359.

Otsuki, T., J.S. Wilson and M. Sewadeh. 2001a. What price precaution? European harmonization of aflation regulations and African groundnut exports. Eur. Rev. Agric. Econ. 28: 263–283.

Otsuki, T., J.S. Wilson and M. Sewadeh. 2001b. Saving two in a billion: Quantifying the trade effect of European food safety standards on African exports. Food Policy 26: 495–514.

Otsuki, T. 2017. Consumer's Valuation of Food Safety Regulations: An Application of Conjoint Analysis. Mimeo. Osaka University, Osaka.

Santos Silva, J.M.C. and S. Tenreyro. 2006. The log of gravity. Rev. Econ. Stat. 88: 641–658.

Wei, G., J. Huang and J. Yang. 2012. The impacts of food safety standards on China's tea exports. China Econ. Rev. 23: 253–264.

Wieck, C., S.W. Schlüter and W. Britz. 2012. Assessment of the impact of avian influenza–related regulatory policies on poultry meat trade and welfare. World Econ. 35: 1037–1052.

Wilson, J.S. 2002. Standards, regulation, and trade. pp. 428–438. *In*: Hoekman, B., A. Matoo and P. English (eds.). Development, Trade and the WTO: A Handbook. The World Bank. Washington D.C.

Wilson, J.S. and T. Otsuki. 2003. Food safety and trade: Winners and losers in a non-harmonized World. Journal of Economic Integration 18: 266–287.

Wilson, J.S., T. Otsuki and B. Majumdar. 2003. Balancing food safety and risk: Do drug residue limits affect international trade in beef? J. Int. Trade Econ. Dev. 12: 377–402.

Winchester, N., M.L. Rau, C. Goetz, B. Larue, T. Otsuki, K. Shutes, C. Wieck, H.L. Burnquist, M.J.P. Souza and R.N. Faria. 2012. The impact of regulatory heterogeneity on agri-food trade. World Econ. 35: 973–993.

Xiong, B. and J. Beghin. 2012. Does European aflatoxin regulation hurt groundnut exporters from Africa? Eur. Rev. Agric. Econ. 39: 589–609.

Xiong, B. and J. Beghin. 2014. Disentangling demand-enhancing and trade-cost effects of maximum residue regulations. Econ. Inq. 52: 1–35.

8

Toward a Win-Win Integration of Agriculture and the Food Sector
Perspectives from the Mekong Region

Manabu Fujimura

Introduction

International trade in agriculture and food can play an important role in economic integration toward a win-win outcome in which less-developed economies realize their potential by joining cross-border value chains with regional as well as international players in the global market. This chapter focuses on the Mekong region where Cambodia, Laos, Myanmar, and Vietnam (CLMV)—relatively less-developed countries in Southeast Asia—are presented with new opportunities in this context. Growth of these economies through win-win linkages enhances their understanding of the benefits of producing high-value-added crops and stimulates institutional developments such as food safety laws, standards of good agricultural practice, and geographical indication. Such institutional development would in turn create a virtuous circle whereby the upgrading of agriculture and food sectors in these economies brings higher income to local farmers, and then contributes to further economic growth. With careful management of cross-border value chains, we argue that food credence attributes such as organic farming and geographic indication that are specific to the less-developed economies can be mobilized to enhance economic development toward a win-win outcome.

Department of Economics, Aoyama Gakuin University, Room 816, Building 8, 4-4-25 Shibuya, Shibuya, Tokyo, Japan, 150-8366.
E-mail: manabu@cc.aoyama.ac.jp

In this chapter, we first present the forms of agricultural integration observed in the Mekong region and discuss cross-border value chains that are emerging therein. We then discuss country-specific experiences and case studies of emerging models that combine external knowhow and existing comparative advantages in the region. Finally, challenges in further pursuing these opportunities are discussed.

Agricultural Integration and Cross-Border Value Chains Involving the Mekong Region

A type of agricultural integration unique to the geographically contiguous Mekong region can be termed "agricultural corridor". A corridor can be defined as a space of linear nature connecting large agglomerations (economic nodes) across a geographic area through a number of transport routes. Along these routes, the corridor links large urban centers (with high economic density) and other smaller nodes (intermediate cities and towns) that may exist in between the land surrounding the corridor (Nogales 2014). A corridor would normally develop progressively from a transport corridor through to a logistic variant, then on to trade, and finally to an economic corridor that involves various activities such as cross-border investments (Fig. 1). Some empirical studies in the Mekong region indicate that development of transport infrastructure appears to lead to such progressive evolution (e.g., Fujimura and Edmonds 2008, Fujimura 2017). The agriculture and food sector would be an essential part of such corridor development, connecting less-developed economies with their neighbors and beyond.

Agribusiness is one of the most commonly prioritized sectors that is expected to act as an engine for growth in developing country corridors. Governments see "agricultural corridors" as a source of foreign exchange earnings beyond food security, whereas foreign investors see it as an opportunity for high returns. Recently, host governments have become keen on fostering collaborative and inclusive business models, by which foreign investors seek to promote involvement by smallholders and local investors. However, bespoke regulatory frameworks are needed to ensure the sustainability and success of agricultural corridors. The debate is still open in terms of the best practices for incorporating agricultural development strategies into spatial development programs, because of the specificities of particular agricultural products in question (Nogales 2014).

Components of agricultural corridors include cross-border agricultural clusters and special economic zones (SEZs). They have become prominent features of regional development in the Greater Mekong Sub-region (or GMS, comprising Cambodia, Laos, Myanmar, Vietnam, Thailand, Yunnan Province, and the Guangxi Autonomous Region of China). Private firms participate actively, especially in ensuring reliable

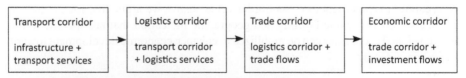

Fig. 1. Evolutionary development of corridors.
Source: Modified from Nogales (2014) p.9.

input supplies by relocating close to the border or entering into contract farming to guarantee supply availability (Nogales 2014).

Some cross-border SEZs are established along the routes of economic corridors linking neighboring countries (see Fig. 2 for the network of economic corridors in the GMS). One example on the Laos-Vietnam border is the Lao Bao Special Economic and Commercial Area, which includes coffee and fruit processing (canned and bottled juice beverages) (ADB 2010a). Along the East-West Economic Corridor (EWEC), the agro-industry is growing fast in several segments. These include the Nghe An Economic Zone in Vietnam, with major new private and foreign investments in beer, sugar, and milk factories: the Hoi An and Chu Lai Open Economic Zone in Tam Ky in Vietnam, with the presence of some agri-food processing firms such as PepsiCo and the Uni-President China food production company (Ishida 2012).

The most common factor in agricultural corridors, regardless of their locational characteristics (SEZ, cluster, or non-cluster), is cross-border contract farming. Chinese and Thai firms are increasingly engaging in cross-border contract farming with farmers in CLMV countries. Thailand has been actively pursuing contract farming as a tool for regional economic integration, building upon the Declaration of the 2005 Ayeyawady-Chao Phraya-Mekong Economic Cooperation Strategy (ACMECS) Summit. In the declaration, Thailand pledged tariff-free imports of all approved agricultural products derived under contract farming in ACMECS countries. Table 1 illustrates subsectors and value chains in GMS corridors where cross-border contract farming agreements abound.

It is worth noting that corridors exhibit heterogeneous climatic and topograhic characteristics, and thus different combinations of comparative advantages along their routes. For example, the North-South Economic Corridor provides good opportunities for investing in agriculture and the agro-industry, including food processing, non-food

Table 1. Contract farming experiences in GMS.

Corridor	Participating countries	Subsectors and value chains
North-South Economic Corridor	Laos	Rubber, tea, maize (northern Laos/Yunnan province of China)
	Myanmar	Cassava, sugar (Guangxi Region of China/northern Vietnam)
	China	Sugarcane, maize, watermelon, banana, cabbage, etc. (northern Laos/northern Thailand)
	Thailand	Biofuel from cassava, jatropha and sugarcane (China) Biofuel from jatropha and oil palm (Thailand) Bioethanol from maize, cassava and sweet sorghum (Myanmar)
East-West Economic Corridor	Laos	Sugar, beer, milk, and other beverages (Vietnam)
	Myanmar Vietnam Thailand	Pinewood oil, beer, and sugar (Laos)
		Rice (Thailand and Vietnam)
		Organic food (Laos, Myanmar)
Southern Economic Corridor	Cambodia Vietnam Thailand	Maize (Cambodia and Thailand)
		Ethanol from cassava and sugarcane
		Rubber, rice, pluses, fruit, and vegetables

Source: Modified from Nogales (2014), p.115.

Fig. 2. Economic corridors in the GMS.
Source: ADB (2012a) p.11.

agro-based industries (e.g., forest products and bioenergy industries), agricultural
machinery and equipment, and cottage industries linked to community tourism
(ADB 2010b). The Southern Economic Corridor offers good potential for investing
in the production and processing of commercial and industrial food crops, as well
as ethanol production from cassava and sugarcane (ADB 2010a). Similarly, The

East-West Economic Corridor prioritizes support to agriculture-based processing activities through cross-border contract farming in Laos' Savannakhet Province, and rice processing in Vietnam and Thailand.

Although CLMV economies can naturally exploit opportunities for agricultural integration within the Mekong region due to its geographical proximity, further opportunities lie in their integration with external players in the global market—or inter-regional value chains involving both advanced and developing economies. Firms in advanced economies including Japan have significant experience and accumulated knowledge and knowhow in adding value to agricultural products. They are keen to apply their expertise to emerging opportunities by collaborating with agricultural firms and farmers in developing economies. In so doing, however, it is essential to identify and package food credence attributes into value added which are specific to the producers.

Emergence of cross-border agricultural integration creates positive feedbacks for the participating countries in terms of understanding the benefits of high-value-added crops and stimulating institutional development such as food safety laws, agricultural standards, certification, and geographical indication (GI).

For example, Cambodia put into effect food safety legislation covering imported and domestic retail foods as well as street vendor foods beginning in 2016. Myanmar has started regulating unsafe food processing factories and promoting good agricultural practice (GAP) and aims at gaining international certification in this area. Vietnam amended its food safety law in July 2016 and now imposes heavy penalties on serious violations such as food poisoning. The AEON group (a large Japanese retailer), operating in Vietnam, has started labeling the origins of their meat, fish, and vegetable products as well as the usage of pesticides.[1] In May 2016, the Lao government developed standards for GAP and Organic Agriculture, and established the Lao Certification Body and Clean Agriculture Farm Models for GAP and organic agriculture.[2]

Certification of organic agriculture is very important in international trade for developing economies such as CLMV. The information attached to the produce generates value, reflected as a price premium for the certified products. Growth of the market for organic products in the Asian region is expected to be dramatic in the future as indicated by the rapid increase in land being certified as organic, which increased about 10-fold from 300,000 ha in 2001 to 3.2 million ha in 2012. China saw its organic market quadruple between 2007 and 2012 (Setboonsarng and Markandya 2015).

In upscaling organic farming in developing economies, its cost-benefit effectiveness is an issue. The costs of adopting organic agriculture include training, organizing smallholder groups, subsidies on inputs, inspection, and certification. In the four case studies from Thailand, China, and Sri Lanka that were examined by Markandya et al. (2015), the costs of inspection and certification were as little as 3%, but as much as 57% of total costs. Sri Lanka was on the high end, reflecting its high internal control costs. In China and Thailand, these costs were much more modest. From the Thai experience, certification costs did not exceed 5% when organic projects

---

[1] Nikkei News (in Japanese), 23 August 2016.
[2] Vientiane Times, 24 May 2016.

had grown to an optimum size. Most cases in their survey revealed the total production cost in organic agriculture to be lower than in conventional agriculture, whereas the former is more labor-intensive than the latter. They compared the costs of poverty reduction through the adoption of organic agriculture in China and Thailand with the cost of achieving the Millennium Development Goal (MDG) of halving the percentage of households in poverty (around $554–$880 per head) and make a qualified claim that the former would be about one-twentieth of the latter. Although the comparison is rough, it seems to indicate that organic agriculture can contribute to poverty reduction in a reasonably cost-effective way.

Interest in GI has grown worldwide and can promote traditional production methods and certification of products which are indigenous to developing countries. In the last decade, over 150 GIs have been registered in Southeast Asia alone. Thailand currently has 64 GI registrations, Vietnam 41, Malaysia 35, Indonesia 26, and so on. Some examples include Lamphun Brocade silk from Thailand, Kampot Pepper from Cambodia, and Van Yen Cinnamon from Vietnam. In order to receive a GI certification on a product, farmers must meet specification standards on the origin, production, and control mechanisms of the product.[3]

It is not widely known but coffee is one of Laos' top income earners among all exports of agricultural products. Currently contributing to the promotion of Lao coffee are internationally recognized standards such as Fair Trade and organic-produced certifications. GI would also help promoting Lao coffee. Coffee produced in the Bolaven plateau is on the priority list for Laos' first GI applications.[4]

In the following section, we will discuss country experiences and case studies of agricultural corridors in the Mekong region as well as emerging business models that combine external knowhow and existing comparative advantages in the region.

Country Experiences and Case Studies

Cambodia

As noted above, pepper is one of Cambodia's major agricultural exports. Large pepper farms are in Kampot and Kep provinces. Prior to 2013, close to 32 hectares of land were home to pepper plants. In 2014, the cultivated area had reached more than 90 hectares, largely due to the presence of new foreign investors. Companies from Japan, India, and a conglomerate of investors from Singapore, Malaysia, Hong Kong, and China are pouring money into pepper plantations, mostly in joint ventures with local partners. By virtue of GI, exports and prices are on the rise. The "Kampot pepper" GI brand covers a very specific area: five of the eight districts in Kampot and one district in Kep.[5]

Hironobu Kurata, a Japanese volunteer-turned entrepreneur arrived in Cambodia in 1994, leased a land of about 6 ha from a local traditional pepper-producing farm in Koh Kong Province in 1997, and started pepper production for the Japanese market. The farm's location is suitable for pepper production with abundant rainfall,

---

[3] Vientiane Times, 28 October 2016.
[4] Vientiane Times, 22 October 2015.
[5] Phnom Penh Post, 21 February 2014.

good soil drainage, and short daylight hours. The farm hires 14 local staff. Drawing on local traditional knowhow, Kurata aims at establishing a "Made in Cambodia" brand. The farm sends its harvest to a processing factory in Phnom Penh, where peppers are selected and packaged in exportable form. In 2006, under the technical assistance of IFC (International Financial Corporation) GTZ (Deutsche Gesellschaft fur Technische Zusammenarbeit), Kurata helped the country establish the Cambodia Organic Agriculture Association and chairs it currently. In 2011, "Kurata Pepper" gained a certificate for organic crops in Japan. Kurata pepper is now sold at about 60 contracted food stores in Japan, whereas the product is exported to Europe under the "Kampot Pepper" brand as it is so designated under European GI.[6]

Within the Mekong region, Charoen Pokphand Foods (CPF), one of the multinational arms of CP Group, a large Thai conglomerate, has invested about US$8 million in Cambodia's Pailin Province to develop a silo and drying plant for maize. The purpose is to ensure high-quality raw material to serve its feed production for both the domestic market and exports. Since 1995, CP has invested a total of about US$100 million in Cambodia, including feed production, livestock farming, and food processing such as sausage-making and slaughter houses.[7]

Cambodian products enjoy duty-free access to European markets under the Everything But Arms (EBA) Agreement. An example taking advantage of this status is Confirel Company Ltd. It was established as a social enterprise to contribute to the sustainable development of rural Cambodia by providing farmers with regular income based on palm sugar. In 2004, the company received organic certification for palm sugar from ECOCERT, an organic certification organization based in France. Confirel purchases palm sap and palm sugar from farmers at a premium price under a contractual farming arrangement. In 2009, about 500 households were under contract with Confirel in the Kampong Speu province. Kampong Speu Palm Sugar, along with Kampot Pepper, was among the first Cambodian products to earn GI status. The collection and processing of its sugar have traditionally been an important source of income for many rural households. However, these traditional livelihood activities are poorly paid and, as a result, have recently come under increasing pressure from rural–urban migration, as well as illegal logging. Against this backdrop, Confirel initiated the production of organic palm products for export. The company provides technicians who train farmers in sap collection and sugar production in accordance with organic standards. The sap is treated and manufactured into palm sugar, palm wine, or palm vinegar. Confirel has developed new techniques for manufacturing products from palm sap, allowing the company to produce crystallized fruits, fruit jams, palm wines, and cocktails (ADB 2012b, pp. 1–8).

In summary, Cambodia shows high potential vis-a-vis forming an inter-regional value chain with Japanese and European firms in specialized crops such as pepper. Preferential market access in Europe is an advantage. Cambodia also provides an example of forming an agricultural corridor with Thai agribusinesses.

---

[6] Kurata's presentation at an investment seminar convened by the Japan-ASEAN Center, Tokyo, on 13 June 2017.

[7] The Nation, 4 March 2013.

Laos

Although Laos is a mountainous land-locked country, it is endowed with vast untapped natural resources including land. Particularly, its southern provinces possess large potential for agricultural products. For example, the Bolaven Plateau that lies across Salavan, Sekong, Champasak, and Attapeu Provinces is considered a base of high-value organic crops, especially coffee and vegetables. Coffee beans produced there are exported mainly to Europe, Japan, and Vietnam, whereas vegetables are exported mainly to Thailand.

Coffee planting in Laos originated in the 1920s when French colonizers brought in Arabica species and started coffee production. As of 2013, the coffee-planted area is about 77,000 ha countrywide but 70,000 ha are concentrated on the Bolaven Plateau with its elevation at about 1200 m and suitable climate for coffee. Most coffee production is undertaken by smallholder farmers. In recent years, investors from Vietnam, Thailand, Singapore, China, and so on, have invested in large-scale plantation in the plateau.[8]

The Bolaven Plateau Coffee Producers' Cooperative, established in 2007, has been exporting more than 1000 tons of organic coffee per year. France is the top buyer accounting for some 800 tons per year. The members of the cooperative comprise 1391 families from three provinces, including 826 families engaged in the organic production process. There are also 56 producers' groups, including 36 groups focused on organic production. The current area dedicated to growing the cooperative's organic coffee is around 3000 ha. Many foreign countries especially in Europe are interested in Lao organic coffee and have offered a better price than what is received for normal coffee.[9]

Coffee has been Laos' top agricultural export but it has yet to make substantive inroads into Japan, the largest neighbor market in Asia. Consumers there have a low recognition of Lao coffee, despite its high recognition among coffee experts. Although coffee brands such as Dao-Heuang and Sinouk are well known and popular among Japanese tourists visiting Laos, the scale of coffee exports to Japan is very limited compared with coffee from Latin America and Africa. Thus, a further effort for GI is warranted in order to expand exports to this increasingly affluent Asian market. The same goes with Vietnamese coffee as will be discussed below.

There is also huge potential for the production of organic vegetables in Laos, particularly around the Bolaven Plateau. For example, Pakxong Development Enterprise Export-Import Co., Ltd., founded in 2004 by Ms. Inpeng Samuntee, an entrepreneur formerly associated with timber trade, succeeded in establishing export contracts with buyers from Thailand, Europe, and the Middle East. It has a logistics service agreement between Lao and Thai companies, under which refrigerated cargo trucks transport fresh organic vegetables from the Bolaven Plateau to a warehouse at a Thai port, where they are packed and loaded onto cargo planes or ships for export destinations.

Initially the company contracted with about 160 local farm households to produce cabbages and now contracts with over 10,000 households for various vegetables with a total cropping area of about 7500 ha and total exports of 80,000 tons per year

---

[8] JETRO Daily 8 May 2015.
[9] Vientiane Times, 6 December 2016.

to neighboring countries.[10] The company organized a cross-border trade agreement between Thai buyers and Lao sellers, with the endorsement of the commerce authorities of Thailand's Ubon Ratchathanee Province and Laos' Champasak Province (Fig. 3).

The company provides direct loans to its field representatives for the purchase of trucks used for collecting produce and transporting it to the delivery center near the border. Because the soils are quite fertile, vegetables can be grown with a minimum of agrochemicals. When the produce reaches the delivery center near the border, it is unloaded, trimmed, re-packed, weighed, and then loaded onto Thai buyers' trucks. Buyers pay cash for the vegetables at the price agreed with the field representative, who then delivers cash payments to the farmers for the vegetables sourced that day.

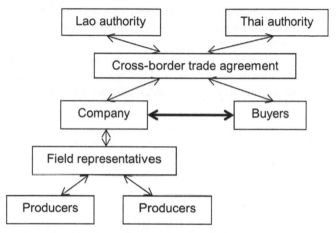

Fig. 3. Stakeholders in a cross-border trade agreement for fresh vegetables.
Source: ADB (2012b) p.10.

Lessons from Ms. Inpeng's case include: (i) some degree of collaboration between producers, exporters, and buyers can result in additional benefits for all stakeholders as opposed to producers independently planning production and engaging in price competition; (ii) contracting for minimum prices and volumes benefits all transacting parties, and provides exporters with sufficient certainty to encourage them to invest in improving the supply chain (ADB 2012b, pp. 9–14).[11]

In Sekong Province, a Japanese agricultural company, Agri-San, has been trying to solidify exports of high-value vegetables to Japanese markets. Taniyama Siam, a Thai subsidiary of Agri-San, had an eye on Laos with its large area of untapped virgin land. Its newly established "Advance Agriculture" exports fresh vegetables, especially okra, from Laos to Japan.[12]

Their Lao operation started land preparations in 2007, built an access road, a packing factory, and so on, from virtually nothing. Their crops are "organically"

---

[10] Vientiane Times, 22 February 2013; Nikkei News (in Japanese) 12 September 2014.

[11] The author met Ms. Inpeng in person in March 2009 and found her exceptionally business-oriented and flexible.

[12] In 2013 Agri-San was purchased by JALUX Fresh Foods. Both Taniyama Siam in Thailand and Advance Agriculture in Laos are now owned by JALUX Asia Ltd. in Thailand.

produced but some pesticides are necessary to keep harmful insects away. Integrated Pest Management (IPM) is adopted. Cut plants are recycled, fermented, and made into organic fertilizer. "Dense planting" occurs to maximize yields. Thorough on-site data collection is carried out: acidity, moisture, sunlight, and so on, are measured daily by local staff and communicated to Taniyama Siam in Thailand and Agri-San in Japan for remote management. Their Thai operation is under contract farming but the Lao operation is completely self-controlled for the time being to secure efficiency and stability. Lao okra is priced more than three times higher than Thai okra. Further, it meets demand in a slack period left by major okra producers in southern Japan such as Kagoshima, Kochi, and Okinawa prefectures.

Figure 4 shows the flow of okra from the farm in Sekong Province to the final consumers in Japan. The whole process takes about three days, which is within a normal perishing limit of okra. Although it was not possible to obtain the exact breakdown of the value chain for this case, the author's interviews with the company in Japan as well as in Thailand provide indicative information in this respect, as shown in Table 2. About half of the wholesale value appears to fall into Lao resources including labor. When we talk about a "win-win" cross-border linkage, this type of information is

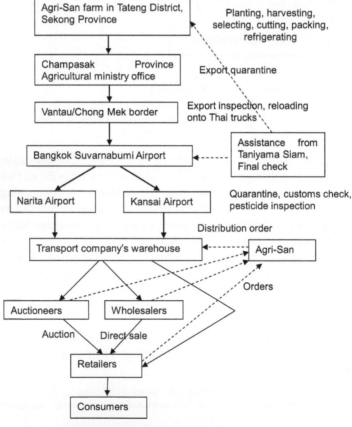

Fig. 4. Flow of okra exports from Laos to Japan.
Source: Author's interview.

Table 2. Value chain of okra export from Laos to Japan.

Value/Cost attribution		Value share
Laos	Seeds, fertilizers, pesticides	52% of wholesale price
	Packing material	
	Labor (about 1/5 of production cost)	
	Other inputs including land	
	Costs at the Lao/Thai border	
Thailand	Transport, logistics, and others	28% of wholesale price
Japan	Quarantine, customs, inspection	20% of wholesale price
	Forwarding	
	Promotion/advertisement	
Wholesale price		100%
Retailer's mark-up		30%–40%

Source: Author's interview.

most valuable. Likewise, when we talk about "fair" trade, as claimed by many "fair traders," often with certification, the ultimate verdict as to whether they are really fair should come down to disclosure of value chains such as this.

Within the Mekong region, again, Thai agri-business has an advantage with its geographical proximity to Laos. Mitr Phol Group, Asia's biggest sugar producer, reportedly earmarked about US$110 million to double production in Laos to 90,000 tons by 2015. Of the total amount, about 70% goes to developing production facilities and the rest to farm plantation. The company had branched into Laos in 2006 with its wholly owned Mitr Lao Sugar Co. It obtained a 40-year concession from the Lao government for a sugarcane plantation area spanning 62,500 rai, or 100 km². Up to 80% of sugar production in Laos is exported. The major sugar makers in Laos are two Thai companies—Mitr Phol and Khon Kaen Sugar Industry Plc—and a Vietnamese firm, Hoang Anh Gia Lai.[13]

In summary, Laos shows high potential in forming agricultural corridors in organic vegetables and coffee with neighboring countries, as well as an inter-regional value chain with Japanese firms.

Myanmar

Myanmar was called the rice basket of the world from the early 1900s to the 1940s encouraged by the British colonial government. However, the country was largely withdrawn from the rice export market from the early 1960s until recently due to its isolation under the "Burmese way of socialism" adopted by the military regime. Myanmar's rice sector was collectivized and sales to the government were forced, with rationing under a controlled price ceiling. In principle, farmers had to surrender all their harvest except that which was used for their own consumption and seeds. As the procurement price was less than half of the black market price, the rice procurement system was effectively a heavy tax on the producers. Although the government came to realize the problem and abolished the system of coerced rice sale rationing in 2003,

---

[13] *Bangkok Post*, 29 May 2015.

it was only in 2012 that the government liberalized rice exports completely for the private sector (Kubo 2013).

With the opening up of the economy since late 2012, Myanmar is now well positioned to regain its role as a rice exporter. Indeed, the country's rice exports rose more than 40% in FY2014-15, with border exports to China being dominant despite Beijing's official ban on imports of Myanmar rice for food safety reasons.[14]

The Myanmar Agribusiness Public Corporation (MAPCO), established in 2012, is trying to modernize the country's rice production system: upgrading rice mills, improving seed quality, providing farm machines to farmers, and so on. Japan's Mitsui Corporation, a *sogoshosha* or large trading company, entered into an alliance with MAPCO with the aim of exporting rice to Japan. Kubota and Yanmar, Japanese manufacturers of agricultural machinery announced plans to establish import bases in Myanmar. Marubeni, another Japanese trading company plans to establish a fertilizer factory in Tilawa Industrial Zone, in a suburb of Yangon. A Singaporean chemical company, Ben Mayer, formalized a joint-venture with MAPCO for fertilizer production in May 2016.[15]

Although rice would be a natural choice for Myanmar's comeback as an exporter of agricultural products, high-value crops are likely to follow as the country, such as Laos discussed above, still possesses large areas of virgin land, ironically as a result of underdevelopment for many decades. For example, a domestic food company named Myanmar Belle entered into alliance with Japanese companies with their know-how in frozen food and plans to export to Japan a variety of vegetables including taro, okra, green beans, and spinach; about 5000 tons per year of these vegetables are produced by local contract farmers. They would be used for ingredients for processed food in Japan. If realized, Myanmar would surpass Vietnam as an exporter of frozen vegetables to Japan.[16]

In summary, the huge potential of Myanmar's agriculture is just starting to unfold. Modernization of its agriculture would greatly help food security for its own people as well as neighboring countries and beyond. Foreign direct investment would flow in to tap the country's virgin land. Therefore, the government needs to carefully formulate investment rules for sustainability. The country shows solid potential in forming an agricultural (rice) corridor with China. It also offers promise of an inter-regional value chain in frozen vegetables for the Japanese market.

Vietnam

Vietnam is also endowed with suitable climate and topography for high-value agricultural products, particularly in its south-central highland areas.

Vietnam is the world's second largest exporter of coffee after Brazil. Dak Lak Province, at 500–1000 m above sea level, is the main producer of Vietnamese coffee as well as rubber and pepper. Like Laos, coffee production in Vietnam originated in 1910–1920s when French colonizers brought Robusta species (as opposed to Arabica

---

[14] Irrawaddy, 8 April 2015.
[15] Nikkei News (in Japanese), 6 July 2016.
[16] Nikkei News, 21 November 2015.

species in Laos due to altitude differences). There were more than 40 plantations in the province by 1960s. About 40% of Robusta coffees marketed around the world originate in Vietnam. They are also used in processed variants such as instant coffee and canned coffee.

However, conventional coffee production in Vietnam depends heavily on chemical fertilizers and water. Although certification has become increasingly popular, excessive application of irrigation and chemical fertilizers have made it difficult for Vietnamese coffee to apply for overseas certification. Ho et al. (2017) examined economic and environmental performance of conventional versus alternative production that is compatible with eco-friendly certification. Their conclusions included: (i) coffee farms could reduce environmental pressures by more than 50% without sacrificing coffee produce revenue; (ii) participation in certification schemes reduces the use of irrigation water and other environmental pressures; (iii) there appears to be a convergence in the level of eco-efficiency between certified and non-certified farms after several years.

Dalat in Lam Dong Province, with its mild climate at 1400–1500 m above the sea level, is renowned for its wine production as well as organic vegetables. Also, endowed with unpolluted soil, Dalat attracts numerous foreign investors, particularly from Japan. A dozen Japanese affiliated firms operate in the Lam Dong Province producing vegetables as well as flowers. Currently Vietnam is at the most advanced stage among CLMV in commercialized organic agriculture. Vietnamese consumers are awakening to the importance of food safety and the Vietnamese government welcomes foreign investors who bring in skills and knowhow on good practice in this area.

An example of Japanese investment in organic vegetables is NICO NICO YASAI ("happy smile vegetables"). The company operates three farms at Buon Ma Thuot (Dak Lak Province), Krong Bong (Khanh Hoa Province), and Mang Den (Kon Tum Province) which serve markets in Ho Chi Minh City, and one farm at Moc Chau (Son La Province) which serves markets in Hanoi.[17]

AEON supermarket in Ho Chi Minh City started to sell "Japanese strawberries" shipped from Dalat. They are priced four times higher than regular local strawberries but are very popular among shoppers due to the "Japan" brand image. They are produced by a Japanese affiliated farm with low usage of chemical fertilizers and pesticides. Similarly, in Datat, a Vietnam-Japan joint venture named An Phu Lacue started producing Japanese premium brand lettuce (supposed to be sweet and crunchy) in 2014, working with local contract farmers. The lettuce is retailed at local Co.opmart, Big C, Metro, and so on, under the "Asagiri" brand. As the quality of the lettuce attracted many local farmers, the company established the product standard and now purchases lettuce from non-contract farmers as well. The company plans to export lettuce to Hong Kong and Singapore in the near future. Although Japan imports a variety of vegetables from China, food safety is becoming an issue for Chinese-made ingredients among Japanese restaurants and ramen shops, and so on. Traders dealing with imported vegetables now look for alternative sources to diversify risk. Dalat-made vegetables are promising candidates. With an improved cold chain infrastructure in the future, export of this fresh produce to Japan, as well as to neighboring markets such as China and Thailand, would emerge as new opportunities.[18]

---

[17] Home page for NICO NICO YASAI: http://niconicoyasai.com/ja/home-page/
[18] Nikkei News, 5 March 2015.

A unique example of Japanese high-tech investment is Rrfarn Green Farm Co., Ltd., a "vegetable factory" established by an auto parts company at Lon Hau Industrial Zone in Long An Province of Mekong Delta. Its main products include organically grown frill lettuce using artificial lighting in a closed sterilized space. It is aimed at markets in Southeast Asia.[19] Similarly, Fujitsu and AEON Agri collaborate in operating an IT-based fruit and vegetable factory (e.g., producing high-sugar content tomatoes) in Ha Nam Province in the north, with an area of 1000 m² set aside for pilot farming. Fujitsu administers the data collection system and AEON AGri analyzes the data and supervises local farmers remotely from Japan. Fujitsu has formed an alliance with PFT, Vietnam's fastest growing IT firm, and intends to market its "crowed-based" farm management support system, under which temperature, moisture, and other conditions are digitally controlled and monitored.[20]

With its fertile Mekong Delta in the south, Vietnam is one of the major exporters of rice. Angimex-Kitoku, a joint venture between An Giang Province's trading corporation and Japan's rice wholesaler Kitoku Shinryo, established rice mills in the province and exports locally produced *japonica* rice to Singapore, Malaysia, Indonesia, as well as the USA and the EU. The company further plans to produce Japanese brand "Koshihikari" with its contract farmers and export it to Japanese retailers and restaurants.[21]

Local efforts for strengthening high-value agricultural products also deserve attention. For example, GI is gaining recognition as a way of adding value for export-oriented produce. Phu Quoc Nuoc Mam (fish sauce made on Phu Quoc Island in the south) is a case in point (JETRO 2017). Nuoc Mam is made from only fish and salt with no additives or preservatives. Although Thai Nam Pla is processed quickly using concrete tanks, Nuoc Mam is produced naturally in a wooden barrel taking a long period of time and is considered to have a unique flavor and superior nutritional content. Regular Nuoc Mam uses two to three kinds of fish for ingredients but Phu Quoc Nuoc Mam uses six kinds of fish caught in the Gulf of Thailand which is an area endowed with rich seaweed to sustain the fish. The caught fish are placed in salt in the fishing vessels and once at the ports they are immediately transported to the island's processors. Freshness and mild odor are the distinctive features of Phu Quoc Nuoc Mam. It took Phu Quoc Nuoc Mam Association 11 years to gain GI status through trial and error. Finally in 2012, the association gained GI status (called "PDO"). Maintenance of the GI status requires adherence to the traditional production method and adequate hygiene protocols.

Social enterprise is an alternative way of promoting high-value niche products for poor farmers. In 2007, the International Market Development and Investment Joint Stock Company (MDI) was founded as an outgrowth of a personal commitment to social entrepreneurship and ecological production (ADB 2012b, pp. 23–28). MDI deals in fair trade agricultural and handicraft products that are produced in an environmentally responsible manner. It obtained certification from FLO Cert GmbH for export and processing of fair trade tea and became the first fair trade certified exporter and processor of cashewnuts in Vietnam. MDI works exclusively with small-

---

[19] Homepage for Rrfarn Green Farm: http://www.rrfarn.com/
[20] Nikkei News, 19 January 2016; 25 February 2016.
[21] Nikkei News, 17 October 2015.

scale farmers and also works closely with producer groups set up by associates. These producer groups share two characteristics: (i) they are located in remote and poor areas of the country, and (ii) the vast majority of their members belong to ethnic minorities.

MDI's domestic sales are mainly carried out in Hanoi, where it maintains its own retail outlet while part of their products are exported, including loose-leaf tea, teabags, green coffee beans, roasted coffee, and roasted cashewnuts. Although MDI initially focused on European export markets such as Denmark, Germany, and the Netherlands, export markets within Asia—particularly Hong Kong, China, and potentially Japan—are becoming increasingly important. MDI provides interest-free advances to producers for financing farm inputs and they are deducted from purchase payments. MDI employs its own extension staff. Some specialists from research institutes and provincial agricultural extension services are hired to provide specific training, for example in terms of compost-making.

In summary, Vietnam provides numerous examples of inter-regional value chain formation for the Japanese markets in various agricultural products. Among CLMV, Vietnam has, thus, far gained the most experience in aiming at global high end markets through establishing GIs.

Conclusions and Challenges

The country experiences and case studies discussed herein indicate the potential and promise for less-developed economies in the Mekong region to increase agricultural income by forming agricultural corridors in the region as well as inter-regional value chains in the global market. Contract farming is the most common form of regional integration, whereas configuring and maintaining inter-regional value chains requires external investors to bring their production and marketing expertise which has already been well-tested in the global market. In either context, food credence attributes such as organic farming and geographic indication that are specific to the Mekong region play key roles. However, each economy has unique constraints in terms of geography, human resources, technology, institutions, and so on. This final section focuses on these aspects.

For Cambodia, its limited market and underdeveloped logistics infrastructure do not provide incentives for establishing local certification bodies. Laos is also hindered by a limited market size and, additionally, the geographical disadvantage of being land-locked leading to high transport costs. Myanmar, with its long period of economic isolation, also suffers from underdeveloped transport and logistics infrastructure as well as a lack of market-based institutions in all sectors. Although the country is on the right track to undo decades of stagnation and decline, it will take some time to meet the dual challenges of stabilizing civilian control of the country and preparing market-based institutions for sustainable economic growth. Vietnam has been best positioned among CLMV to attract investments toward win-win agricultural linkages in global value chains. Vietnam is endowed with a reasonable amount of agricultural land as well as abundant labor, but its virgin land for organic farming is limited. Its domestic GAP (VietGAP introduced in 2008) covers only 0.2% of vegetable farms nationwide.[22] It may well be the case that those Vietnamese entrepreneurs who have been successful so far in high-value crops, bring their experiences and knowhow to bear on Cambodia and Laos in order to further advance regional integration.

In the following, specific challenges associated with the country experiences delineated in the previous sections are discussed.

In the case of Kurata Pepper in Cambodia, Mr. Kurata notes the country's almost complete lack of mid- and downstream clusters in the agricultural sector, such as processing, packing, and logistics specific to agriculture. That part of the sector is currently dependent on external investors and traders. To break this sheer constraint, he recommends that the government and donors coordinate in setting up an agricultural special economic zone where public markets, refrigerated warehouses, rental processing factories, inspection facilities, export procedures, and so on, are all offered.

In the case of Confirel, the limited size of the domestic market does not provide incentives for establishing a local certification body or for foreign certification companies to establish offices in Cambodia. Confirel depends on using inspectors from importing countries. This has been a major barrier to exporting to countries with high food quality standards. Consequently, the company's production of palm products continues to be subsidized by its more profitable business activity of importing wine and food products from France into Cambodia. The public sector can significantly lower the cost of certification and increase participation by local producers in export markets by facilitating establishment of a national certification body.

In the case of okra production by Agri-San in Laos, the company points out the cumbersome administrative procedures in cross-border transport. For example, Champasak's provincial office in Pakse, in charge of plant quarantine, does not reasonably cater to the requirements of private sector businesses. Transshipment between Laos and Thailand involves a change of vehicles and drivers, adding to transport costs. There is much room for improvement in terms of the Lao government's awareness and pro-active facilitation of cross-border movement of goods and people. In addition, improvement in cold chain logistics is essential to realize Lao's export potential for high-value vegetables.

For Myanmar, its duty-free access to the EU market under the EBA policy has been a particular boon to its garment industry so far but has also opened the door to the rice trade. Only Cambodia and Myanmar enjoy this duty-free access among regional rice exporters. But it is still difficult to attract European buyers, as the domestic industry has trouble meeting their high standards. To accelerate developments and improvements in its business environment, Myanmar should engage foreign investors with knowhow and experience in order to exploit opportunities like the EBA.

In the case of Vietnamese coffee, challenges include the following: (i) Brand recognition is low and processing for higher value is limited. Most of the exported coffees are in raw beans and their export prices are low. Efforts are required to promote domestic firms' capacity to process coffee and brand marketing. Further collaboration with Japanese firms with processing technology and local producers is warranted; (ii) International coffee prices are unstable and often move downward suddenly. Even when they move upward, export prices of Vietnamese coffee often do not keep up with

---

[22] JETRO Daily, 1 May 2015.

them due to inadequate communications infrastructure. In 2015, Dak Lak Province set up Buon Ma Thout Coffee & Commodity Exchange to facilitate online international transactions and to align export prices with international trends. Also, in the same year, Lam Dong Province collaborated with Japan's UCC Coffee to convene the "Vietnam Coffee Competition" in Datat and invited baristas from inside and outside the country to evaluate local coffees; (iii) climate change appears to bring increasingly worrisome impacts such as a decrease in water availability in the dry season (December to March) on the one hand and intensified rainfall washing soil down slopes in the wet season (April to November). Foreign NGOs provide technical support to local cooperatives in mitigating such effects via water-saving production methods, underground water management, on-farm planting, and grafting.[23]

In the case of Phu Quoc Nuoc Mam, its producers face two major challenges. One is a dwindling production level, from 30 million liters in 2012 to 20 million liters in 2016. This downward trend can be explained in terms of dwindling fish catches around the island due to overharvesting and the difficulty of expanding production capacity because of the traditional process required by GI certification protocols. Phu Quoc Nuoc Mam Association note that "Some fishing members voluntarily constrain the harvesting time but improved management requires government intervention on the fishing cycle as well as environmental regulation." Another challenge is the lack of capacity for marketing overseas. Most of the producers are small-scale and do not have resources for marketing. This has made the brand vulnerable to piracy. Some self-defensive measures are necessary as well as government support for brand promotion overseas (JETRO 2017).

In the case of MDI, the social enterprise, a major problem is achieving consistency in the quality and quantity of output. The company has learned that it must reduce its reliance on European markets and focus additionally on Asian markets such as Hong Kong, China, Japan, as well as neighboring countries in Southeast Asia. Although markets for organic products are emerging in Asian countries, "fair trade" remains broadly unfamiliar to their consumers. Expanding markets require collaboration between the government, international agencies, and NGOs in facilitating eco-friendly commercial activities and their marketing.

In Vietnam, the application of VietGAP covers production of vegetables, fruit, tea leaves, rice, and coffee. But, its requirements include analyses of soil and water quality and associated facilities. Such requirements are often beyond the resources available to local smallholders. Neither are such practices well recognized among local consumers. In the meantime, the government introduced BasicGAP, a simplified version of VietGAP, in order to accommodate the constraints faced by smallholders. It reduces safety standard procedures for vegetable production, for example, from 65 in VietGAP down to 25, and downgrades many of them from "required" to "desirable" in order to qualify for BasicGAP.[24]

Finally, key factors to the emerging cross-border business models are their up-scalability, replicability, and sustainability. The extent of up-scalability and replicability depends on public and private sector coordination and non-profit sector assistance.

---

[23] JETRO Daily, 8 April 2015.
[24] JETRO Daily, 1 May 2015.

In particular, constraints for smallholders in terms of quality control, price volatility, and brand recognition need to be addressed. The extent to which these emerging business models are sustainable depends on an equitable share of value accruing to all stakeholders from investors, producers, traders, marketers, and down to consumers.

References

Asian Development Bank (ADB). 2010a. Toward Sustainable and Balanced Development. Strategy and Action Plan for the Greater Mekong Subregion – Southern Economic Corridor. Manila.

ADB. 2010b. Toward Sustainable and Balanced Development. Strategy and Action Plan for the Greater Mekong Subregion – North-South Economic Corridor. Manila.

ADB. 2012a. GMS Economic Cooperation Program: Overview. Manila.

ADB. 2012b. Case Studies on Cross-border Ecotrade. Manila.

Fujimura, M. and C. Edmonds. 2008. Road infrastructure and regional economic integration. Evidence from the Mekong. *In*: Brooks, D.H. and J. Menon (eds.). Infrastructure and Trade in Asia. Edward Elgar. Asian Development Bank Institute, Tokyo.

Fujimura, M. 2017. Evaluation of cross-border transport infrastructure in GMS: Three approaches. ADBI Working Paper No. 771.

Ho, T.Q., V.N. Hoang and C. Wilson. 2017. Eco-efficiency analysis of sustainability-certified coffee production: Evidence from Vietnam. Paper presented at International Symposium on food credence attributes: How can we design policies to meet consumer demand, held at Aoyama Gakuin University, Tokyo.

Ishida, M. 2012. Emerging Economic Corridors in the Mekong Region. BRC Research Report 8. Bangkok Research Center. Bangkok, IDE-JETRO.

Japan External Trade Organization (JETRO). 2017. Phu Quoc Nuoc Mam: registering EU's GI over 11 years. (in Japanese) JETRO Sensor, April 2017 Issue, pp. 96–97.

Kubo, K. 2013. Rice yield gap between Myanmar and Vietnam: A matter of price policy or pubic investment in technology? ASIAN J. Agr. Dev. 10(1): 1–24.

Markandya, A., S. Setboonsarng, Q.Y. Hui, R. Songkranok and A. Stefan. 2015. Chapter 2: The costs of achieving the MDGs by adopting organic agriculture. *In*: Markandya, A. and S. Setboonsarng (eds.). Organic Agriculture and Post-2015 Development Goals: Building on the Comparative Advantage of Poor Farmers, ADB, Manila.

Nogales, E.G. 2014. Making Economic Corridors Work for the Agricultural Sector, FAO Agribusiness and Food Industries Series 4, Rome.

Setboonsarng, S. and A. Markandya. 2015. Organic agriculture, poverty reduction, climate change, and MDGs. *In*: Markandya, A. and S. Setboonsarng (eds.). Organic Agriculture and Post-2015 Development Goals: Building on the Comparative Advantage of Poor Farmers, ADB, Manila.

9

Factors Influencing Farmers' Demand for Agricultural Biodiversity

*Muditha Karunarathna** and *Clevo Wilson*#

Introduction

Agricultural biodiversity[1] is of fundamental significance to human societies, providing socio-cultural, economic and environmental benefits (for example, see, Mozumder and Berrens 2007, Tisdell 2014). It is essential to food security and poverty alleviation in rural economies. The conservation and sustainable use of all aspects of agricultural biodiversity may present opportunities for enhancing soil fertility, naturally controlling pests, reducing the use of pesticides while increasing yields and incomes (Brock and Xepapadeas 2003). As well, diversified agricultural production offers opportunities to expand new markets and further increase the level of food security for rural households (Birol 2004, Ceroni et al. 2005, Ho et al. 2017). The underlying causes for the loss of agricultural biodiversity are extremely complex. They are closely related to increasing food demand, growing market pressure, agricultural development policies, demographic, economic and social factors (Mozumder and Berrens 2007).

* Senior Lecturer, Department of Economics and Statistics, Faculty of Arts, University of Peradeniya, Peradeniya, Sri Lanka.
  E-mail: muditha1974@yahoo.com
Professor, School of Economics and Finance, QUT Business School, Queensland University of Technology, GPO Box 2434, Brisbane, QLD, 4001, Australia.
Corresponding author: clevo.wilson@qut.edu.au

[1] FAO, (1999) defined agricultural biodiversity as the variety and variability of animals, plants and micro-organisms that are used directly or indirectly for food and agriculture including crops, livestock, forestry and fisheries.

Many agricultural practices such as reliance on monoculture, exotic/cross breeds, high yielding varieties, mechanization, and misuse of agricultural chemicals have caused long term negative impacts on agricultural biodiversity at all levels. Such loss of biodiversity may be accompanied by the loss of cultural diversity for traditional communities and their impoverishment (Franks 1999).

Using agricultural biodiversity sustainably can provide environmental, economic and socio-cultural benefits on national, regional and global scales (Hengsdijk et al. 2007). Furthermore, agricultural biodiversity contributes to the provision of wildlife habitats, protecting watersheds, and reducing the use of harmful chemicals (Gauchan et al. 2005). Therefore, understanding the underlying causes of degradation of agricultural biodiversity would assist global efforts to develop badly needed more effective sustainable development policies.

This chapter aims at identifying the determinants of crop and livestock variety diversity (richness in crop/animal varieties/breeds) using a farm household model to predict farmer demand for crop and livestock variety. Small-scale farms data drawn from Sri Lankan farm households is collected to observe which are most likely to sustain agricultural biodiversity. The findings are designed to assist in the formulation of agri-environmental policies which promote biodiversity in family farm management practices. Section 2 discusses the relevant literature in this area. Section 3 explains theoretical aspects of deriving demand for agricultural biodiversity. Section 4 identifies the empirical model specification and relevant variables. Section 5 provides descriptive statistics of the relevant variables while subsequent sections show the results of different models of estimation. The final section summarizes the results of key findings of the study.

Literature Review on Demand for Agricultural Biodiversity

Several studies have used econometric models to identify the determinants of diversity in livestock and crops in developing or transitional economies. Detailed case studies, conducted in Peru (potatoes), Turkey (wheat), and Mexico (maize), have sought to identify some of the important factors that positively and negatively affect the conservation of agricultural biodiversity (Brush et al. 1992, Meng 1997, Van Dusen 2000, Smale et al. 2002). However, most of these studies (Brush et al. 1992, Franks 1999) on *in situ* conservation of agricultural biodiversity on farms concentrate on diversity within a single crop or animal breed. When analyzing the multiple benefits of the farms under semi-subsistent rural areas, concentration on variety diversity is more important than considering a single crop.

According to Fafchamps (1992) crop diversity may be particularly important for farmers with limited opportunities to trade and participate in markets. He identified agro-ecological heterogeneity and imperfect markets with high transaction costs in rural areas as contributing factors to the demand for agricultural biodiversity. Brock and Xepapadeas (2003) developed a conceptual framework for valuing biodiversity from an economic perspective. Bellon (2004) argued that crop diversity maintained by farming households' results from the interplay between a demand and a supply for this diversity. According to them demand interventions should increase the value of crop diversity for farmers or decrease the farm-level opportunity costs of maintaining it, while supply interventions should decrease the costs of accessing diversity. Bunning and Hill (1996) identified different roles and responsibilities of women in conserving crop diversity.

Meng (1997) investigated the diversity of traditional varieties of wheat on Turkish farms. He analyzed the impacts of a combination of factors, including missing markets, farmers' attitudes towards risk and environmental constraints on wheat diversity outcomes. According to this study, regional effects, off-farm income and distance from markets significantly explain diversity of traditional varieties of wheat on Turkish farms. In his discussion of the value of plant genetic resources for food and agriculture in the United Kingdom, Franks (1999) found that the UK's agri-environmental conservation schemes do not prioritize the conservation of genetic diversity of agricultural crops.

According to Van Dusen (2000) agro-ecological and market characteristics significantly affect the levels of diversity maintained by households. He developed a theoretical model in which a household's decision to plant a *milpa* variety is linked to household specific, agro-ecological, and market variables. The results from the Poisson regressions showed that a range of household, village, environmental, and market conditions affect the diversity outcomes. Market integration, measured by distance to a regional market, use of hired labour, and international migration, were found to negatively affect diversity outcomes. Agro-ecological conditions, measured by the number of plots, plots with different slopes, and the high altitude region, were all found to positively increase agricultural biodiversity in the study area.

Li-zhi (2003) investigated genetic erosion of rice and its possible impacts on the Chinese economy which he found could significantly affect the future yield of any crop in China. Meanwhile Scarpa et al. (2003) showed that for Creole pigs in Mexico, the respondent's age, years of schooling, size of the household and number of economically active members of the household were important factors in explaining breed trait preferences. Accordingly, younger, less educated and lower income households placed relatively higher values on the attributes of indigenous piglets compared to exotics and their crosses. A farm household model was used to identify the factors affecting inter and intra crop species diversity of cereal crops in the northern Ethiopian highlands by Benin et al. (2003). This study found that a combination of factors related to the agro-ecology of a community, its access to markets, and the characteristics of its households and farms significantly affected both inter-specific and intra-specific diversity of cereal crops.

An empirical approach was employed to understand the determinants of farmers' access to and use of, crop genetic resources by Van Dusen (2005). He also investigated the impacts of farmer behaviour on crop populations. Gauchan et al. (2005) investigated the socioeconomic, market and agro ecological determinants of farmers' maintenance of rice diversity at the household level. They assessed spatial rice diversity at the farm level using household survey data. Winters et al. (2005) studied potato diversity in Peruvian farms finding that diversity increased with the size of the land owned, number of different plots cultivated, and distance to the nearest market and wealth indicators.

Wilson and Tisdell (2006) investigated how specialization of production of commodities in the agriculture sector leads to the concentration of genetic materials. Isakson (2007) investigated how the participation of Guatemalan peasants in the market economy is related to on-farm conservation of crop genetic diversity in three crops: maize, legumes, and squash. He found that participation in markets is not inherently detrimental to the provisioning of crop genetic resources. However, without proper protections in place market participation may unleash processes that contribute to genetic erosion over time. Pascual and Perrings (2007) distinguished between the proximate and fundamental causes of biodiversity loss in terms of decentralized behaviour of farming households. In doing so special attention was paid to the interplay between micro-economic decisions and the macro-economic factors (institutional and market conditions) that determine the effects of government policies.

It is clear that despite the large number of studies on agricultural biodiversity more conceptual and theoretical work is needed to understand the factors influencing farmers' demand for agricultural biodiversity in developing countries. In particular, analysis of direct policy relevant variables affecting demand for crop variety diversity and animal variety diversity is not properly explained in the literature. Moreover, while a wider cross-section of case studies has been conducted on commercially-oriented farming systems, an analysis of subsistence-oriented farming systems is required in order to generalize and validate the empirical findings (see, for example, Ceroni et al. 2005).

Derivation of demand for agricultural biodiversity

In order to estimate demand for agricultural biodiversity we use a basic model developed by Singh et al. (1986), Taylor and Adelman (2003) and Van Dusen and Taylor (2005). A similar model was used by Birol et al. (2005) to analyze four components of agricultural biodiversity found on family farms in Hungary. The utility a household derives from various consumption combinations and levels depends on the preferences of its members. Preferences are in turn shaped by the characteristics of the household, such as the age or education of its members, and wealth (Birol 2004). Choices among goods are constrained by the full income of the household, total time (T) allocated to farm production (F) and leisure (l), and a fixed production technology represented by $G(.)$. Suppose a farm family maximizes his/her utility over consumption of market purchased goods, C_m, leisure, C_l, and owned farm outputs, C_f. The utility is maximized subject to budget, time, and production technology constraints respectively. Household utility is influenced by a vector of household characteristics δ_h. The utility function is assumed to be quasi-concave in choice variables (Birol 2004, Van Dusen and Taylor 2005). The prices of all market purchased goods, inputs and wages are assumed to be exogenous, and production is assumed to be riskless. The model can be written as follows:

$$U = U(C_m, C_l, C_f; \delta_h) \tag{1}$$

Constraints:

$$I = wT + I_e - wF - p_x X - p_m C_m \quad \text{(income constraint)} \tag{2}$$

$$G(Q, F, X; \beta_f) = 0 \qquad\qquad \text{(technology constraint)} \tag{3}$$

$$F + L_d + C_l = T \qquad\qquad \text{(time constraint)} \tag{4}$$

Equation (1) gives the utility function of a representative household, while Eqn. 2 gives the full income constraint. Full income is composed of value of stock of total time owned by the household T, exogenous income I_e, the values of household management input used in the small-scale farm production F, other variable inputs required for production of small-scale farm outputs (X) and market commodities consumed by the farm family, C_m. The household faces a technology constraint (Eqn. 3) for the production technology on the small-scale farm (Eqn. 3). It gives the relationship between farm inputs F, X and all outputs Q, and has the properties of quasi-convexity, increasing in outputs and decreasing in inputs (Taylor and Adelman 2003). The vector, β_f represents the fixed agro-ecological features of the small-scale farm, such as soil quality and land shape. The household also faces a time constraint (Eqn. 4). Labour use in small-scale farm cultivation F is one use of labour which competes with other uses, including off-farm employment L_a and leisure C_l.

The household is driven toward the goal of increasing diverse farming within the family farm because of uncertainty, unreliable or missing markets, as well as the desire to consume fresh food. This phenomenon brings about an additional constraint that induces the household to equate small-scale farm output demand and supply, resulting in an endogenous, shadow price for small-scale farm outputs (Singh et al. 1986, Birol 2004). This can be written as follows:

$$Q_f = C_f(Z) \tag{5}$$

where Q_f and C_f denote the quantity supplied and consumed by small-scale farm produce, and Z is a vector of exogenous characteristics related to availability and access to markets. This equality condition implicitly defines the shadow prices for small-scale farm outputs under missing market, which guides production decisions (Birol 2004). The production and consumption decisions of the household cannot be separated when labor markets, markets for other inputs, or product markets are imperfect. Then, prices are endogenous to the farm household and affected by the costs of transacting in the markets (Taylor and Adelman 2003). The specific characteristics of farm households (represented by vector δ_h) and physical access to markets (represented by vector Z) influence the magnitude of transaction costs and hence, the effective price governing the household's choices (Van Dusen and Taylor 2005). The household maximizes utility subject to constraints explained in Eqs. 2, 3, 4 and 5. This maximization results in the following Lagrangian Eqn. 6:

$$L = U(C_m, C_l, C_f; \delta_h) + \lambda(wT + I_e - wC_l - p_m C_m - wF - p_x X)$$
$$+ \rho[Q_f - C_f(Z)] + \mu G(Q_f, F, X; \beta_f) \tag{6}$$

However, when all relevant markets function perfectly, farm production decisions are made separately from consumption decisions (Birol 2004). In this context, full income in a single decision-making period is composed of the net farm earnings (profits) from crop or livestock production (Q_f) (of which some may be consumed on farm and the surplus sold), income that is exogenous to the season's crop/animal breads and variety choices such as stocks carried over, remittances, pensions, and other transfers from the previous season (I_e). The household maximizes net farm earnings subject to constraints and then allocates these with other income among consumption goods (Smale et al. 2001). Farm production decisions, such as crop/animal breeds and variety choices, are driven by net returns, which are determined only by wage, input and output prices (w, p_x and p_o) and farm physical characteristics (represented by vector β_f).[2] This will only change the full income budget constraint by adding farm profit as an income while market prices will have some role to play in decision making (Singh et al. 1986, Meng et al. 1998, Smale et al. 2001). Accordingly, $Q_f - C_f(Z) > 0$ and additional income constraint can be added to Eqn. 6. It can be given as $[Q_f - C_f(Z)]p_o$, where p_o is the output prices of the commodities that are produced by the small-scale farms and has a market.

Assuming interior solutions, the optimal set of choice variables are given by the solutions to the first order conditions with respect to the choice variables are:

$$\partial L / \partial C_m = \partial U / \partial C_m - \lambda p_m = 0 \tag{7}$$

$$\partial L / \partial C_l = \partial U / \partial C_l - \lambda w = 0 \tag{8}$$

$$\partial L / \partial \lambda = w(T - F - C_l) + I_e - p_x X - p_m C_m = 0 \tag{9}$$

$$\partial L / \partial F = -\lambda w + \mu G_f = 0 \tag{10}$$

$$\partial L / \partial X = -\lambda p_x + \mu G_x = 0 \tag{11}$$

$$\partial L / \partial \mu = G(Q, F, X; \beta_f) = 0 \tag{12}$$

$$\partial L / \partial C_f = \partial U / \partial C_f - \rho = 0 \tag{13}$$

$$\partial L / \partial Q_f = \rho - \mu G_f = 0 \tag{14}$$

Equations 7 and 8 imply the optimal demand for market purchased goods and leisure respectively. These equations show that the marginal utility the household receives from each commodity equals the Lagrange multiplier, λ, times its market price, p_m and w respectively. Equation 9 corresponds to the full income constraint, which ensures that the net full income received is spent. Equations 10 and 11 represent the optimal amount of each input required in the small-scale farm, determined by the equality between the Lagrange multiplier, λ, times the price of the input and its marginal product. Equation 12 ensures being on the transformation function. The optimal demand for small-scale farm output is given by Eqn. 13. This condition implies that the marginal utility obtained from consuming small-scale farm products is equal to its shadow price, ρ. The supply of the small-scale farm output is given by Eqn. 14. This implies that the marginal cost of producing small-scale farm products equals to its shadow price. Substituting for the shadow price ρ in 13 and 14, it can be shown that the marginal utility of small-scale farm outputs is equal to the marginal cost of small-scale farm outputs and to the shadow price (Birol 2004). Similar derivation could be found in the study carried out by Birol (2004) for the estimation of the demand for attributes in home garden in Hungary:

---

[2] When comparing farmers among communities located in a broader geographical area, it can be seen that their decisions are also affected by factors that vary at a regional level but that they themselves cannot influence. These include several fixed factors hypothesized to affect variation in the diversity maintained among regions, such as agro-ecological conditions, infrastructure development, and the ratio of labor to land.

$$\frac{\partial U}{\partial C_f} = \mu G_f = \rho \tag{15}$$

The endogenous shadow price is household-specific, depending on the household characteristics that affect access to markets and consumption demand, such as wealth, education, age, household composition. Agro-ecological features of the small-scale farm such as soil quality or irrigation enter the equation through their effect on supply. Fixed factors related to market transactions costs and observed market prices also influence the shadow prices of small-scale farm outputs (Feder and Umali 1993). The shadow price, ρ, can therefore be expressed as a function of all exogenous prices and household, agro-ecological and market characteristics:

$$\rho = \rho^*(P_m, P_x, w; \delta_h, \beta_f, Z) \tag{16}$$

The general solution to the household maximisation problem yields a set of optimal choices for production, input demand and consumption demand as given in following equations:

$$Q_f = Q_f^*(\rho, p_x, w : \beta_f) \tag{17}$$

$$F = F^*(\rho, p_x, w : \beta_f) \tag{18}$$

$$X = X^*(\rho, p_x, w : \beta_f) \tag{19}$$

$$C_i = C_i^*(\rho, p_m, w : \delta_h) I = m, l, f \tag{20}$$

Equation 17 is the optimal supply of small-scale farm outputs while Eqn. 18 provides the expression for optimal demand of household labour in farm production. Equation 19 gives the optimal demand for all other inputs to small-scale farm production and Equation 20 is the optimal demand for market purchased goods (m), household produced goods (f) and leisure (l).

Substituting these solutions for the shadow price (Eqn. 16) into small-scale farm output production and consumption solutions (Eqs. 17 and 20), the optimal production of small-scale farm outputs is seen to be a function of all exogenous variables:

$$Q_f = Q_f^*(P_m, P_x, w; \delta_h, \beta_f, Z) \tag{21}$$

We assume that the household does not value diversity itself rather than the direct benefits of it. Therefore, diversity is not explicitly in the utility function. The diversity within a given household is the result of the choice of which crops to produce, subject to constraints. This 'diversity outcome' in the constrained case takes the form of a derived demand for a number of varieties resulting from the farmer's utility maximisation subject to income, production, and market constraints. Following Van Dusen and Taylor (2005) the level of agricultural biodiversity maintained on the small-scale farms, which is a direct outcome of the production and consumption choices of the farm household, is a function of all prices, and characteristics of the households, markets, and of the small-scale farm plots. This relationship can be given as shown in Eqn. 22:

$$BD = BD\{Q_f^*(P_m, P_x, w; \delta_h, \beta_f, Z)\} \tag{22}$$

Some of the interesting applied economic analyses of agricultural biodiversity based on either farm household or variety choice models are Brush et al. (1992), Meng (1997), Smale et al. (2001) and Birol (2004). Studies in this area commonly use count data analysis or Logit/Probit models for empirical estimation.

Empirical Model Specification: Relevant Variables and Data

In this study agricultural biodiversity is investigated in terms of crop diversity and livestock diversity. In order to understand the important determinants of variety demands, different types of policy relevant variables are selected for our regression analysis—information based on the pilot survey and that provided by regional agricultural specialists. They are divided into three main categories namely household characteristics, market characteristics and other characteristics (see Table 1).

It is clear that some variables are taken as numbers while other variables are defined as dummy variables. Experience in farming is one of the important variables used in the analysis. Experience of the household head in agricultural activities is expected to have a quadratic relationship with the selection of a diverse farming system (Van Dusen 2000). That is, younger households may be more willing to try out different crops and varieties, while older households with more experience in farming may be more set in their production activities and less likely to try new crops and varieties. Therefore, we hypothesised that demand for the agricultural biodiversity will decrease

Table 1. Definition of potential explanatory variables.

Variables	Definition
Household characteristics	
EXP	Experience of farm decision maker (number of years)
OWN	Household owns a business vehicle or not: dummy-1 if Yes, Otherwise 0
HMP	Household member's participation in agricultural activities (%)
GEN	Decision maker, male or female: dummy-1 if Male, Otherwise 0
INC	Off-farm income of the family (Rs. 000)
SHL	Shared labour (number in the last season)
WLH	Household wealth: dummy-1 if wealthier, Otherwise 0
Market characteristics	
NMA	Number of market access days per week (number)
DIMK	Distance to the nearest market (KM)
DSN	Direct sales or not (intermediary): dummy-1 if Yes, Otherwise 0
PRIF	Price fluctuation of the output (index)[i]
Other characteristics	
AS	Receiving agricultural subsidize: dummy-1 if Yes, Otherwise 0
IOM	Percentage of investment of owned money[ii]

Note: i. Price fluctuation indexes were constructed using average unit price changes over the last two seasons for crops and livestock outputs.
ii. This variable is created by taking the percentage value of own money invested to total farm investment in the last season. Total farm investment includes own money plus borrowing for the last season.

with experience in farming. Owning a business vehicle can have positive correlation with agricultural biodiversity. This is because a business vehicle can help farmers take different products into different markets. Given the limited number of market places and market access days in rural areas in developing countries, business vehicles can be used to sell farm products directly to the market. This will avoid an intermediary transaction.

The variable household member's participation in agricultural activities shows the number of mandates received from members of the family (except household head) for agricultural work during the last season. Participation rate captures family labour availability for farming activities. The number of members in the family is expected to have a positive effect on diversity through its effects on overall labour capacity. Moreover it is clear that as family size increases so does expenditure on food consumption[3] which can be reduced by greater diversity given the contribution of own farm on their food requirement is higher. In terms of gender, women's knowledge in seed selection and management can be expected to contribute towards increased richness. On the other hand, lack of skills—such as in ploughing—may restrict their decisions to grow a number of varieties. Off-farm income is expected to have a negative correlation with agricultural biodiversity, given it is likely to generate managerial and labour constraints. Shared labour by reducing labour constraints is expected to have a positive correlation with diverse farming systems.

To differentiate households according to wealth dummy variables are created by classifying houses as luxury/upper middle class, ordinary and small house/cottage. Household differentiation is also achieved through accounting for presence of a telephone, electricity, piped water, vehicle road to the house, water sealed toilet and attached bathrooms. Thirdly, durable assets are considered. They included vehicles, threshing machines, water pumps and motorcycles. If a household belongs to a luxury/upper middle class or ordinary house and has at least four of the aforementioned facilities with at least two of the asset varieties, that household is identified as wealthy. It is hypothesized that wealth is negatively correlated with agricultural biodiversity given wealth helps reduce the risk of shortages faced by poorer farmers.

Market characteristics, as explained in Table 1, are examined in terms of determinants of agricultural biodiversity. Market infrastructure operates in several ways that may not be dissociable in a given location at one point in time. For example, the more removed a household is from a major market centre, the higher the costs of buying and selling in the market and the more likely that the household relies primarily on its own production for subsistence and on less specialized production activities. On the other hand, as market infrastructure reaches a village, new trade possibilities may emerge, adding crops and production activities to the portfolio of economic activities undertaken by its members. The theory of the household farm predicts that the higher the transaction costs faced by individual households within communities, the more they rely on the diversity of their crop and variety choice to provide the goods they consume Van Dusen (2000).

We hypothesized that the number of market access days has a positive correlation with agricultural biodiversity as it helps minimize the risk of not selling the surplus. A

---

[3] A positive correlation can be expected between agricultural biodiversity and income spent on food consumption as well.

direct sale variable is included to see whether it has some impact on selecting a diverse farming system. It is expected that farmers who sell their output to market directly are more likely to maintain a diverse farming system. A variable to capture the effect of price fluctuation on agricultural biodiversity is used (average output price changes for crops and livestock over the last two cultivation seasons) which is expected to have a positive correlation with agricultural biodiversity.

It is accepted that receiving agricultural subsidies helps reduce financial constraints on farmers and has a positive impact on selecting a more diverse farming system. Own money investment in the farm allows farmers to use their own savings while others commonly borrow money from both formal or informal sources (e.g., village shops).[4] We hypothesized that the percentage of own money contribution to total farm expenditure has a positive correlation with agricultural biodiversity given farmers often borrow money in order to maintain a specialization system. It is noted, however, that although agricultural subsidies are important for determining crop varieties, they are not so for animal varieties given agricultural subsidy policies only focus on the crop sector. The theoretically possible signs for different variables are given in Table 2.

The Poisson model (PM) or negative binomial model (NBM) for count data may be the suitable model(s) for estimating the determinants of farm families' decisions about how many crop and livestock species to cultivate (see, for example, Greene 1997). In

Table 2. Explanatory variables used in the demand model.

Variable	Definitions	CD	AD
Household characteristics			
EXP	Experience of farm decision maker	–	–
OWN	Household owns a business vehicle or not	+	+
HMP	Household member's participation	+	+
GEN	Decision maker, male or female	+/–	+/–
INC	Off-farm income of the family	–	–
SHL	Shared labour	+	+
WLH	Household wealth	–	–
Market characteristics			
NMA	Number of market access days per week	+/–	+/–
DIMK	Distance to the nearest market	+/–	+/–
DSN	Direct sales or not (intermediary)	+/–	+/–
PRIF	Price fluctuation of the output	+	+
Other characteristics			
AS	Receiving agricultural subsidies	–	NA
IOM	Percentage of investment of owned money	+	+

Note: Definitions of all variables are given in Table 1. Expected signs in each variable are provided in this table. As shown in the table, some variables can take positive or negative depending on the situation.

---

[4] In such cases the interest paid can be very high: around 20% per month in most rural areas.

this study it was found that PM provides better estimations than NBM. Therefore, PM is used for the estimation. The empirical model specification is as follows:

$$Y_i = \beta_0 + \beta_1 EXP + \beta_2 OWN + \beta_3 HMP + \beta_4 GEN + \beta_5 INC + \beta_6 SHL + \beta_7 WLH + \beta_8 NMA + \beta_9 DIMK$$
$$+ \beta_{10} DSN + \beta_{11} PRIF + \beta_{12} AS + \beta_{13} IOM + U_i \tag{23}$$

where Y_i is a count-dependent variable representing the diversity indices, namely, crops or livestock. All other independent variables are as explained in Table 2. The predictions based on this econometric model enable us to profile households that are most likely to sustain current levels of crops diversity and animal diversity.

We use primary data along with secondary data for the analysis. Survey data were collected covering approximately 746 farmers in three agricultural districts in Sri Lanka (Anuradhapura—247, Ampara—248 and Kurunegala—251). In August 2010, a pilot survey was conducted in randomly chosen areas of Anuradhapura, Ampara and Kurunegala districts. The main survey was started at the beginning of September 2010 and completed at the end of October 2010. Surveys in all districts were carried out by administering a questionnaire through a face-to-face interview with the head or any other working member of the households. The questionnaire was validated in a pilot survey and in a number of focus group discussions. The data collection was carried out by a trained group of researchers under the close supervision of their search team. Interviews took place in the interviewee's home. The participants were informed about the purpose of the study and their verbal consent obtained. A field supervisor reviewed the quality of the data gathered. It was confirmed that the survey questions were clearly understood by the respondents.

Household Socio-Economic Characteristics

We estimated the diversity regression equations for selecting crop varieties and animal varieties. Most farmers in a given district cultivate or maintain approximately similar crops or livestock. Rice, different types of vegetables and cash crops are found to be the common type of crops that farmers cultivate. Animal breeds include mainly cattle, chickens, goat, pigs and buffalos. Most households cultivated between two and six types of crop varieties. In the case of animals, most households maintained two to three animal varieties in the study areas. On an average, approximately 92% of the sampled respondents' main occupation is farming. Approximately 8% of respondents are employed in the government or private sector for whom agricultural practice is a secondary activity and source of income. Some 32% of households have a secondary income. 23% work as waged labourers on some days in the month.

Directly relevant socio-demographic characteristics of the household that determine the diversity of crops and livestock they grow and raise include the average experience of farming—24, 19 and 26 years for the Ampara, Anuradhapura and Kurunegala samples respectively. Approximately 78% of respondents are male while 22% are female. Household members' participation in agricultural activities is very high in rural communities in Sri Lanka. Average participating rates were 87, 92 and 96% of the total number of households (greater than 14 years old) in Ampara, Anuradhapura and Kurunegala districts respectively. Off-farm income is not significant for most households as their main income is determined by farm output. Earning from waged labourers, small-scale business and government family allowances (*Samurdi* allowance) are among the most common off-farm income sources in rural areas. One of the interesting aspects of rural households is explained by the shared labour. This variable represents the magnitude of the social capital. The average number of shared labour events per season is 12, 21 and 18 for Ampara, Anuradhapura and Kurunegala respectively and which, on average, accounts for approximately 17% of their total labour usage in a given season.

It was found that a significant percentage of families belong to the less wealthy category with only 31, 18 and 22% of respondents identified as wealthy families in Ampara, Anuradhapura and Kurunegala district respectively. The number of market access days in differed between districts varying from 1 to 7 days per week. As well, there are different types of markets including weekly fairs where farmers can sell their products directly and intermediary traders who visit villages. In general, informal discussions with farmers reveal that marketing is the biggest problem for all areas. This is because in some seasons there is no demand for their product while in other seasons they do not get an expected price. It was revealed that one of the main objectives of their agricultural activities is to meet the family food requirement. The marketable surplus of small-scale farms in rural areas is relatively small with most households maintaining a stock until the next harvesting season. Consumption rate of some of the crops and livestock products are as high as 98% of their output with an average rate of 73% for rice and 95% for vegetables. While the distance to the nearest market is relatively higher in the Anuradhapura district sample, average price fluctuations are similar in all three districts. Moreover, a significant difference could not be observed where subsidies were received for cultivating crops in different districts. This might be due to existing government policy which makes any farmer who has their own land eligible for the fertilizer subsidy.

Estimated results of this analysis are reported with their interpretation in the next sections. As we covered three separate districts in our survey, we adopted the following to analyse the data for the models that represents agricultural biodiversity. First, separate regressions were run for district-based data separately. Second, the pool data model was run after combining three data sets together. A dummy variable was included in the pool data model to capture the effects of regional fixed factors for Anuradhapura and Kurunegala, as compared to Ampara. The next section discusses the determinants of crops variety diversity in separate district and pool data models.

Determinants of Crops Variety Demand

The estimated results of the four regression models are reported in Table 3. We report only marginal effects for all regression models. For the crops variety diversity, four Poisson regression models were estimated: three for separate districts data and one for the pool data for all districts.

The results show that experience in farming is highly significant in all models and has a positive coefficient value and not consistent with our initial hypothesis that older farmers with more experience have a better understanding of the benefits of diverse farming systems. Therefore, this variable can serve as a proxy for farmers'

Table 3. Poisson regression results for crops variety model.

Variables	Ampara	Anuradhapura	Kurunegala	Pool data
EXP	0.022(0.004)*	0.016(0.003)*	0.018(0.004)*	0.011(0.002)*
OWN	0.325(0.118)*	0.073(0.129)	0.267(0.118)**	0.208(0.096)**
HMP	0.009(0.002)*	0.006(0.001)*	0.007(0.002)*	0.008(0.001)*
GEN	0.181(0.131)****	0.456(0.121)*	0.235(0.106)**	0.263(0.078)*
INC	−0.016(0.006)**	−0.004(0.005)	−0.002(0.001)**	−0.004(0.002)***
SHL	0.033(0.007)*	0.029(0.008)*	0.019(0.007)*	0.037(0.005)*
WLH	−0.445(0.115)*	−0.222(0.111)**	−0.024(0.068)	−0.126(0.056)**
NMA	0.152(0.036)*	0.086(0.020)*	0.066(0.023)*	0.154(0.019)*
DIMK	−0.124(0.032)*	−0.094(0.022)*	−0.105(0.025)*	−0.097(0.015)*
DSN	0.350(0.112)*	0.647(0.129)*	0.495(0.110)*	0.387(0.068)*
PRIF	0.008(0.002)*	0.002(0.001)***	0.007(0.001)*	0.004(0.001)*
AS	−0.216(0.153)****	−0.444(0.162)*	−0.3840.108)*	−0.405(0.094)*
IOM	0.021(0.004)*	0.003(0.001)***	0.009(0.002)*	0.010(0.001)*
Anuradhapura	-	-	-	0.813(0.141)*
Kurunegala	-	-	-	0.286(0.114)**
N	248	247	251	746
Pseudo R^2	0.207	0.181	0.256	0.208
Wald chi^2 (13)	634.37	1206.75	1696.93	2469.98

Note: i. Definitions of the variables used in the regression analysis are shown in Table 1. In the pool data analysis, Ampara is used as the base district when creating dummy variables.
ii. Standard errors are shown in brackets. *, **, *** and **** denote the significance of the variables at the 1%, 5%, 10% and 20% levels respectively.
iii. Marginal effects on the count dependent variable are reported in this table. These coefficients indicate how a one unit change in an independent variable alters the count dependent variable.

age. Further implied is that human capital and access to information are favourable for growing a wider range of crop varieties in rural areas in Sri Lanka.

Owning a business vehicle is not significant in the Anuradhapura sample but is significant for the other three models. This implies having a business vehicle may help reduce market transaction costs for selling any farm surplus.

The household members' agricultural labour participation variable is highly significant in all models. This implies that a diverse farming system requires more labour time and is thus consistent with the theory. As hypothesized, households headed by men grow more diverse varieties. This can be associated with the skill or requirement for frequent manual work for cultivating more varieties. The influence of this variable is uniform and significant across all models.

The result of this study shows that off-farm income has a significant negative effect on crop variety diversity which can be attributed to farmers purchasing most food consumption needs from the market and given the incentive to maintain a diverse farming system is less as it needs a relatively higher amount of labour. However, the literature related to the effect of off-farm income and crop diversity shows contradictory findings (Bellon and Taylor 1993, Meng 1997).

Shared labour—an important part of social capital in rural areas in Sri Lanka—shows a significant positive correlation with crop variety diversity. The coefficient of household wealth is negative and significant indicating that the greater the wealth (and therefore decreased risk aversion), the less likely it is that a diverse set of crops is planted (Van Dusen 2000). Wealth may also be a proxy for networks, information, and access to outside market opportunities in the presence of various kinds of market imperfections.

The relationship between markets and the conservation of agricultural biodiversity is complex. To examine the issue we have included four market-related variables in the study. They are the number of market access days per week, distance to the nearest market, direct sales or not (intermediary) and price fluctuation of the output in the previous season.

As expected, increased market access had a positive and significant coefficient in all four models. The distance of the household farm to the nearest market, which is a major component of the cost of engaging in market transactions related to seed, labour, other inputs, and farm produce, is as hypothesized, is shown to negatively affect crop diversity. This is due to the higher transaction cost of market access, which limits interaction with the market.

The price fluctuation—a proxy for risk of future return of farm output—is, as expected, positively related to variety demand. The higher the market price fluctuation, the higher the likelihood that a household will cultivate more crops and minimize the risk. Receiving agricultural subsidies is also significant and negatively related to crop diversity in all models.

Percentage of own money invested in agricultural activities over the last season is positively related to variety selection—the higher the expenditure the greater the number of varieties. Finally, the pool data results show that heterogeneity among districts is significant. This is expected as the three selected districts were based on differences in the country's agricultural biodiversity.

In summary, markets and a number of other characteristics have a significant impact on crop diversity levels across small-scale farms in Sri Lanka. In the next section, we investigate variables which determine animal variety diversity.

Determinants of Livestock Variety Demand

We investigate the determinants of livestock variety demand in separate district data and pool data. All variables included in the crop variety model are retained except agricultural subsidy. This is because there are no direct subsidies for this farming activity. The estimated results are shown in Table 4. They show that experience in agricultural activities is significant at 1% in Anuradhapura and for the pool data model and at 5% level for the Ampara and Kurunegala districts. Thus, farmers who have more experience in farming are likely to maintain a diverse livestock farming system. Owning a business vehicle is not a significant determinant of livestock varieties in Ampara and Anuradhapura samples and only weakly significant for Kurunegala. The pool data model is significant at 1%. This is because livestock farms are mainly maintained for family consumption purposes in Sri Lanka's rural areas.

Table 4. Poisson regression results for animal variety model.

Variables	Ampara	Anuradhapura	Kurunegala	Pool data
EXP	0.003(0.001)**	0.018(0.003)*	0.008(0.003)**	0.010(0.001)*
OWN	0.023(0.035)	0.079(0.155)	0.195(0.124)****	0.085(0.054)****
HMP	0.004(0.001)*	0.003(0.002)****	0.004(0.001)*	0.004(0.001)*
GEN	−0.014(0.035)	−0.182(0.118)****	−0.056(0.103)	−0.095(0.047)**
INC	−0.006(0.001)*	−0.019(0.003)*	−0.016(0.002)*	−0.014(0.001)*
SHL	0.008(0.003)**	0.019(0.010)**	0.026(0.006)*	0.021(0.004)*
WLH	−0.096(0.054)***	−0.366(0.133)*	−0.591(0.129)*	−0.410(0.055)*
NMA	−0.117(0.022)*	−0.104(0.021)*	−0.027(0.025)	−0.063(0.011)*
DIMK	0.017(0.001)**	0.032(0.024)****	0.042(0.016)**	0.029(0.012)*
DSN	0.103(0.037)*	0.162(0.120)****	0.131(0.123)	0.082(0.051)***
PRIF	0.001(0.000)**	0.003(0.001)*	0.002(0.001)***	0.002(0.000)*
IOM	0.002(0.000)*	0.009(0.001)*	0.001(0.001)*	0.001(0.001)*
Anuradhapura	-	-	-	0.614(0.091)*
Kurunegala	-	-	-	0.402(0.080)*
N	241	243	242	726
Pseudo R^2	0.443	0.297	0.378	0.362
Wald chi^2 (12)	290.09	254.81	321.11	757.38

Note: i. Definitions of the variables used in the regression analysis are shown in Table 1. In the pool data analysis, Ampara is used as the base district when creating dummy variables.
ii. Standard errors are shown in brackets. *, **, *** and **** denote the significance of the variables at the 1%, 5%, 10% and 20% levels respectively.
iii. Marginal effects on the count dependent variable are reported in the table. These coefficients indicate how a one unit change in an independent variable alters the count dependent variable.

The household members' participation variable is highly significant in most of the models indicating the important role of family members in farm laboring. The gender variable is not significant in the Ampara and Kurunegala samples although significant at 20% and 5% levels for Anuradhapura. The negative coefficient implies households headed by women raise more diverse animal varieties, which can be explained by women's capacity to manage a small-scale business from home.

The results show off-farm income has a significant negative effect on animal variety demand. As for crop diversity, when off-farm income is higher, farmers can be assumed to purchase most food needs from the market and rely less on farm diversity. As well, off-farm employment reduces on farm labour availability. Shared labour has a significant positive correlation with animal variety diversity. The coefficient for household wealth is negative and significant. That is, the greater the wealth of the household, the less likely the household is to have a diverse set of animals.

The coefficient for the number of market access days per week variable is significant at 1% in all models except Kurunegala, and has a negative sign. This result shows that livestock diversity decreases when market access days increase. Furthermore, the distance to the nearest market variable shows that households further from markets are shown to have more diversity as distance to nearest market increases. Therefore, households closer to the market will select more marketed items, providing evidence of market participation. Thus subsistence dependent households located away from the market are more likely to maintain a diverse livestock system for their own consumption.

The variable 'direct sales or not' is weakly significant in the Anuradhapura sample and not significant for Kurunegala. Price fluctuation of output, as noted, can be taken as a proxy for risk of future return of output and, as expected, is positively related to variety demand. As hypothesized, percentage of own money invested in farm activities over the last season is also positively related for all models to stock variety. Pool data results show that heterogeneity of animal varieties among districts is significant.

Overall, the results show that for small-scale farms, markets and a number of other characteristics have a significant impact on variation in livestock diversity levels. In the next section the main conclusions drawn from this study are set out.

Key Findings

We find that the key variables promoting agricultural biodiversity are farm household and market characteristics. Other characteristics include percentage of own savings invested in agriculture. In particular, the centrality of markets in shaping diversity does not suggest a trade-off between development and diversity. This is because, as integration with outside markets increases, the level of crop and animal diversity on farms could also be increased. However, there are exceptions. Furthermore, in the study we find that households with more experience, more labour availability and more food required for consumption grow more diverse crops or livestock because they have the resources to do so. The reasons for this were discussed.

Off-farm income, wealth and agricultural subsidies were shown to be negatively related with agricultural biodiversity for small-scale farms. Also of significance is that output price fluctuation is one of the important variables producing significant results in all models.

In each statistical analysis conducted, whether descriptive or econometric, regional heterogeneity has emerged. Hence, formulation of agri-environmental policies and programmes supporting agricultural biodiversity in Sri Lanka will need to take account of this. Equally, policies and programmes affecting the wealth, education or labour participation of family members, or the formation of food markets within settlements, will influence their choices.

Such findings are clearly relevant to future policy formulations aimed at increasing crop and livestock diversity. That is, the results of the models provide a means of identifying the characteristics of families that are most likely to sustain agricultural biodiversity. Profiles can, therefore, be created to design targeted, least cost, incentivized mechanisms to support conservation as part of national environmental programmes.

References

Bardsley, D. 2003. Risk alleviation via *in situ* agrobiodiversity conservation: Drawing from experiences in Switzerland, Turkey and Nepal. Agric. Ecosyst. Environ. 99: 149–157.

Bellon, M.R. and J.E. Taylor. 1993. Folk soil taxonomy and the partial adoption of new seed varieties. Econ. Dev. Cult. Change. 41(4): 763–786.

Bellon, M.R. 2004. Conceptualizing interventions to support on-farm genetic resource conservation. World. Dev. 32(1): 159–172.

Benin, S., B. Gebremedhin, M. Smale, J. Pender and S. Ehui. 2003. Determinants of cereal diversity in communities and on household farms of the northern Ethiopian highlands. Environment and Production Technology Division, Discussion Paper No. 105, International Food Policy Research Institute, Washington DC.

Birol, E. 2004. Valuing agricultural biodiversity in home gardens in Hungary: An application of stated and revealed preference methods. PhD Thesis. University College London, UK.

Birol, E., G. Bela and M. Smale. 2005. The role of home gardens in promoting multi-functional agriculture in Hungary. Euro. Choices. 4(3): 14–26.

Brock, W.A. and A. Xepapadeas. 2003. Valuing biodiversity from an economic perspective: A unified economic, ecological and genetic approach. Am. Econ. Rev. 93(5): 597–614.

Brush, S.B., J.E. Taylor and M. Bellon. 1992. Technology adoption and biological diversity in Andean potato agriculture. J. Dev. Econ. 39(2): 365–387.

Bunning, S. and C. Hill. 1996. Farmers' Rights in the Conservation and Use of Plant Genetic Resources: A Gender Perspective. Rome, FAO.

Ceroni, M., S. Liu and R. Costanza. 2005. The ecological and economic roles of biodiversity in agroecosystems. *In*: Jarvis, D.I., C. Padoch and D. Cooper (eds.). Managing Biodiversity in Agroecosystems. Columbia University Press, NY, USA.

Di Falco, S. and C. Perrings. 2003. Crop genetic diversity, productivity and stability of agroecosystems. A theoretical and empirical investigation. Scott. J. Polit. Econ. 50(2): 207–216.

Fafchamps, M. 1992. Cash crop production, food price volatility and rural market integration in the third world. Am. J. Agric. Econ. 74(1): 90–99.

FAO. 1999. The Strategic Framework for FAO 2000–2015. Food and Agriculture Organisation of the United Nations. Rome.

Feder, G. and D. Umali. 1993. The adoption of agricultural innovations: a review. Technol. Forecast. Soc. Change. 43(3-4): 215–239.

Franks, J.R. 1999. *In situ* conservation of plant genetic resources for food and agriculture: a UK perspective. Land. Use. Policy. 16(2): 81–91.

Gauchan, D., M. Smale, N. Maxted, M. Cole, B.R. Sthapit, D. Jarvis and M.P. Upadhyay. 2005. Socioeconomic and agroecological determinants of conserving diversity on-farm: the case of rice genetic resources in Nepal. Nepal Agricultural Resource Journal 6: 89–98.

Greene, W.H. 1997. Econometric Analysis (3rd Edn.), New Jersey: Prentice Hall.

Grootendorst, P.V. 1995. A comparison of alternative models of prescription drug utilisation. Health. Econ. 4(3): 183–198.

Heisey, P.W., M. Smale, D. Byerlee and E. Souza. 1997. Wheat rusts and the costs of genetic diversity in the Punjab of Pakistan. Am. J. Agric. Econ. 79(3): 726–737.

Hengsdijk, H., W. Guanghuo, M.M. Van den Berg, W. Jiangdi, J. Wolf, L. Changhe, R.P. Roetter and H.V. Keulen. 2007. Poverty and biodiversity trade-offs in rural development: a case study for Pujiang county, China. Agric. Syst. 94(3): 851–886.

Hilbe, J.M. 2005. Negative Binomial Regression. New York: Cambridge University Press.

Hilbe, J.M. 2011. Negative Binomial Regression (2nd Edn.), New York: Cambridge University Press.

Ho, T.Q., V.N. Hoang, C. Wilson and T.T. Nguyen. 2017. Which farming systems are efficient for Vietnamese coffee farmers? Econ. Anal. Policy. 56: 114–125.

Isakson, S.R. 2007. Uprooting diversity? Peasant farmers' market engagements and the on-farm conservation of crop genetic resources in the Guatemalan highlands. Working Paper 122, Political Economy Research Institute, University of Massachusetts Amherst.

Li-zhi, G. 2003. The conservation of Chinese rice biodiversity: genetic erosion, ethnobotany and prospects. Genet. Resour. Crop. Evol. 50(1): 17–32.

Maikhuri, R.K., K.S. Rao and R.L. Senwal. 2001. Changing scenario of Himalayan agroecosystems: loss of agrobiodiversity, an indicator of environmental change in Central Himalaya, India. Environmentalist 21(1): 23–39.

Meng, E.C.H. 1997. Land Allocation Decisions and *In Situ* Conservation of Crop Genetic Resources: The Case of Wheat Landraces in Turkey. Unpublished Doctoral Dissertation, University of California at Davis, California, USA.

Meng, E.C.H., M. Smale, M. Bellon and D. Grimanelli. 1998. Definition and measurement of crop diversity for economic analysis. *In*: Smale, M. (ed.). Farmers, Gene Banks and Crop Breeding: Economic

Analyses of Diversity in Wheat, Maize and Rice. Dordrecht and Mexico, D.F.: Kluwer and International Maize and Wheat Improvement Centre.

Mozumder, P. and R.P. Berrens. 2007. Inorganic fertilizer use and biodiversity risk: An empirical investigation. Ecol. Econ. 62(3-4): 538–543.

Nagarajan, L., M. Smale and P. Glewwe. 2007. Determinants of millet diversity at the household-farm and village-community levels in the dry lands of India: The role of local seed systems. Agric. Econ. 36(2): 157–167.

Pascual, U. and C. Perrings. 2007. Developing incentives and economic mechanisms for *in situ* biodiversity conservation in agricultural landscapes. Agric. Ecosyst. Environ. 121(3): 256–268.

Scarpa, R., A.G. Drucker, S. Anderson, N. Ferraes-Ehuan, V. Gomez, C.R. Risopatron and O. Rubio-Leonel. 2003. Valuing genetic resources in peasant economies: the case of 'hairless' Creole pigs in Yucatan. Ecol. Econ. 45(3): 427–443.

Singh, K., P. Ram, V. Singh and S.K. Kothari. 1986. Effect of dates of planting and nipping on herb and oil yield of Mentha arvensis. Indian Journal of Agronomy 31: 128–130.

Smale, M., M. Bellon and A. Aguirre. 2001. Maize diversity, variety attributes, and farmers' choices in Southeastern Guanajuato, Mexico. Econ. Dev. Cult. Change. 50(1): 201–225.

Smale, M., E. Meng, J.P. Brennan and R. Hu. 2002. Determinants of spatial diversity in modern wheat: Examples from Australia and China. Agric. Econ. 28(1): 13–26.

Taylor, E. and I. Adelman. 2003. Agricultural household models: genesis, evolution and extensions. Rev. Econ. Househ. 1(1): 33–58.

Tisdell, C. 2014. Ecosystems functions and genetic diversity: TEEB raises challenges for the economics discipline. Econ. Anal. Policy. 44(1): 14–20.

Van Dusen, E. 2000. *In Situ* Conservation of Crop Genetic Resources in Mexican Milpa Systems. Unpublished Doctoral Dissertation, University of California, Davis, USA.

Van Dusen, E. 2005. Understanding the factors driving on farm crop genetic diversity: empirical evidence from Mexico. pp. 127–145. *In*: Cooper, J., L.M. Lipper and D. Zilberman (eds.). Agricultural Biodiversity and Biotechnology in Economic Development. Springer: US.

Van Dusen, E. and J.E. Taylor. 2005. Missing markets and crop diversity: Evidence from Mexico. Env. Dev. Econ. 10(4): 513–531.

Van Dusen, E., D. Gauchan and M. Smale. 2005. On-farm conservation of rice biodiversity in Nepal. Paper Presented at the American Agricultural Economics Association Annual Meeting, Providence, Rhode Island, USA.

Wilson, C. and C. Tisdell. 2006. Globalization, concentration of genetic material and their implication for sustainable development. pp. 251–262. *In*: Aurifeille, J., S. Svizzero and C. Tisdell. (eds.). Leading Economic and Managerial Issues Involving Globalisation. Nova Science Publishers, Inc., New York.

Winkelmann, R. 2008. Econometric Analysis of Count Data (5th Edn.): Springer.

Winters, P., L.H. Hintze and O. Ortiz. 2005. Rural development and the diversity of potatoes on farms in Cajamarca. pp. 146–161. *In*: Smale, M. (ed.). Valuing Crop Biodiversity: On-Farm Genetic Resources and Economic Change. CABI Publishing, Wallingford, UK.

PART III
Culture and Ethics

10

Culture and Ethics Concerning Food Attributes

Ian Werkheiser

Introduction: Consuming Ethics, Consuming Culture

There is little question among marketers as well as academics that consumers have preferences for certain attributes in the products they consume rather than simply preferences in products themselves, and that these preferences are expressed in purchasing decisions when those attributes can be determined (e.g., Ford et al. 1988). These include credence attributes: those attributes which cannot be determined by the consumer except through trusting signs, labels, etc. (*ibid.*). It is unsurprising that this behavior holds for food as it does for other commodities; indeed it may be particularly true for food, which is not simply an object we buy, but one we take into our own bodies and think about through complex cultural and personal frameworks, and which embodies the entire food system (Poppe and Kjærnes 2003). It would be a mistake, however, to see food credence attributes, consumers' perception of these, and their resulting behavior, purely in terms of individual, subjective preferences and purchasing decisions. To see them as purely subjective would be to miss out on a number of important ethical considerations. To see them as purely individual would be to miss out on a number of important cultural interactions. Other chapters in this section explore particular examples of these sometimes overlooked ethical and cultural factors; this chapter seeks instead to take a more general survey of that terrain and look for potential hazards and stumbling blocks, as well as resources, for those thinking about consumer preference and food credence attributes.

Department of Philosophy, University of Texas Rio Grande Valley, ELABS Building, Room 342, 1201 West University Drive Edinburg, TX 78539, USA.
E-mail: ian.werkheiser@utrgv.edu

In Section one, this chapter examines some of the effects of food credence attributes and their signals on consumers. This includes ethical issues around honesty in signals such as fraudulent, misleading, or confusing signals. It also includes ethical-epistemic issues around the clarity and understandability of these signals. Those issues of signaling are sometimes overlooked, at least in their full implications and impact, but they are better represented in the literature than the final topic in that section: some of the concerns that arise in the use of food credence attributes and signals to shape behavior. In Section two, this chapter examines some of the effect of food credence attributes and their signals on others in the food system beside consumers. The first topic in that section looks at the effects of food credence attributes and particularly official signals of them on cultural continuance. The second topic is the effects of food credence attributes and their signals on the food system itself, and the ethical issues that arise with those effects. Ultimately, this chapter argues that the promotion of food credence attributes should only be done with much more careful consideration of their perils and power than is usually done by marketers and policy makers.

Ethical and cultural issues surrounding food credence attributes can be divided into two areas: those issues surrounding the effects of the use of food credence attributes for consumers on the one hand, and the issues surrounding the effects of food credence attributes for other actors in the food system and the environment on the other. Let us examine each set in turn.

Effects of the Use of Food Credence Attributes for Consumers

One important ethical issue within this first area is honesty in signals of food credence attributes. Of course, if labels or other signals to consumers contain outright lies, it is an obvious breach of ethics as well as (often) the law. The hazards of such lies have been known for some time (Nelson 1970, Darby and Karni 1973). One such hazard is that they can impede agential autonomy for the consumers to be able to pursue projects that are important to them; to pick one example, if being vegan is important to someone's identity and eating exclusively vegan food is a vital part of their life projects, then a product which falsely claims to be vegan harms that identity and pursuit of their life projects.

Dishonest food credence attributes can also undermine any individual or societal goods one might associate with a free market of informed actors. To the extent that market forces can lead to improved safety, better products, more environmental processes, higher salary, or whatever other improvements one might attribute to market forces (obviously these are all debatable), these goods are much more difficult to achieve if the market is distorted with false signals (Golan et al. 2001).

These dishonest signals also risk eroding trust in food attribute signals and in societal institutions, with far-reaching negative effects. To pick only one of the many dangers in eroded trust in these institutions, it is much easier for spurious health claims to proliferate if people do not pay attention to credence attributes on packages as the sign for what food contains and can do. Most countries and international trade institutions have recognized some combination of the above reasons as over-determinative against false signaling of food attributes, and thus have laws against them (e.g., Cheftel 2005, Endres and Johnson 2011). Yet despite this, such false signals are unfortunately common (to pick only some quite recent cases, see, e.g., Olmstead 2016, Whoriskey 2017).

Though perhaps less obviously than outright lies, misleading or confusing signals can also have very similar negative effects and should likewise be viewed as ethical hazards. In most legal systems, misleading consumers through signals is seen as quite close to an outright lie both as a violation of norms and as a violation of the law. The line between signals which some people might misinterpret and signals which are actively misleading are unclear and often decided only through case-by-case decisions and litigation (Negowetti 2014). Signals which merely confuse consumers (words like "natural", "healthy", "sustainable", "ethically sourced", and so on are particularly egregious and well-known examples of terms which provide much more confusion than clarity for consumers) should also give us pause. In addition to less extreme versions of the hazards consumers face from lies and deception, confusion also highlights epistemic concerns we should worry about as ethical issues.

A perhaps crude way to put this is to imagine food signals of various kinds as moving between two poles on a continuum. At one extreme, food credence attributes are very easy for consumers to read and understand (examples toward this pole might include a skull-and-crossbones image, or a "Mr. Yuck" face, or the "traffic lights" food labels in the U.K.). However, signals at this pole achieve clarity and understandability at the cost of accuracy and nuance. Consumers may very well understand that a clear signal means "don't eat this," but many foods so labeled could easily be parts of a healthy lifestyle if they were eaten in moderation or were eaten by people with a high level of activity, for example. Similarly, a consumer can easily understand a green light to eat food, but this signal glosses over the quantity, or what is in the rest of the diet the person creating that label imagines the consumer eats (they must make those kinds of assumptions, or very few foods indeed would be seen as acceptable, since they do not on their own provide all the vitamins and other nutrients the consumer needs). As an example of these kinds of assumptions and glosses, it is helpful to come back to the "traffic lights" system of food credence attribute signals in the U.K. This is a clear labeling system with the advantage that it is quite easy for consumers to understand. So much so that despite an initial rejection by the EU, it seems to be gaining momentum across Europe (Michalopoulos 2017). Some in Italy are upset that many of the traditional foods in an Italian diet are "red lights". This despite that fact that Italian people in general are healthier than people in the U.K. and indeed much of Northern Europe. The assumptions being made about a consumer's activity level, portion control, and other factors playing into total health in this case are being shaped in ways that are representative of some groups but not others (Davies 2013, Michail 2016) (instances of how this plays out not for the consumer but for the ethnicity or the culture, see the next section of this chapter).

Further, while clear signs which tell consumers to avoid a given product are relatively easy to create, it is much more difficult to create a clear sign with other messages. One example of this is the debate in the U.S. about whether or not to label products that contain genetically modified ingredients. Advocates of labels are right that there are quite a few people in the U.S. who wish to have that information in order to avoid GMO foods, and they may well have a claim to that information regardless of the scientific consensus on the hazard or lack thereof in GMO foods. This is because even in the absence of any demonstrated risk, there is a general principle that we ought to defer to consumer sovereignty and autonomy, and enable consumers to make the decisions which they see as most in keeping with their view of the good life and their values. If they dislike GMOs for aesthetic reasons, religious reasons, or other similarly "unscientific" reasons, many have argued that they still ought to have the ability to make those decisions (see Thompson 1997, Streiffer and Rubel 2004).

All that said, however, it is also the case that opponents of GMO labels are right that such labels may well carry a stigma which could lead people who would not otherwise care about GMO to avoid the product under an assumption that any product with a "may contain X" label is, all things being equal, less preferred than a product without such a label, and the "X" in question is presumptively bad. Even more neutral labels, such as QR codes to explain what ingredients are genetically modified, if not present on all products, can easily be viewed as one of the "do not eat" symbols like the skull and crossbones discussed earlier. Indeed, an advertising campaign around QR codes made that visual connection, calling it the "Mark of Monsanto" (see Mercola 2001 for an example of this connection being drawn). Conversely, it is also possible that clear signals will sometimes lead to perverse effects, where consumers prefer those foods labeled in some way as unhealthy or otherwise "bad" to those labeled as "good". This is a well studied (if still debated) phenomenon in the case of health warnings on cigarette packs (Erceg-Hurn and Steed 2011), and there is also suggestive research that it is present in the context of food labels (e.g., Wansink and Chandon 2006). The conclusion we can draw from these cases is that the clearer a signal is, the more it elides, and the more likely consumers are to see it as a binary decision on whether food is "good" or "bad", and how that decision plays out is not as predictable as we might think.

At the other pole of the continuum, food credence attributes can be quite nuanced, with lots of information consumers can use to improve their own judgment about whether or not to eat a given product (examples toward this pole might include full nutritional information, a complete ingredients list, or a transparent accounting of product sourcing). However, signals at this pole achieve nuance and accuracy at the cost of clarity and understandability. The kind of complete information consumers could truly use often does not fit on a package and must instead be available only with additional work such as scanning a QR code or following a URL, thereby placing an additional burden on the consumer, and often making an informed decision at the point of sale impossible or at least extremely difficult, due to information overload and decision fatigue (Malhotra 1982, Verbeke 2005, Wansink et al. 2004).

Furthermore, in order to truly be well informed, consumers would need to not only have access to complete information, but also have access to background knowledge necessary to put the information they are getting into context. Without sufficient background knowledge, it is quite possible that a given consumer will ignore the information available, or use the mere presence of information in a heuristic (as discussed above in the case of GMO QR codes). Such background information is not evenly distributed between communities in a given society. Rather, individuals with more education and familial wealth, we can presume, are more likely to have the required background information to make sense of any food credence attributes of which they are made aware, and to be more comfortable with doing the research required to gain that information. This could lead to the further exacerbation of class differences in nutrition and diet-related health. Yet requiring or encouraging producers to provide not only information about the product but also background information about the significance of that information is a burden on them, and creating that education is quite challenging to do well. Companies would have strong incentives to use any such explanation to promote their own product compared to competitors by emphasizing the importance of metrics by which they do better than other products in the field, and downplaying the significance of ways in which they fall short. Such explanations are also likely to greatly exacerbate the problems of overload and fatigue mentioned above, and would again presumably be more sought after and better understood by those from privileged backgrounds. All this means that the ethical dimensions of epistemic challenges in food labelling are difficult problems, and one which extend far beyond merely making sure everything on a label is true.

While honesty and clarity in signals present a host of serious problems, there is at least fairly broad consensus on the underlying ethical issues at play and the need for action, though with often quite serious disagreement on what form exactly that action should take. There is less widespread acknowledgment of ethical concerns around the use of food credence attributes to modify behavior or achieve other positive ends. This is a common goal of the use of these signals, over and above merely satisfying consumer preferences (Golan et al. 2001, Thaler and Sunstein 2008). Yet it is not the simple moral good and efficacious lever that it may first appear to be. One concern is that such social engineering also limits agential autonomy in much the way lying does. Consumers who are having their behavior shaped via food credence attributes are in some ways having their selves "Fragmented" (Bovens 2009) in that their intentions and plans for themselves, perhaps even their identities, are being combined with the intentions of others. Another concern is that there is a possibility people used to being "nudged" in this way are losing the practical ability to maintain and pursue their own projects. If these signals purport to show consumers what they need to know, they are not practicing making their own decisions about what characteristics of the food they should value and how they should evaluate them in light of those values (Furedi 2011). This concern can be avoided, perhaps by biting the bullet and saying that limiting agential autonomy and eroding the practical ability of individuals to be rational agents is acceptable in order to achieve positive social ends. It is also possible to reject the premises of the criticism, by arguing that it is possible to actually enhance autonomy via credence attributes (e.g., Del Savio and Schmietow 2012), though any plan to do so is likely to run into some of the pitfalls we have been discussing in terms of providing background information to consumers.

Another concern with using food attributes to modify consumer behavior as a way of intervening in serious social problems is that it runs the risk of misattributing the causes of the problem onto consumer choice. One example of this can be found in food credence attributes used to intervene in obesity. There is a growing body of research on "obesogenic environments", which is taken to include factors such as urban planning which discourages walking and other exercise, the prevalence of advertising for unhealthy foods, higher access to unhealthy foods than healthy foods, but also the possibility of environmental pollutants which change people's metabolism (e.g., Swinburn et al. 1999, Egger and Swinburn 1997).

The causative power of obesogenic environments makes it less likely that food credence attribute signals to inform consumers will have much of an effect on their weight. At the same time, however, they are quite likely to have an effect on the consumers' sense of identity and their self-esteem. If food credence attributes are used in a way which implies that an individual consumer's choice is responsible for their weight issues, it may well stigmatize them further in the eyes of society (lowering the chance that society will help them, in addition to increasing their feelings of marginalization) and may lead overweight consumers to blame themselves entirely for their weight issues, in a way which can lead to depression and a lower sense of self-worth (O'Connor 2011, Schwartz and Puhl 2003). Disgust is a powerful and effective emotion to evoke as a means of motivating behavioral change, but the effects of that disgust on people when viewed by others or when they view themselves is a significant harm (Puhl and Brownell 2003). This is particularly the case for women (see Lieberman et al. 2012). How much worse, then, to enact a regime of behavioral modification with the known harms to self-esteem, if it is doomed to failure due to the absence of environmental factors. A more effective and less cruel policy intervention would take an "ecological approach" which engages with the wide range of factors surrounding weight gain and try to make the environment in which people live more leptogenic (e.g., Sacks et al. 2007, Schwartz and Puhl 2003). This is not to say that food credence attributes couldn't be a part of such a holistic approach; indeed they may well be. Issues only arise when food attributes and consumer choice are elevated above their proper role.

As we have seen, there are serious ethical concerns about the effects on consumers of food credence attributes, whether they arise naturally or are created by businesses or the government. In the next section, we will look at the ways in which food credence attributes affect other actors in the modern food system.

Effects of Food Credence Attributes for Others

Possible benefits of consumers using food credence attributes for the rest of the food system and the supporting environment are fairly well known, particularly in regards to attributes like "organic", "fair trade", "shade grown", and so on. Less well-known, however, are possible risks to the food system and actors in it. As in the previous section on risk to consumers, this chapter will take a survey approach by lightly touching on some of the more important concerns in this area, and leave the other chapters in this section of the book to more deeply explore particular cases.

One concern in this area is the ways in which particular food credence attributes can silence cultural diversity and harm cultural continuance. For example, which foods are marked as "healthy" in some sense in the minds of consumers is not merely a neutral project of weighing nutritional information. Rather, such decisions are filtered through various cultural and social lenses. The end result is that few if any of the food credence attributes of health in the U.S.—from labels and packaging, to graphic representations of a healthy diet for children (the so-called "food pyramid" or the more modern "healthy plate"), to presence in the media, to layout in grocery stores, and so on—are likely to apply to the cuisine of the immigrants from many different countries. The same can be said of other attributes, such as "proper", "delicious", or even in some cases "edible". It is true that in addition to dominant cultural assumptions and attributes, sometimes particular attributes of some minority communities and cultures in a given society are included, such as when food is labeled as kosher or halal in societies which are not majority Jewish or Muslim. However, even when such exceptions are carved out, this is usually only done for a few of the least marginalized minority cultures. If it is a voluntary commercial attribute, it is more likely to be created for communities with significant purchasing power in the market. If it is a mandatory governmental attribute, we can assume that it is more likely to be created for communities with significant political power. In the U.S., kosher labels are more common than halal labels (at least in most of the country and for national products), and both are far more common than foods with the Hindu Pramãn Standard, or foods labeled as Ital and therefore appropriate for Rastafari diets.

This puts members of marginalized cultures in a difficult position. To what extent do they take on board dominant cultural attributes' significance, and to what extent do they resist cultural hegemony and try to maintain their traditional understandings of food? The cost of such resistance can include ridicule, further marginalization, and even jail time if traditional understanding of what is "proper", and what is "edible" are sufficiently different from the dominant cultural assumptions (Stercho 2006). On the other hand, the cost of fitting in can in extremis be the disappearance of one's culture (e.g., Mittendorf 2017). This is a fraught balancing act for anyone in that situation, but it is particularly troubling for parents, or for everyone in a community which is connected well enough to be concerned for the transmission of their culture to future generations. Deciding whether to give children the ability to fit in with the dominant cultural institutions and mores or to maintain knowledge into future generations is always difficult, but it is particularly sharp in the context of food, due to food's deep phenomenological connection with being a member of a community or not (Leichter 2017). While this section has been focusing here on quite marginalized, minority cultures in larger, dominant societies, we see this same phenomenon playing out on other scales as well. Recall the discussion in the previous section about the controversy around the "traffic light" system of food labels in the U.K. and increasingly in the rest of Europe. Though Italy is in a much less vulnerable position than the cultures we have been discussing, they nevertheless have a similar problem when their culture is not represented in the food credence attributes of the wider European society.

Another concern brings us back to the use of food credence attributes to make positive change. In the last section we looked at concerns around modifying individual behaviors. Here we will look instead at the effects of such interventions on other actors in the food system and on the wider environments in which the food system is embedded.

One concern along these lines is the ways in which labels can distort the behavior of food system actors trying to achieve those labels. Such distortion can have quite perverse consequences which can cause surprising side effects, actually work against the goals of the label in the first place, and harm those actors. For example, the "fair trade" label has become quite a popular food credence attribute in recent years, in large part because it is understood to have a very beneficial effect on multiple actors in the food system, particularly workers harvesting ingredients. Yet as Sarah Besky shows in her book *The Darjeeling Distinction: Labor and Justice on Fair-Trade Tea Plantations in India*, in the case of Indian tea plantations the label can have quite the opposite effect of the one for which consumers and the certifiers are presumably working. Besky speaks about how many of the benefits provided by fair trade certification are not helpful to the workers. For example, she argues that loans given to tea pickers for purchasing cows, with the goal of them selling milk and reduce their poverty, are nearly useless for people working all day in the fields away from where the cow would be, on a plantation too far away from any neighboring community which might purchase their milk. She also discusses how those benefits are often given in lieu of traditional or legal obligation the plantation owner has to the workers. For example, she discusses how many of the plantation owners use Fair Trade subsidies to pay for temples or churches rather than doing it themselves, as plantation owners are supposed to under Indian law. Finally, she discusses how the desire for Fair Trade certification has led some plantations to put the workers into even worse working conditions than normal, such as increased working hours spent improving the plantation, in order to meet the standards before an inspection (Besky 2014). As another example, organic certification in the US is such an onerous project that many small-scale farmers find it prohibitively difficult to achieve, requiring additional paperwork and convoluted rules which increase their time and costs. Yet if they do not achieve that certification, some of those farmers worry that they will be severely limited in their ability to reach the kinds of customers willing to pay a premium for their produce (Piso et al. 2016).

In both of these examples, we see the concern that food credence attributes, particularly labels, are sometimes simply too blunt an instrument to achieve positive social change on their own. Another reason to suspect this is the ways in which such attributes can be easily hijacked by bad actors. Many signals of food credence attributes, particularly labels, can become for consumers mere status symbols to be purchased on the cheap, and for food sellers a goal to be achieved by the letter of the law while violating the spirit of the attribute if it would increase profit margins to do so. This can lead to the attribute losing all its meaning, and falling far short of any promise it had for social change. One such food credence attribute which is facing this kind of threat is food understood to be "local". Early advocates of local food saw it as an avenue to challenge the hegemony of the capitalist industrial food system, and a proxy for a host of positive values of social justice and environmentalism. However, as it has grown in popularity it is now at risk of becoming one consumer choice among many for those that can afford it, and indeed a way of maintaining class differences. It is also at risk of becoming a shield for food sellers to use when avoiding more stringent goals of worker welfare or true environmental sustainability (Werkheiser and Noll 2014). When we hope for a proxy attribute or goal to stand in for a whole host of values, we cannot be surprised when the proxy itself is maximized in ways that avoid achieving, or even work against, the values for which it was originally meant to stand (Toadvine and Morar 2015).

The example of "local" as a food credence attribute also points to a third concern with using these attributes to create social change: possible perverse effects where the attributes are used to increase social problems. This is similar to our earlier discussion of the ways in which food credence attributes meant to help address problems that consumers have (e.g., labelling food as "healthy" in order to help people eat nutritiously) can lead to perverse effects (people using "healthy" as a sign that something doesn't taste good, and therefore avoiding it). In the case of the use of "local", there are at least two possible ways in which this attribute can increase social problems. One of these is that local food may reduce cash flow from wealthy nations to poor ones whose economies are largely dependent on agricultural exports (Navin 2014). If the decrease in support of poor nations' economies via food imports were compensated with other forms of developmental aid that might mitigate the harm, but this is rarely brought up as a necessary component of a move toward local food by advocates of the social effects of that food credence attribute (*ibid.*). Another related problem is that local food can be used as a proxy for jingoistic local control (Fairbairn 2012). In this way, it can lead to a further marginalization of already vulnerable groups, such as when the food ways and agricultural knowledge of immigrant and displaced communities are marginalized as being insufficiently local (Alkon and Mares 2012). At such times, the focus on local as a food attribute can actually worsen social relations between marginalized and dominant communities, and can worsen the marginalized communities' access to culturally appropriate food; presumably outcomes that would be abhorrent to many of the proponents of local food.

There is a final concern with the use of the markets to address profound ethical and justice-based issues, namely that it runs the risk of reducing those important issues to mere preference. Even granting the possibility that signals could achieve positive social ends, say that the use of "Union Made" or "Fair Trade" labels could improve workers' treatment, this avoids the issue of workers' *rights* in a way that could be pernicious. If workers have a *right* to fair treatment, on some ethical theories it is a harm to their dignity to reduce that fair treatment to one option among many for distant consumers to decide with their dollars. Similar concerns can be raised for other food credence attributes, including environmental preservation, protection of endangered species, animal cruelty, and so on. If something is a right demanded by morality or justice, it is precarious to leave it to the whims of consumer preference (particularly preferences which are manipulable by advertising and other effects). Additionally, it is also a diminution of those issues which can be insulting or even harmful (again, on some ethical theories, such as deontological ones). If the only reason you are not being abused by your employer because of your age, race, or gender is because their good treatment of you is a successful marketing strategy, you might well feel that this is not a secure enough footing for that fair treatment to allow you to make future plans. You might also feel a markedly worse relationship to your employer and to your society than if such fair treatment was guaranteed by law.

Conclusion

The above chapter has focused on concerns around the use of food credence attributes, but there is another term which is often closely associated to food credence attributes and which is worth examining briefly here, namely the idea of consumer *preference*. The idea in much of the work on food attributes is that consumers use those attributes to satisfy their preferences. This is true, nearly a tautology, but it can also be misleading if we use the simple word "preference" to obscure the vast range of motivation in human activities around food. For example, as was mentioned earlier in this chapter, food attributes have a profoundly cultural component. For marginalized communities, equating efforts to preserve their culture into future generations via teaching their children traditional foodways, with an individual consumer's preference for a particular flavor of soda, seems like a category error. As another example, we have discussed how people sometimes respond to food credence attributes in ways which are opposite of the intent of the signal (e.g., understanding health labels to mean "not tasty"). It seems misleading to call this a mere preference, say for unhealthy foods. This is because we are discussing the decisions made by people who have poor diets which include far too much of foods which we know to have quasi-addictive properties and which are marketed to them by a multi-billion dollar industry. Such consumers are acting irrationally, but predictably so, given what we know about the ways their bodies and minds are being manipulated. Seeing that as simple preference lets marketing of unhealthy foods and the obesogenic environment off the hook in a way which is quite pernicious. These critiques are not a call to abolish talk of preferences from discussions of consumer perception of food attributes. Rather, they are a call to be more mindful of the term's connotations and limitations.

More generally, this chapter is in no way trying to argue that we—policy makers, activists, academics, food producers, food sellers, and consumers—should not pursue the use of food credence attributes. For one thing, this would probably prove impossible. Consumers will naturally look for signs suggestive of particular qualities in a product they are considering buying. This is perhaps particularly true for food, given that we have an evolutionary predilection to look carefully at food for signs that it might be inedible. Thus consumers will generate, particularly in the world of social media, their own food credence attributes, making abandoning them impossible. Additionally, despite the concerns described above, it is quite possible that we will be able to use food credence attributes to increase consumer autonomy and agency, benefit other actors in the food system, and make the food system itself more sustainable. In order to do that successfully, what this chapter and the following chapters in this section show is that we must be cautious when promoting or signaling these attributes. They are powerful tools, but ones that are too often viewed in simplistic, optimistic terms, and thus misapplied.

References

Alkon, A.H. and T.M. Mares. 2012. Food sovereignty in US food movements: radical visions and neoliberal constraints. Agr. Hum. Val. 29: 347–359.

Besky, S. 2014. The Darjeeling Distinction: Labor and Justice on Fair-Trade Tea Plantations in India (California Studies in Food and Culture). University of California Press, Berkeley. USA.

Bovens, L. 2009. The ethics of the nudge. pp. 207–27. *In:* Yahoff-Grüne, T. and S.O. Hansson (eds.). Preference Change. Springer, Dordrecht. Germany.

Cheftel, J.C. 2005. Food and nutrition labelling in the European Union. Food Chem. 93: 531–550.

Darby, M.R. and E. Karni. 1973. Free competition and the optimal amount of fraud. J. Law Econ. 16: 67–88.

Davies, L. 2013. Italy claims 'traffic-light' labelling unfair on Mediterranean food. The Guardian. 21 October 2013.

Del Savio, L. and B. Schmietow. 2012. Environmental footprint of foods: the duty to inform. J. Agr. Environ. Ethics 26: 787–796.

Egger, G. and B. Swinburn. 1997. An "ecological" approach to the obesity epidemic. BMJ 315: 477–483.

Endres, B. and N.R. Johnson. 2011. United States food law update: the FDA Food Safety Modernization Act, obesity and deceptive labelling enforcement. J. Food Law Policy 7: 135.

Erceg-Hurn, D.M. and L.G. Steed. 2011. Does exposure to cigarette health warnings elicit psychological reactance in smokers? J. Appl. Social Psychol. 41: 219–237.

Fairbairn, M. 2012. Framing transformation: the counter-hegemonic potential of food sovereignty in the US context. Agr. Hum. Val. 29: 217–230.

Ford, G.T., D.B. Smith and J.L. Swasy. 1988. An empirical test of the search, experience, and credence attributes framework. Adv. Consum. Research 15: 239–244.

Furedi, F. 2011. Defending moral autonomy against an army of nudgers. Spiked. <http://tinyurl.com/6kafka> (accessed on 6 October 2017).

Golan, E., F. Kuchler, L. Mitchell, C. Greene and A. Jessup. 2001. Economics of food labeling. J. Consum. Policy 24: 117–84.

Leichter, D.J. 2017. Edible justice: between food justice and the culinary imaginary. pp. 13–30. *In:* Werkheiser, I. and Z. Piso (eds.). Food Justice in US and Global Contexts. Springer, New York. USA.

Lieberman, D.L., J.M. Tybur and J.D. Latner. 2012. Disgust sensitivity, obesity stigma, and gender: contamination psychology predicts weight bias for women, not men. Obesity 20: 1803–14.

Malhotra, N.K. 1982. Information load and consumer decision making. J. Consum. Res. 8: 419–30.

Mercola, J.M. 2001. The chilling history of Monsanto's rise to power. Bibliotecapleyades. <https://tinyurl.com/y93552h6> (accessed on 6 October 2017).

Michail, N. 2016. Italy raises red flag once more over UK's traffic light label. Food Navigator. <https://tinyurl.com/yautxm9b> (accessed on 6 October 2017).

Michalopolous, S. 2017. 'Traffic light' food labels gain momentum across Europe. Euractiv. <https://tinyurl.com/y9vgy8nn> (accessed on 6 October 2017).

Mittendorf, R.W. 2017. Religious slaughter in Europe: balancing animal welfare and religious freedom in a liberal democracy. pp. 285–98. *In:* Werkheiser, I. and Z. Piso (eds.). Food Justice in US and Global Contexts. Springer, New York. USA.

Navin, M. 2014. Local food and international ethics. J. Agr. Environ. Ethics 27: 349–368.

Negowetti, N.E. 2014. Food Labeling Litigation: Exposing Gaps in the FDA's Resources and Regulatory Authority. Governance Studies at Brookings, Washington, DC. <http://www.brookings.edu/> (accessed on 6 October 2017).

Nelson, P. 1970. Information and consumer behavior. J. Polit. Econ. 78: 311–29.

O'Connor, L.K. 2011. Weight-based stigma and deficit thinking about obesity in schools: how neoliberal conceptions of obesity are contributing to weight-based stigma. M.A. thesis, Ontario Institute for Studies in Education, University of Toronto (St. George Campus).

Olmstead, L. 2016. Real Food/Fake Food: Why You Don't Know What You're Eating and What You Can Do About It. Algonquin Books, New York. USA.

Piso, Z., I. Werkheiser, S. Noll and C. Leshko. 2016. Sustainability of what? Recognizing the diverse values that sustainable agriculture works to sustain. Environ. Val. 25: 195–214.

Poppe, C. and U. Kjaernes. 2003. Trust in food in Europe: a comparative analysis. Professional Report no. 5. National Institute for Consumer Research, Oslo. <www.trustinfood.org/SEARCH /BASIS/tif0/all/publics/DDD/24.pdf> (accessed on September 2011).

Puhl, R.M. and K.D. Brownell. 2003. Psychosocial origins of obesity stigma: Toward changing a powerful and pervasive bias. Obesity Rev. 4: 213–227.

Sachs, G., B.A. Swinburn and M.A. Lawrence. 2008. A systematic policy approach to changing the food system and physical activity environments to prevent obesity. Aust. NZ Health Policy 5: 13.

Schwartz, M.B. and R. Puhl. 2003. Childhood obesity: a societal problem to solve. Obesity Rev. 4: 57–71.

Stercho, A.M. 2006. The importance of place-based fisheries to the Karuk Tribe of California: A socioeconomic study. Doctoral dissertation, Humboldt State University, Germany.

Streiffer, R. and A. Rubel. 2004. Democratic principles and mandatory labeling of genetically engineered food. Public Aff. Quart. 18: 223–249.

Swinburn, B., G. Egger and F. Raza. 1999. Dissecting obesogenic environments: the development and application of a framework for identifying and prioritizing environmental interventions for obesity. Preventative Med. 29: 563–570.

Thaler, R. and C. Sunstein. 2008. Nudge: Improving Decisions about Health, Wealth, and Happiness. Yale University Press, New Haven. USA.

Thompson, P.B. 1997. Food biotechnology's challenge to cultural integrity and individual consent. Hastings Center Report 27: 34–39.

Toadvine, T. and N. Morar. 2015. Biodiversity at twenty-five years: revolution or red herring? Ethics, Policy Environ. 18: 16–29.

Verbeke, W. 2005. Agriculture and the food industry in the information age. Eur. Rev. Agr. Econ. 32: 347–368.

Wansink, B. and P. Chandon. 2006. Can "Low-fat" nutrition labels lead to obesity? J. Market. Res. 43: 605–617.

Wansinsk, B., S.T. Sonka and C.M. Hasler. 2004. Front-label health claims: when less is more. Food Policy 29: 659–667.

Werkheiser, I. and S. Noll. 2014. From food justice to a tool of the status quo: three sub-movements within local food. J. Agr. Environ. Ethics 27: 201–210.

Whoriskey, P. 2017. Why your 'organic' milk may not be organic. *Washington Post*, 1 May, 2017.

11

Food Credence Attributes, Multi-Criteria Analysis and the Ethics of Food Choice

Mario Giampietro

Introduction

Over the last years the quality of food products has gained much attention among consumers in developed societies (Kasriel-Alexander 2015). This attention has been accompanied by a growing interest in finding "sustainable" alternatives to conventional techniques of food production, such as organic food (Canavari et al. 2007, Stolz et al. 2011, IFOAM 2016), and alternative methods of food distribution (local production) to replace the long or global food chains that tend to compress producer profit (Du Puis and Goodman 2005, Martinez 2010, Morgan 2010, Murdoch et al. 2000). This growing change in consumer attitude reflects a more general concern about the lack of sustainability of the existing structure and operation of the modern food systems (Blay-Palmer 2008, Porter and Kramer 2011).

However, in practice, the definition and analysis of food system performance and quality of food products have resulted to be elusive concepts. In fact, the relations between food system activities, human health, environment, economy and cultural identity of a given society are extremely complex and impossible to represent in terms of a finite set of attributes that can be measured and modelled (Tansey and Worsley 1995, Sobal et al. 1998, Mc Michael 2000, Vorley 2003, Giampietro 2004, Sonnino and Marsden 2006, Ericksen 2008, Tansey and Rajotte 2008, Lang et al. 2009, Patel 2014, Tilman and Clark 2014, Lang and Heisman 2015). The more so as the globalization

Institut de Ciència i Tecnologia Ambientals, Universitat Autònoma de Barcelona, 08193 Bellaterra, Spain; ICREA, Pg. Lluís Companys 23, 08010 Barcelona, Spain.
E-mail: mario.giampietro@uab.cat

of the food system has become almost inevitable (even in local production systems, most inputs such as energy and fertilizer, are imported).

Recent experiences done in the European research project GLAMUR (www. glamur.eu), which had the goal of comparing the performance of long and short food chains (Brunori et al. 2016), have clearly shown that the complexity of food systems makes it impossible to define a 'one size fits all' set of attributes to define the quality of food products (Gamboa et al. 2016). Dealing with complex issues, such as the production and consumption of food in a modern urban society, it is simply impossible to identify a set of attributes (what is called in common jargon "the framing of a multi-criteria analysis") capable of determining a substantive ranking that indicates what is the "best" (or the "optimal") choice to make. Indeed, several studies on consumer attitude have shown that the definition of quality of a food product reflects not only the information available but also the normative values of the consumer (Tregear and Ness 2005, Paloviita 2010, Giampietro and Bukkens 2015, Robinson and Carson 2015). For this reason there is also an unavoidable ethical dimension in any food choice, especially related to sustainability (the sustainability of our body, of our family, of our landscape, of our community).

The term 'food credence attribute' has been exactly proposed to indicate that most of the 'qualities' of a food product cannot possibly be properly evaluated by the consumer. Indeed, consumers have to rely on generic perceptions of or beliefs about the food product and its associated food system at the moment of buying. This situation entails the risk of poor choices due to (sometimes purposeful) lack of information and possible manipulation of the perceptions and choices of the consumers through marketing campaigns.

In the following section, I will illustrate the general problems faced when trying to characterize the performance of the modern food system and the resulting quality of its food products. I will show that these problems are rooted in the co-existence of many different dimensions and scales of analysis that have to be considered and represented simultaneously according to the many non-equivalent perceptions and normative values of the various social actors involved. In the next section, I will then frame this discussion in more technical terms: What are the problems faced when using multi-criteria analysis to rank options when dealing with complex food choices? Last but not the least, I will argue that (i) moral choices have to play a key role in the redefinition of a more sustainable food system, and (ii) that food should not be considered a mere commodity, but something special associated with the definition of our very identity.

The Complexity of the Concepts of Food System and Food Quality

Can we define a set of food attributes that captures the 'true' quality of food products?

In order to characterize the performance of food systems we have to consider both (i) the attributes that are directly relevant for the consumers themselves, and (ii) the attributes that are relevant for assessing the environmental impact generate by the food system and the stability of ecological services both locally and globally (indirect impacts on current and future food consumers). While the former attributes are already difficult to define and measure—e.g., the nutritional value of the food supply, food safety, food security, food sovereignty, food convenience, food taste, symbolic meaning of food consumption, preservation of cultural identity—the latter are even more so, as they not only involve the impact of food production in a given country but also the impact associated with trade in a globalized economy (e.g., externalization). Indeed, dealing with the quantitative characterization of the performance of food systems we inevitably face an information space that is difficult to tame: (i) the definition of what is relevant depends on who will use the information; (ii) the representation of cause and effect of our choices is affected by a large dose of uncertainty (Giampietro and Bukkens 2015).

From a technical perspective we can re-state this problem as: Any quantitative analysis aimed at assessing the 'performance' of food systems faces an epistemological challenge that is common to all purposive quantitative analyses attempting to assess the performance of complex systems (Giampietro et al. 2006). Indeed, the problem faced by the scientist in generating an integrated assessment of the food system is similar to the problem faced by the consumer in buying food at the supermarket, that is, the problem of 'simplification' of complexity. Simplifying an enormous information space into a simple protocol to be used for selecting food items generates the phenomenon of hypocognition—i.e., the unavoidable negligence of relevant aspects of the complex system under analysis that were not included in the chosen frame used for the choice (Lakoff 2010).

A practical example: are more efficient methods of producing food an 'improvement'?

Scientific literature on technological progress in agriculture tells us about the impressive achievements of science-driven innovations in the field of production of food and fibers. For example, in advanced production systems less than 1.6 kg of feed input is required to produce 1 kg of poultry meat (output) in 35 days (Poultryhub. org 2009), whereas chicken raised in the traditional way may require up to 8 kg of feed per kg of poultry meat produced (van Eekeren et al. 2006). This represents an increase in efficiency in poultry production of 5 times! Clearly, this progress was not obtained while all other elements of the production process remained equal; increased efficiency was obtained by moving chicken production from open-air yards to animal factories. *"With the use of incubator, chicken are no longer produced in parallel as in the old system dictated by nature . . . "Chicken farm" has thus become a misnomer: the situation calls for replacing it by "chicken factory". Because of the new system a pound of chicken sells in the United States for less than a pound of any other kind of meat, while in the rest of the world, where the old system still prevails, chicken continues to be "the Sunday dinner"* (Georgescu-Roegen 1971, p.253).

An even more spectacular result has been obtained with milk productivity of cows. The net milk produced per cow per day (yearly average) differs a lot among countries and production systems: It is below 1 litre in pastoralist societies, less than 5 litres in pre-industrial farms, more than 50 litres in modern "milk factories", and it can reach peaks of more than 75 litres in cases where production is boosted with hormones (Giampietro and Bukkens 2015).

In these examples of chicken and milk production, we can readily calculate a number, i.e., a quantitative index characterizing, respectively, an increase in efficiency (an output/input ratio of chicken meat produced per unit of feed) and an increase in productivity (a given pace of milk flow per cow per day). But how useful and effective are these numerical indices in defining whether or not we are witnessing an *improvement* in the food system? Should we consider the chicken or the liter of milk 'better' than before because the process of production is more efficient?

When studying the performance of food production systems, we must be extremely careful and always look for an integrated set of indicators that allows us to better understand the pros and cons (in terms of side effects) of the technical change(s) we want to introduce. In other words, we must guarantee the quality of the indications given by any single numerical index of performance, by checking the existence of other criteria of performance neglected in our choice. For instance, how can we know whether one kilogram of poultry meat produced in a distant chicken factory is *the same* as one kilogram of poultry produced in a farm backyard in a nearby village in the traditional way? What set of criteria (indicators of performance) could we use to answer this question? Similarly, the same search for side effects should be done when assessing the consequences of moving from an average production level of a few liters of milk per day by the family cow milked by hand in the backyard to an average production level of over 50 liters of milk per day per cow achieved in high-tech, large-scale dairy farms. If we are unable to account for all the side effects entailed by this 'progress', then how reasonable is it to assume that a liter of milk is *the same* regardless of the system used to produce it? What is the meaning of the concept of *sameness* in this discussion?

Do not make the mistake of thinking that this abstract discussion is irrelevant for practical purposes. For example, the implementation of the controversial *principle of substantial equivalence* did cause heated ideological debates (Giampietro 2002). This latter concept has played an especially important role in the regulation of genetically modified organisms (GMOs) in food production and the use of bovine growth hormone (BGH) to enhance milk productivity in cows (Giampietro 2002).

In real life, innocent questions like whether a tomato produced from genetically modified plants is *the same* as a tomato produced from traditional and local varieties, or whether the milk produced by cows injected with a synthetic hormone is the same as milk produced by traditional cows, have led to ferocious ideological battles. Because of the high stakes involved, every single term used has been fiercely debated and it has been necessary to get into philosophical questions such as:

- Is it possible by only comparing a few selected chemical substances, chosen from among thousands, to declare the substantive equivalence of two batches of milk?
- How reasonable is the claim of reductionism that it is only the chemical composition of the edible parts of a food product that matters?

The avoidance of this type of philosophical questions has been instrumental in the past, in defining and implementing the concept of substantial equivalence and in the decision of the U.S. Food and Drug Administration (FDA) to not require labelling for milk produced with BGH or food products obtained from genetically modified plants. Thus epistemological discussions like those above may result crucial for the future survival of small local organic farmers and dairy-farms and in relation to the health of our children.

We may ask ourselves then: How defendable is the use of reductionism in making these decisions? Imagine that by considering just a few chemical elements, such as the amino acids making up meat protein, a scientist concludes that pork is substantially equivalent to beef. Given this conclusion, should we no longer label meat as pork or beef simply because they were considered equivalent according to some arbitrary reductionist criterion? Should then Jews and Muslims no longer have the option to avoid eating pork because this type of consideration (religion) is 'unscientific' and does not merit labeling (Giampietro 2002)?

This example raises two important questions:

1. *What* criteria should be used for defining the 'performance' and 'improvements' in the food system and the 'quality' of food?
2. *Who* should choose these criteria and *how*?

The questions directly relate to the use of multi-criteria analysis and the ranking of the quality of food products.

The Problems in Using Multi-Criteria Analysis to Rank Food Choices

On complexity and relevance

By definition a complex phenomenon is a phenomenon which can and should be perceived and represented using simultaneously several narratives, dimensions, and scales of analysis (Simon 1962, 1976, Rosen 1977, 2000, Salthe 1985, Ahl and Allen 1996, O'Connor et al. 1996, Cillier 1998, Funtowicz et al. 1999, Allen et al. 2001, Giampietro 2003). This conceptualization of complexity entails that the definition of *what is complex* for quantitative analysis depends on a preliminary definition of *what is relevant* for those that will use the analysis. This means that when coming to food choices only the consumers can decide what is relevant for them.

The technical problems with multi-criteria analysis

Multi-criteria analysis (MCA) has been developed about two decades ago with the aim of avoiding the pitfalls of reductionism—in particular the simplifications of Cost-Benefit Analysis—in relation to the quantitative assessment of different types of costs and benefits (Zeleny 1982, Bana e Costa 1990, Nijkamp et al. 1990, Vincke 1992, Roy 1996, Beinat and Nijkamp 1998, Jansen and Munda 1999, Hayashi 2000, Bell et al. 2001, Belton and Stewart 2002). An overview of the state of the art in this field has been provided by Figueira et al. (2005).

The final output of a process of MCA is a quantitative or qualitative representation of the problem in the form of either an impact matrix or a graphic representation in the form of a performance space (e.g., a radar diagram with multiple indicators) that is then used to generate a ranking over a set of options. An example of such an impact matrix is given in Fig. 1. In this example the rows of the matrix are represented by 12 attributes of performance divided over four criteria of performance (three attributes for each indicator): (i) impact on the economy; (ii) impact on health; (iii) impact on food security; (iv) impact on the environment. The values assigned to these attributes are used to characterize three alternatives—the columns of the matrix—represented in this example by: (i) high-tech milk production; (ii) organic milk production; and (iii) subsistence milk production. In this example the characterization is not done with quantitative data but qualitative judgment. These qualitative judgments can either be based on qualitative indicators or quantitative indicators.

A multi-criteria characterization based on an integrated set of indicators of the type illustrated in Fig. 1 has a main problem in that it represents a static output that is semantically closed. Indeed, it is necessarily based on a closed and finite information space determined by a series of pre-analytical choices of narratives, encoding variables, targets, sampling and measurement schemes, done in the pre-analytical phase. However, this analysis should be contextualized both in space and time in order to be effective in determining an informed choice.

An effective system of characterization of performance should be open to feed-backs within a process of deliberation. On the contrary, the organization of the information space illustrated in Fig. 1 limits the focus of the analysis in relation to both: (i) *How* to carry out the analysis—because it already provides a finite set of relevant attributes for consideration (relevant to whom?); and (ii) *What* is the option to be analyzed in terms of performance—because it provides only a finite set of possible choices.

In conclusion, it is impossible to formalize—"once and for all" and "one size fits all"—the characterization of the performance of a food product in general terms. A generic characterization of the pros and cons of a "lamb chop" or "the content of

DIMENSIONS CRITERIA	RELEVANT ATTRIBUTES	CONSIDERED ALTERNATIVES		
		High tech commodity for the market	Organic quality product for the market	Household subsistence
Economic	final price	GOOD	MORE OR LESS	GOOD
	production cost	MORE OR LESS	MORE OR LESS	GOOD
	rural development	BAD	GOOD	BAD
Health	food safety	GOOD (?)	GOOD	BAD
	quality of milk	MORE OR LESS	GOOD	MORE OR LESS
	risk protection	MORE OR LESS	MORE OR LESS	BAD
Food Security	reliable supply	GOOD	MORE OR LESS	MORE OR LESS
	available subsidies	GOOD	MORE OR LESS	BAD
	convenience	GOOD	BAD	BAD
Environment	GHG emission	BAD	GOOD	GOOD
	N leakages	BAD	GOOD	GOOD
	deforestation (feed)	BAD	MORE OR LESS	GOOD

Fig. 1. Example of multi-criteria impact matrix (from Giampietro and Bukkens 2015).

a box of cereals" is certainly useful to help a potential buyer in choosing the desired food product when in the supermarket but only to a certain extent.

The Ethical Dimension of Food Choices

The change of meaning of food production in modern society and the paradigm of industrial agriculture

The paradigm of industrial agriculture can be defined as the existence of an uncontested consensus over the idea that a massive use of technology (capital) and fossil energy in agriculture is justified in order to achieve two key objectives: Boost the productivity of labor in the agricultural sector and boost the productivity of land in production (Giampietro 2009). The undiscussed acceptance of the industrial paradigm of agriculture is the result of one of the most extraordinary successes of technological progress in the history of humankind. In the last century, the world population has tripled, growing from 2 billion at the beginning of 1900 to more than 6 billion at the beginning of 2000. The increase in the *pace* of population growth has been even more impressive: the world population increased from 3.5 billion at the beginning of the 1970s to 6.5 billion in 2005. That is, in only 35 years the population increased more than it did in the previous 35 thousand years! The paradigm of industrial agriculture made it possible to cope with the increased demand for food coupled to this spectacular population growth. *"Between 1960 and 2000, the demand for ecosystem services grew significantly as world population doubled to 6 billion people and the global economy increased more than sixfold. To meet this demand, food production increased by roughly two-and-half times* (Millennium Ecosystem Assessment (MEA) 2005, p.5).

Progress in agriculture not only increased the ability to produce more food per hectare, but also more food per hour of work in agriculture. In the industrial and post-industrial eras there are no countries that have a GDP of over 10,000 USD p.c. and, at the same time, more than 5% of the work force engaged in agriculture. In the richest countries, the percentage of farmers in the work force is invariably lower than 2% (Arizpe et al. 2011). This dramatic change in the performance of food production has gone hand in hand with a change in the meaning of agriculture and its associated role in society. This change can explain the progressive unsustainability of agriculture that is felt on both sides of the interface: (i) social side—the progressive disappearance of rural communities; and (ii) ecological side—the progressive increase in stress placed on ecosystems and agroecosystems. We can describe the evolution in the meaning of 'agriculture', taking place over the past centuries, as follows:

Meaning #1: Agriculture is an activity supporting the existence and economic development of rural societies. This meaning was uncontested in pre-industrial times. Indeed, preindustrial society was fully aware of the importance of maintaining and reproducing rural communities and the natural resources—including ecosystem services—on which these rural communities relied.

Meaning #2: Agriculture is an activity having the goal of feeding the cities. After the industrial revolution we had the explosion of the industrial paradigm of agriculture. Huge injections of capital and fossil-energy-based inputs made it possible to replace many of the required inputs previously provided by the agro-ecosystem. Machine power, fertilizers, pesticides, improved seeds and irrigation eliminated many of the limits on agricultural productivity associated with the low pace of conversion of natural processes (Giampietro 2003). But the industrial paradigm of agriculture also had its side-effects: it triggered a profound socio-economic transformation consisting in a systemic elimination of farmers and a consequent destabilization of rural communities. Willard Cochrane (1958) neatly described the mechanism, which he called the "agricultural technology treadmill", through which the market mechanism effectively eliminated farmers. This mechanism works in 5 steps, iterated in time:

1. Many small farms all produce the same product. Because not one of them can affect the price, all will produce as much as possible against the going price;
2. A new technology enables innovators to capture a windfall profit: *'innovation'*;
3. After some time, others follow: *'diffusion of innovations'*;
4. Increasing production and/or efficiency drives down prices. Those who have not yet adopted the new technology must now do so lest they lose income: *'price squeeze'*;
5. Those who are too old, sick, poor or indebted to innovate eventually have to leave the scene. Their resources are absorbed by those who make the windfall profits. This is called *'scale enlargement'*, but would better be called *'farmers elimination'*.

Meaning #3: In a globalized word powered by financial operations food is just another commodity, handled by agro-businesses. In the last decade (post-industrial era) the meaning of food production has changed again. Globalization and financialization of the economy have led to a total loss of contact with biophysical feed-backs. For example, according to the report of the McKinsey Global Institute (2015), since 2007 the global debt has increased by 57 trillion USD. This means that the increase in debt has outpaced world GDP growth and, considering the biophysical roots of the economy, that the world did not have any economic growth since 2007. In this situation, whenever it is convenient, virtual flows of money (debts and quantitative easing) are used to liquidate the biophysical assets of society (structural elements associated with technology and biophysical processes). In other words, the world economy entered in the era of *Ponzi scheme economics*.

To give an example of the nature of the implications of this change let us consider the extraordinary economic performance of Dutch agriculture in the last decade. As shown in Figs. 2 and 3, in 2014 Dutch agriculture was the number one exporter in Europe and the second exporter in the world in terms of economic value. According to this economic ranking the Netherlands is exporting more than Canada and Argentina summed together. Looking at the biophysical roots of these money flows—the availability of arable land—we see that these exports are based on less than 1 million hectares of arable land (available in the Netherlands) versus almost 74 million hectares of Canada and Argentina combined—the ratio is 1/82!

Obviously Dutch agriculture represents an economic sector that is completely open and maximizes the pace and density of imported and exported flows through a continuous liquidation of ecological funds (e.g., area of natural habitat) and, most importantly, through the use of virtual land, water and labor (embodied in imports) thus externalizing any negative impacts on the environment. But what has this brilliant

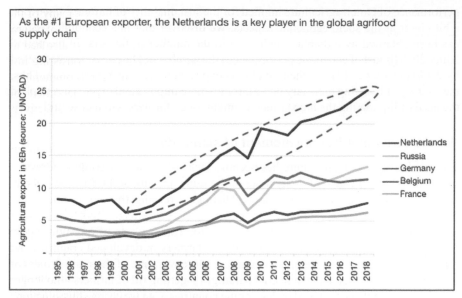

Fig. 2. Ranking of agricultural export in Europe (source: https://www.pwc.nl/nl/assets/documents/pwc-megatrends-impacting-the-dutch-agrifood-industry.pdf).

Rank	Country	Value of Food Exports (US Dollars	Arable land hectares	
			p.c.	total
1	United States	$149,122,000,000.00		
2	Netherlands	$92,845,387,781.00	0.06	0.9 M
3	Germany	$86,826,895,514.00		
4	Brazil	$78,819,969,000.00		
5	France	$74,287,121,198.00		
6	China	$63,490,864,000.00		
7	Spain	$50,960,954,460.00		
8	Canada	$49,490,302,612.00	1.25	45.7
9	Belgium	$43,904,482,740.00		↓
				73.6
10	Italy	$43,756,176,567.00		↑
11	Argentina	$37,171,872,677.00	0.93	27.9

Fig. 3. Ranking of agricultural export (in economic value) in the world, 2014 (Based on data from the US Department of Commerce's International Trade Administration; Source: https://www.worldatlas.com/articles/the-american-food-giant-the-largest-exporter-of-food-in-the-world.html).

performance of Dutch agriculture implied for the sustainability of farmers? If we look a bit closer at the socio-economic aspects, we discover that this economic success has been obtained by a dramatic reduction in the number of farmers. In the period 2000–2010 (boost of economic performance) the number of Dutch farmers decreased by 23% (Eurostat 2017) and their number is expected to still further decrease. Thus, the strategy of boosting monetary flows through more imports and exports has implied the need of liquidating the human fund element—i.e., farmers—in this sector!

For the consumer it is the moment of stand-up!

Technical progress in agriculture has become locked-in in a 'Concorde syndrome': innovations have the goal of doing more of the same, in spite of the fact that nobody is happy with the actual performance. Through stressing the environment, the paradigm of industrial agriculture has resulted in: (i) a surplus of food commodities without a demand in developed countries and a too-expensive-to-afford food supply for developing countries; and (ii) land without farmers in developed countries and farmers without land in developing countries. Moreover, one of the main drivers of technological progress in the industrial paradigm—getting rid of farmers—no longer makes sense in either developed or developing countries. Can we change this situation? Can consumers play an active role in creating the conditions for a different path of development of the food system?

Reflecting on the consequences of the change in the meaning of food production (as discussed in the previous section) we can start from an analysis of the resulting change of meaning of 'food' for both consumers and producers. In this reflection we have to consider the existence of two different meanings (narratives) about food: (i) food is something 'special' associated with the identity of both producers and consumers; (ii) food is just another commodity. Then in relation to these two narratives we can try to answer the following questions:

1. What is special about 'food' and what are the implications for our choices?
2. What are the relevant attributes of food commodities and what are the implications of defining a finite set of quality indicators for 'food'?

Narrative #1: Food is 'special' because of: (i) the history of both the consumer and the producer; and (ii) the place of both consumption and production. So any type of food always reflects the 'special' characteristics of the identity of the consumer and the producer, the 'special' characteristics of the place in which the food product is produced and consumed. History and place matter and, therefore, the identity of a product helps to preserve the identity of the consumer, the producer, and the time and location-specific process of production. This implies that consumers, when choosing food items, express their identity (taste, symbolic role) and moral values. The consumers are reinforcing or weakening the processes of production. In a world in which food products have a clear identity, producers can take pride in their work because it is appreciated and makes it possible to make a decent living.

Narrative #2: Food is just a mere commodity whose 'quality' can be assessed using a standard set of attributes of performance, such as (i) reliability in supply; (ii) safety; (iii) nutrient composition; (iv) affordability (price); (v) taste; (vi) convenience. In this narrative the history and the place of either production or consumption do not matter. As a consequence the identity of the consumer and the producer is not explicitly associated with food choices. It is the combination of indicators of performance over the set of criteria and the relative weighting factors that is supposed to determine the quality of what to buy.

In the second narrative the identity of the producer is irrelevant in relation to the decisions on how to generate the supply of food products. Farmers are told what to do, how and when. The supply of food commodities has to be reliable and programmable. Prices and subsidies define the option space of producers. As a result, within an international context determined by subsidies and regulations, farmers are never playing a level field. Moreover, the necessity of complying with international quality standards forces them to hide special characteristics of their products that may not be compatible with selected markets and regulations.

Conclusion

Looking at existing trends of technological development of agricultural production systems, we can predict that the 'more-of-the-same solution' will be unlikely to last very long. Especially when considering rural development there are two magic words to take into account in the discussion, that is, (i) multi-functionality, and (ii) reconnection with urban societies. These two terms imply (i) the need of diversifying the socio-economic activities taking place in the rural landscape—not only farming!, and (ii) the need of re-integrating urban and rural population—not only considering urban elites! The sooner we start discussing and learning how to work on these two objectives the better.

Reducing the importance of economic narratives in the framing of the issue of rural development and sustainability is of paramount importance. When dealing with the sustainability of the food system and the quality of food making the 'right' choices requires wisdom, visions, values, and above all that we care for the farmers, the environment, our health, our culture, our identity, the future of our children. A boost in productivity or a high return on the investment in agricultural production (or agribusiness) within the 'business as usual' narrative of progress will not do it.

Three main points should be driven home from this chapter:

1. Food is not, cannot, and should not be a mere 'economic commodity';
2. The services of (agro)ecosystems cannot be replaced, therefore they should not be monetarized. They must be protected and respected;
3. The separation of our society into a second-rate rural community, supposed to merely produce food commodities to feed the cities, and an urban elite taking care of business is naïve, immoral, and unsustainable.

References

Ahl, V. and T.F.H. Allen. 1996. Hierarchy Theory. Columbia University Press.
Allen, T.F.H., J.A. Tainter, J.C. Pires and T.W. Hoekstra. 2001. Dragnet ecology, "Just the facts Ma'am": the privilege of science in a post-modern world. Bioscience 51: 475–485.

Arizpe-Ramos, N., M. Giampietro and J. Ramos-Martin. 2011. Food security and fossil energy dependence: An international comparison of the use of fossil energy in agriculture (1991–2003). Critical Reviews in Plant Sciences 30: 45–63.

Bana e Costa, C.A. (ed.). 1990. Readings in Multiple Criteria Decision Aid. Springer-Verlag, Berlin.

Beinat, E. and P. Nijkamp (eds.). 1998. Multicriteria Evaluation in Land-use Management: Methodologies and Case Studies. Kluwer, Dordrecht.

Bell, D.E., B. Hobbs, E.M. Elliot, H. Ellis and Z. Robinson. 2001. An evaluation of multi-criteria methods in integrated assessment of climate policy. J. Multi-Criteria Decision Anal. 10: 229–256.

Belton, V. and T.J. Stewart. 2002. Multiple Criteria Decision Analysis: An Integrated Approach. Kluwer, Dordrecht.

Benessia, A., S. Funtowicz, M. Giampietro, A. Guimarães Pereira, J. Ravetz, A. Saltelli, R. Strand and J.P. van der Sluijs. 2016. Science on the Verge – Amazon Book, in the Series "The Rightful Place of Science" Consortium for Science, Policy & Outcomes Tempe, AZ and Washington, DC.

Blay-Palmer, A. 2008. Food Fears: From Industrial to Sustainable Food Systems. Ashgate Publishing, Ltd.: Farnham, UK.

Box, G.E.P. 1979. Robustness in the strategy of scientific model building. pp. 201–236. *In*: Launer, R.L. and G.N. Wilkinson (eds.). Robustness in Statistics. Academic Press, New York.

Brunori, G., F. Galli, D. Barjolle, R. van Broekhuizen, L. Colombo, M. Giampietro, J. Kirwan, T. Lang, E. Mathijs, D. Maye, K. de Roest, C. Rougoor, J. Schwarz, E. Schmitt, J. Smith, Z. Stojanovic, T. Tisenkopfs and J.M. Touzard. 2016. Are local food chains more sustainable than global food chains? Considerations for Assessment. Sustainability 8: 449.

Canavari, M. and K.D. Olson (eds.). 2007. Organic Food. Consumers' Choices and Farmers' Opportunities; Springer: New York, NY, USA.

Cillier, P. 1998. Complexity and Postomdernism: understanding complex systems. Routledge, London, 156 pp.

Cochrane, W. 1958. Farm Prices: Myth and Reality. Minneapolis, MN: University of Minnesota Press.

Du Puis, E.M. and D. Goodman. 2005. Should we go "home" to eat? Toward a reflexive politics of localism. J. Rural Stud. 21: 359–371.

Ericksen, P.J. 2008 Conceptualizing food systems for global environmental change research. Glob. Environ. Chang. 18: 234–245.

Eurostat. 2017. <http://ec.europa.eu/eurostat/statistics-explained/index.php/Agricultural_census_in_the_Netherlands> (accessed on 12 October 2017).

Feyerabend, P. (1975, 2010). Against Method, Verso publisher, London.

Figueira, J., S. Greco and M. Ehrgott (eds.). 2005. Multiple-criteria Decision Analysis: State of the Art Surveys. Springer International Series in Operations Research and Management Science, New York.

Funtowicz, S., J. Martinez-Alier, G. Munda and J. Ravetz. 1999. Information tools for environmental policy under conditions of complexity. European Environmental Agency, Experts' Corner. Environmental Issues Series, No. 9.

Gamboa, G., Z. Kovacic, M. Di Masso, S. Mingorría, T. Gomiero, M. Rivera-Ferré and M. Giampietro. 2016. The complexity of food systems: Defining relevant attributes and indicators for the evaluation of food supply chains in Spain. Sustainability 8: 515.

Georgescu-Roegen, N. 1971. The Entropy Law and the Economic Process. Harvard University Press, Cambridge, MA.

Giampietro, M. 2002. The precautionary principle and ecological hazards of genetically modified organisms. AMBIO 31: 466–470.

Giampietro, M. 2004. Multi-Scale Integrated Analysis of Agro-ecosystems. CRC Press, Boca Raton, FL, 472 pp.

Giampietro, M., T.F.H. Allen and K. Mayumi. 2006. The epistemological predicament associated with purposive quantitative analysis. Ecol. Compl. 3: 307–327.

Giampietro, M. 2009. The future of agriculture: GMOs and the agonizing paradigm of industrial agriculture. pp. 83–104. *In*: Guimaraes Pereira, A. and S. Funtowicz (eds.). Science for Policy: Challenges and Opportunities. Oxford University Press, New Delhi.

Giampietro, M. and S.G.F. Bukkens. 2015. Quality assurance of knowledge claims in governance for sustainability: transcending the duality of passion vs. Reason. Int. J. Sustainable Dev. 18: 282–309.

Guimarães Pereira, Â. and S. Funtowicz. (eds.). 2015. Science, Philosophy and Sustainability: The end of the Cartesian dream. Routledge series: Explorations in Sustainability and Governance.

Hayashi, K. 2000. Multi-criteria analysis for agriculture resource management: a critical survey and future perspectives. Eur. J. Operational Res. 122: 486–500.

IFOAM (International Movement of Organic Agriculture Movements). 2016. Consolidated Annual Report of the IFOAM Action Group. 2016. <http://www.ifoam.bio/sites/default/files/ annual_report_2016. pdf> (accessed on 30 August 2017).

Janssen, R. and G. Munda. 1999. Multi-criteria methods for quantitative, qualitative and fuzzy evaluation problems. pp. 837–852. *In*: van den Bergh, J. (ed.). Handbook of Environmental and Resource Economics. Edward Elgar, Cheltenham, UK.

Lakoff, G. 2010. Why it matters how we frame the environment. J. Env. Comm. 4: 70–81.

Lang, T., D Barling and M. Caraher. 2009. Food Policy: Integrating Health, Environment and Society; Earthscan: London, UK.

Lang, T. and M. Heisman. 2015. Food Wars: The Global Battle for Mouths, Minds and Markets, 2nd ed.; Earthscan: London, UK.

Leslie, I. 2016. The sugar conspiracy. The Guardian [April 7].

Kasriel-Alexander, D. 2015. Top 10 Global Consumer Trends for 2016; Euromonitor International: London, UK.

Martinez, S. 2010. Local Food Systems; Concepts, Impacts, and Issues; Diane Publishing: Collingdale, PA, USA.

McKinsey Global Institute. 2015. Debt and (not much) Deleveraging (February). <https://www.mckinsey. com/global-themes/employment-and-growth/debt-and-not-much-deleveraging> (accessed on 12 October 2017).

Mc Michael, P. 2000. The power of food. Agric. Hum. Values 17: 21–33.

Millennium Ecosystem Assessment. 2005. Ecosystems and Human Well-Being: Synthesis. <https://www. millenniumassessment.org/documents/document.356.aspx.pdf> (accessed on 12 October 2017).

Mirowski, P. 2013. Never Let a Serious Crisis Go to Waste: How Neoliberalism Survived the Financial Meltdown. Brooklyn: Verso Books.

Morgan, K. 2010. Local and green, global and fair: The ethical foodscape and the politics of care. Environment and Planing A 42: 1852–1867.

Munda, G. 2008. Social Multi-Criteria Evaluation for a Sustainable Economy. Springer-Verlag, Berlin, 210 pp.

Murdoch, J., T. Marsden and J. Banks. 2000. Quality, nature, and embeddedness: Some theoretical considerations in the context of the food sector. Econ. Geogr. 76: 107–125.

Nijkamp, P., P. Rietveld and H. Voogd. 1990. Multicriteria Evaluation in Physical Planning. North-Holland, Amsterdam.

Nijkamp, P. and H. Ouwersloot. 1997. Multidimensional sustainability analysis: the flag model. pp. 255–277. *In*: van den Bergh, C.J.M. and M.W. Hofkes (eds.). Theory and Implementation of Economic Models for Sustainable Development. Kluwer, the Netherlands.

O' Connor, M., S. Faucheux, G. Froger, S.O. Funtowicz and G. Munda. 1996. Emergent complexity and procedural rationality: Post-normal science for sustainability. pp. 223–248. *In*: Costanza, R., O. Segura and J. Martinez-Alier (eds.). Getting Down to Earth: Practical Applications of Ecological Economics. Island Press/ISEE, Washington, DC.

Paloviita, A. 2010. Consumers' sustainability perceptions of the supply chain of locally produced food. Sustainability 2: 1492–1509.

Patel, R. 2013. Stuffed and Starved: From Farm to Fork the Hidden Battle for the World Food System; Portobello Books: London, UK.

Polimeni, J., K. Mayumi, M. Giampietro and B. Alcott. 2008. Jevons' Paradox: The Myth of Resource Efficiency Improvements. Earthscan Research Edition, London, 192 pp.

Porter, M.E. and M.R. Kramer. 2011. The big idea: Creating shared value. Harv. Bus. Rev. 89: 2.

Rayner, S. 2012. Uncomfortable knowledge: the social construction of ignorance in science and environmental policy discourses. Econ. and Soc. 41: 107–125.

Robinson, G.M. and D.A. Carson (eds.). 2015. Handbook on the Globalization of Agriculture. Edward Elgar: Cheltenham, UK.

Rosen, R. 1977. Complexity as a system property. International Journal of General Systems 3: 227–232.

Rosen, R. 2000. Essays on Life Itself. Columbia University Press, New York, 361 pp.

Roy, B. 1996. Multicriteria Methodology for Decision Analysis. Kluwer, Dordrecht.

Salthe, S.N. 1985. Evolving Hierarchical Systems: Their Structure and Representation. Columbia University Press, New York.

Simon, H.A. 1962. The architecture of complexity. Proc. Amer. Philos. Soc. 106: 467–482.

Simon, H.A. 1976. From substantive to procedural rationality. *In*: Latsis, J.S. (ed.). Methods and Appraisal in Economics. Cambridge University Press, Cambridge.

Sobal, J., L.K. Khan and C.A. Bisogni. 1998. Conceptual model of the food and nutrition system. Soc. Sci. Med. 47: 853–863.

Sonnino, R. and T. Marsden. 2006. Beyond the divide: Rethinking relationships between alternative and conventional food networks in Europe. J. Econ. Geogr. 6: 181–199.

Stiglitz, J.E. 2011. Rethinking macroeconomics: what failed, and how to repair it. J. Eur. Econ. Assoc. 9: 591–645.

Stolz, M., M. Stolze, U. Hamm, M. Janssen and M. Ruto. 2011. Consumer attitudes towards organic versus conventional food with specific quality attributes. NJAS Wagening. J. Life Sci. 58: 67–72.

Tansey, G. and T. Worsley. 1995. The Food System: A Guide; Earthscan: London, UK.

Tansey, G. and T. Rajotte. 2008. The Future Control of Food; Earthscan: London, UK.

Teicholz, N. 2015. The big fat surprise: Why butter, meat and cheese belong in a healthy diet. Simon & Schuster.

Tilman, D. and M. Clark. 2014. Global diets link environmental sustainability and human health. Nature 515: 518–522.

Tregear, A. and M. Ness. 2005. Discriminant analysis of consumer interest in buying locally produced foods. J. Market. Manage. 21: 19–35.

van Eekeren, N., A. Maas, H.W. Saatkamp and M. Verschuur. 2006. Small-Scale Chicken Production (4th revised edition) Agromisa Foundation and CTA, Wageningen. <http://www.scribd.com/SmallScale-Chicken-Production/d/8422107> (accessed on 12 October 2017).

Vincke, P. 1992. Multicriteria Decision Aid. Wiley, New York.

Vorley, B. 2003. Food, Inc. Corporate Concentration from Farm to Consumer; UK Food Group, International Institute for Environment and Development: London, UK.

Zeleny, M. 1982. Multiple-Criteria Decision-Making. McGraw Hill, New York.

Zellmer, A.J., T.F.H. Allen and K. Kesseboehmer. 2006. The nature of ecological complexity: a protocol for building the narrative. Ecol. Compl. 3: 171–182.

12

Farm Animal Welfare and Consumers[#]

Carmen Hubbard,[1,*] *Beth Clark*[2] *and Laura Foster*[3]

"... the question is not, Can they reason? nor, Can they talk? but, Can they suffer? Why should the law refuse its protection to any sensitive being?... The time will come when humanity will extend its mantle over everything which breathes..."

Jeremy Bentham (1748–1832) on the suffering
of non-human animals
Introduction to the Principles of Morals and Legislation[$]

Introduction

Defining animal welfare

Animal welfare is an important topic within the global debate on sustainability and food security but usually overlooked (Parente and van de Weerd 2012). Research into (farm) animal welfare within the natural and social sciences has flourished and the study of animal welfare is now recognized as a science in its own right (Veissier and Miele

---

[1] School of Natural and Environmental Sciences, Newcastle University; Room 3.11 Agriculture, School of Natural and Environmental Sciences, Newcastle upon Tyne NE1 7RU.
[2] School of Natural and Environmental Sciences, Newcastle University.
   E-mail: beth.clark@ncl.ac.uk
[3] Newcastle University/Consultant.
   E-mail: laurafosteris@gmail.com
* Corresponding author: carmen.hubbard@newcastle.ac.uk

[#] Authorship is shared equally in this chapter.
[$] https://www.utilitarianism.com/jeremybentham.html, last accessed 5 February 2017.

2014). However, what constitutes good animal welfare and how it is measured remain highly challenging and controversial from both scientific and ethical perspectives (Buller and Morris 2003). That is mainly because animal welfare *per se* is a very complex and multidimensional concept (e.g., Hubbard et al. 2007, Fraser and Weary 2004, Scott et al. 2001, Lund and Röcklinsberg 2001, Duncan and Fraser 1997). It is generally related to the well-being or the quality of life of animals (Scott et al. 2001), but the difficulty is that it means different things to different people (Veissier et al. 2016). Some argue that animal welfare is primarily about the physical and mental health of the animal (Webster 2001), so that levels of injuries, incidence of disease, and physical comfort are important. Others argue that welfare is about the subjective feelings of the animal, and their positive or negative experiences (Simonsen 1996). Duncan (2005) highlights the evolution of two schools of thought within the scientific community, the "biological functioning" school, that addresses the first argument, and the "feelings" school that focuses on the later. However, it has been widely accepted that the concept should incorporate the biological functions and the psychological health and feelings but also other characteristics such as the animals' natural behaviour (Hewson 2003). Mellor (2016:14) also stresses that different definitions emphasize different elements. An increase in societal concerns regarding how animals are treated, led Fisher (2009:71) to point out that the understandings of animal welfare are diverse, ranging *"from the absence of suffering, to where the animal is in a state of complete mental and physical health and in harmony with its environment, to an economic or socio-political understanding reflecting human preferences"*. Tuttyens et al. (2010) also point out that finding a universal definition of what constitutes good welfare that will satisfy all those interested in/concerned with animal welfare (i.e., stakeholders) is challenging, and broader, more generalized aspects are likely to meet the expectations of the majority. Hence, two fundamental concepts are thought to be central to good welfare for the general public, i.e., naturalness and humane treatment (Clark et al. 2016). Nonetheless, both pose an interesting challenge for intensive livestock production, as contemporary farming systems inevitably struggle to provide one, or both, of these concepts in the trade-off between productivity and animal welfare. Although modern agri-food production delivers benefits which the general public recognize and enjoy, such as food affordability (Heng et al. 2013), quality and safety (Boogaard et al. 2011), there is a need to demonstrate that both productivity and animal welfare are taken into consideration, especially given the rise in the extensive production systems promoted as acceptable alternatives.

Despite the lack of a universally endorsed definition (Mellor 2016), there are some that are recognized amongst the scientific community. For example, according to the U.K.'s Farm Animal Welfare Committee (FAWC), animal welfare *"concerns both physical and mental health, which is largely determined by the skills of the stockman, the system of husbandry and the suitability of the genotype for the environment"* (FAWC 2009:3). The definition of the World Organisation for Animal Health (OIE) has been endorsed by over 170 countries worldwide. It stipulates that *"animal welfare means how an animal is coping with the conditions in which it lives. An animal is in a good state of welfare if (as indicated by scientific evidence) it is healthy, comfortable, well nourished, safe, able to express innate behaviour, and if it is not suffering from unpleasant states such as pain, fear and distress. ... Animal welfare refers to the state of the animal; the treatment that an animal receives is covered by other terms such as animal care, animal husbandry and humane treatment"* (OIE 2008).[1]

The complexity of the concept also gives rise to disagreement about the most suitable ways to assess welfare. However, it is generally agreed that to be effective, assessment of animal welfare should consider not only physical and mental aspects but also animal behavioural needs and *"recognition of emotional experiences, feelings or affective states"* (Mellor 2016:14). Good animal welfare also *"requires disease prevention and veterinary treatment, appropriate shelter, management, nutrition, humane handling and humane slaughter/killing"* (OIE 2008). At the farm level, however, despite the existence of a variety of approaches to assessing animals' well-being, the evaluation and monitoring of animal welfare is rather complicated, given the heterogeneity of farmed species and the large number of welfare criteria that should be considered (Hubbard et al. 2007). Moreover, the assessment of animal welfare is often 'entrusted' to the latitude of human perceptions and reactions to husbandry conditions. Given the different viewpoints about animal welfare, it is not surprising that there can be also disagreement about how to effectively assess (farm) animal welfare. Indeed, previous animal welfare assessment has relied mainly on resource and management-based parameters or "input assessments", but in the last decade "the approach to [quantitative assessment of] animal welfare has radically changed" with the focus shifted towards the use of so-called "animal-based measures" or "outcome assessment" (EFSA 2015:5). As yet there is no gold standard measure of animal welfare, and whilst there is consensus that animal-based measures are essential in any assessment scheme, it is difficult to argue that only such outcome measures should be used since evaluation of resource inputs (e.g., space, group size and environmental enrichment) can also help to assess behavioural requirements (FAWC 2009). Furthermore, animal-based measures can be laborious and therefore costly, or in some cases may be impractical, e.g., for disease precaution reasons, if the number of visitors to farms needs to be restricted (FAWC 2009). At the farm level the link between the farm productivity and profitability and animal welfare raises also other issues, such as socio-economic issues. Sorensen et al. (2001:11) note that *"what is good for animal welfare ... is not always economical"* and *"failure of viewing animal welfare in the context of other goals can ultimately have a negative effect on the animals"*. Moreover, citizens and consumers' awareness and concerns regarding the welfare of farm animals have increased significantly in recent years. This adds pressure not only on livestock producers and the industry as a whole (e.g., transporters, slaughterhouses, distributors and retailers) but on governments which are called upon to take their role as guardians or custodians of farm animal welfare more seriously. However, different countries have different ways of addressing the issue of farm animal welfare. All of the above make animal welfare a very complex issue both from a scientific assessment and a moral point of view.

---

[1] International Terrestrial Animal Health Code, Chapter 7.1., http://web.oie.int/eng/normes/ mcode/ en_chapitre_1.7.1.htm, last accessed on 25 August 2017.

But Do Consumers Really Care about Farm Animal Welfare?

The citizen—consumer gap

There are signs that societies all over the world are increasingly acknowledging the relative importance of animal welfare, and public concerns regarding how animals are reared and treated are growing. For example, a survey conducted by Ipsos MORI for Compassion in World Farming and the International Fund for Animal Welfare revealed that out of approximately 2,000 people interviewed in the U.K. and 3,000 people from China, Vietnam and South Korea (1,000 from each country) respectively, the majority (over 90% in each of these countries) believe that humans have a moral duty to minimize the suffering of animals as much as possible (MORI 2005).[2] More recently, in a Gallup survey (2015), a third of US respondents said they wanted animals to have the same rights as people, representing a 25% increase since 2008 (Gallup 2015). The latest Eurobarometer (2016) on attitudes of Europeans towards animal welfare (which surveys regularly around 28,000 people across the EU member states) also highlights the importance of protection of welfare for farmed animals. Indeed, EU leads globally in the development of international conventions for the protection of animals. The first major EU act related to farm animal welfare goes back to 1976; the European Convention for the Protection of Animals kept for Farming Purposes (Council Directive 98/58/EC) provides a common legislative framework for all EU member states. More importantly, the Treaty of Amsterdam (1999) recognized "*animals as sentient beings*" and defined animal welfare as an important component of farming practices. A decade later, the Treaty of Lisbon (2009) reinforced these issues and paved the way for the adoption of the EU 'Strategy for the Protection and Welfare of Animals for 2012 to 2015' (European Commission 2016). Under the principle "*Everyone is responsible*", this strategy sets out the foundation for improving animal welfare standards and ensures that these standards are applied across all EU countries (*ibid.*).

And yet, intensive production systems that compromise animal welfare both biologically and psychologically, continue to supply the majority of the world's animal proteins (e.g., two-thirds of the world's poultry meat and eggs, and half of the world's pork) (Park and Singer 2012). While there are undoubtedly groups of actively concerned and welfare-conscious consumers, the majority of people do not claim to take farm animal welfare into consideration at the point of purchase (Clark et al. 2017). Given the growing public concern (European Commission 2016) it would be expected that an increasing proportion of the public are switching to animal goods produced under high(er) welfare production systems. However, research indicates that attitudes towards farm animal welfare do not always translate into purchase decisions (Harper and Henson 2001, Toma et al. 2011), as indicated by the discrepancy between reported willingness-to-pay and existing sales and market share information (Baltzar 2004, Grunert 2006). It should be noted that consumers may not be as rational in their choices as we are led to believe (Lusk 2014), and attitudes and preferences towards different foods are complex (Shephard and Sparks 1994, Bellisle 2006), and influenced by numerous different factors including, potentially, a sensory component, an affective

---

[2] https://www.ipsos.com/ipsos-mori/en-uk/asian-nations-share-british-concern-animals-0, last accessed on 1 June 2017.

component, a cognitive component, a behavioural component Knox (2000), and habit (Grunert 2005), depending on the context of the decision. Although previous reviews of the literature have indicated a growing public concern towards farm animal welfare (Clark et al. 2017, 2016), including a high degree of concern when directly questioned (Johansson-Stenman 2006, Vanhonacker et al. 2007), yet when asked spontaneously, farm animal welfare is not a priority product attribute for most consumers when shopping (Harper and Henson 2001).

In this context, some scholars stress the existence of a so-called citizen-consumer gap, or attitude-behavioural intention gap. This refers to the difference between citizens' ethical views of farm animal welfare, and their behaviour as consumers, i.e., the consumption of animal products derived from low(er) animal welfare sources, despite a belief that farm animal welfare is an important social issue (Vermeir and Verbeke 2006, de Bakker and Dagevos 2012).

The gap between people's views and behaviours in relation to farm animal welfare is also well documented and reflected in the EU's Eurobarometer studies. The 2015 survey revealed that 94% of (about 28,000) respondents believed it is important to protect the welfare of farmed animals, with 82% believing farm animal welfare should be better protected than it is currently. Overall, EU citizens' perceptions of the importance of farm animal welfare have gone up by 5% since 2006. However, the same study revealed that almost half (47%) rarely or never looked for farm animal welfare-friendly labelling on animal products, with only 12% believing the issue to be a matter for consumers (European Commission 2016).

Boogaard et al. (2011) report that whilst the public indicate that they are concerned, they also acknowledge that their purchasing behaviour is supporting the production systems in place, hence evoking a likely cognitive dissonance. Given the negative cognitive and affective implications that this may have on individuals, they are likely to adopt one of two mechanisms to deal with this (Ong et al. 2017), either confronting the consideration and changing their intentions to purchase intensively produced products, or more likely, having a confirmatory bias that enables them to continue with animal product consumption. The latter approach highlights the coping mechanisms and dissociation strategies used by consumers to continue with animal product consumption without accompanying feelings of guilt (Clark et al. 2016). Another means to consider the attitude-behaviour gap would be to study moral self-regulatory processes, such as moral disengagement and meat attachment (Graça et al. 2016). This may prevent individuals from fully expressing their preferences as they may not wish to think about the ethical and moral implications of their dietary practices to enable them to continue with meat consumption, also known as the meat paradox (Loughnan et al. 2014). This can apply at the point of purchase and consumption.

Given that there is a market for animal welfare friendly products, albeit a niche one, these dissonance reducing strategies, including moral disengagement, do not occur for all individuals. Previous studies have indicated that there are certain groups of the population that are more concerned with animal welfare than others, with women, younger people and those who had spent longer in education exhibiting high levels of concern and negative attitudes towards modern production systems (Clark et al. 2017, 2016). There are also individuals for whom concerns over farm animal welfare are strongly tied up with concerns for their own wellbeing, i.e., a human rather than an animal perspective. Indeed, it could be said this forms a 'warm glow' effect or a so-called form of 'impure altruism' (Johansson-Stenman 2006), whereby human concerns make humans feel good, however, these do not have any impact on the improvement of the welfare conditions of the animals under consideration (Lusk and Norwood 2011). Public concerns over food safety also contribute to this. Links between farm animal welfare and human health have been established (Pinar 2006), particularly from a consumer perspective (Hall and Sandilands 2007, Bennett et al. 2012). Still, there is a growing disconnection between animal products and the animals that produce them, as the majority of the public remains distanced from the modern industrial animal production (Spooner et al. 2014). Hence, a growing body of knowledge highlights the lack of public familiarity and knowledge surrounding modern animal production (Clark et al. 2016), which can result in a number of misconceptions.

But What Does Fuel the Gap between Citizens and Consumers?

Global consumerism and intensification in commercial farming

The citizen-consumer gap has been described in different studies, with explanation for the phenomenon linked mainly to a rise in consumerism, usually described as the '*enemy of the citizen*' (Bauman 2009), and population growth. These two major factors are predicted to increase global demand for animal proteins by 70% by 2050 (FAO 2017). In the face of an increase in consumer demand for animal products, individuals are perceived to have minimal influence on the food supply chain, which in general focuses on driving down costs to increase retailer profitability (Napolitano et al. 2010). In a global market for animal products, the absence of international legislation means that standards vary from region to region and traceability can be problematic (Park and Springer 2012). As such, the issue of animal welfare in intensive farming systems has been articulated as a 'vicious circle', i.e., the pressure exerted on producers by retailers to reduce margins and increase profits leads to further intensification and its associated welfare consequences, as for example, lameness in broiler chicken production and the reduced lifespan of dairy cattle (Napolitano et al. 2010). While some scholars are sceptical about the potential for meaningful consumer-citizenship, believing it to be a myth in a consumerist society (Carrigan and Attalla 2001), others argue that individuals' choices are hybrid; that consumers have the potential to affect social change by acting in accordance with ethical views, as well as selfish motivations such as price and convenience (De Bakker and Dagevos 2011).

Consumers' willingness-to-pay

Consumers' willingness-to-pay has dominated research around the potential for people to opt for higher welfare animal products and therefore close the citizen-consumer gap (Harvey and Hubbard 2013). A meta-analysis of willingness-to-pay (WTP) for traceable meat attributes, comprising studies from 23 developed countries, found that animal welfare was one of the top five factors, with respondents claiming to be prepared to pay 14% above base price (Cicia and Colantuoni 2010). However, the relations between consumers' willingness-to-pay and animal welfare is undeniably linked to other factors, e.g., taste, quality, animal species, and disposable income. Napolitano et al. (2008, 2010) found that consumers were more willing to pay for yoghurt that claimed to be from high(er) welfare sources, provided that the sensory experience was also superior. Similarly, consumers assimilated their liking of organic beef to their expectations for higher quality. While these studies show a preference for high(er) welfare standards, according to participants' understanding of the term, and support the evidence of growing concern for animal welfare, they do not explore the perceptions of the participants, or whether accurate information about welfare standards affects participants' liking for the products. Clark et al. (2017) show that consumers are more willing to pay for the high(er) welfare products of species such as chickens, which have featured more in the media. This species bias was also found in a study of the impact of knowledge of farm animal welfare on the attitudes and behaviours of young people, who felt that animals with perceived higher levels of cognition and relation to humans had more capacity for suffering (Jamieson et al. 2012). This demonstrates that information about animals and their welfare has the potential to heavily influence thoughts and actions, and that consumer decisions could be affected by assumed or ill-informed views on farm animal welfare gathered from inaccurate or sensationalized sources such as the media. A critical omission from the willingness-to-pay research, therefore, is an examination of people's understanding of farm animal welfare, which may or may not be accurate or extensive, and whether information can have disruptive effects on their beliefs and behaviours.

Behaviour guided by assumptions and inferences

In addressing the inconsistencies between people's ethical views about animals and their actions in relation to food, Kjaernes (2012) argues that the choices people make are often based on conflated assumptions about what is normal or 'good enough' for themselves, the environment, and farm animals; for example, an inference that organic food scores more highly on all counts. Another factor is the backdrop of what is perceived to be appropriate in a cultural context. By conforming to social norms, such as consumer buying behaviour, people remove the need to be constantly reconsidering the morality of their choices (Kjaernes and Lavik 2007). From an institutional perspective, consumer decisions are reinforced by the (farm) assurances provided through independent bodies and accreditation schemes designed to strengthen trust between producers and consumers (*ibid.*). However, the literature in general does not test whether consumer expectations of these assurance schemes match the quality standards they offer, rather the research suggests consumers make assumptions in this regard to support their decisions (Kjaernes and Lavik 2007, Napolitano et al. 2010). While studies reveal that consumers link benefits such as human health, nutrition and food safety and environmental standards to high(er) farm animal welfare (Clark et al. 2017, Bennett et al. 2012, Kjaernes and Lavik 2007), they do not explore whether these perceptions are grounded in accurate information, and/or if more information of farming processes and animal welfare standards would help to increase levels of willingness-to-pay.

Lack of accountability

Focus group research conducted in 2007 by the EU Welfare Quality project[3] found that, while the majority of U.K. participants stated a concern for farm animal welfare, there was little desire to gather information about welfare standards, or be responsible, ultimately, for making ethical purchase decisions. Feedback suggested an active consumer disassociation between livestock and meat products and that people wanted to be 'forced' to buy more ethical products through enforcement of high(er) welfare standards onto producers and retailers (Evans and Miele 2007).

Across the EU, 83% of citizens believe that public authorities should have sole or joint responsibility for ensuring welfare standards (European Commission 2016), demonstrating a clear expectation that government should be guaranteeing welfare standards on consumers' behalf. The lack of consumer feedback on the U.K. government's consultation on reforming statutory animal welfare codes (only three individuals contributed, DEFRA 2015) indicates a low level of engagement in the policy-level decisions made about animal welfare practice, either through lack of awareness or other barriers, amounting to a lack of accountability, in real terms, for animal welfare practices.

While accepting that '*animal welfare regulation*' is a public good, Harvey and Hubbard (2013a), contest that the issue of animal welfare is a private good, and that individuals must accept responsibility for the decisions they make in relation to animal welfare if the citizen-consumer gap is to be addressed. A key feature of the citizen-consumer gap in relation to animal welfare, they claim, is the free-rider deficit, i.e., the lack of consumer support for a switch to high(er) welfare products based on individuals' perceptions that their action will not make a difference, leading to an overall worse standard of animal welfare than is collectively desired. Hence, there is a need for government intervention (*ibid.*). Public information is described as a critical foundation phase in the maturation of animal welfare issues in society (Harvey and Hubbard 2013b).

What Does Influence People's Behaviour towards Animal Welfare?

Public perceptions of farm animal welfare are recognized as a key driver behind farm animal welfare policy and practice, with food production and supply chains responding to increasing consumer demands for higher welfare standards through product differentiation and marketing (Harvey and Hubbard 2013b). Despite the above-mentioned evidence of a growing social concern for farm animal welfare, studies show that consumers retain poor levels of knowledge of welfare practices in food production systems (Cornish et al. 2016, Clark et al. 2017). Media has been often cited as a primary source of information that affects purchasing decisions and there is a view that information for consumers should be simplified in order to guide more ethical purchases (Evans and Miele 2007). For example, a study in the U.S.

---

[3] Welfare Quality®: Science and society improving animal welfare in the food quality chain EU funded project FOOD-CT-2004-506508; http://www.welfarequality.net/everyone.

found that consumers are highly influenced by media in their perceptions of farming environments, especially the semiotic imagery present in consumer advertising and fictional portrayals of farms, and that they retain unrealistic or inaccurate views of aspects of livestock farming, e.g., housing systems and their welfare implications (Rumble and Buck 2013). Similarly, 87% of people across EU member states felt campaign information to be a good way of influencing children and young people in respect of farm animal welfare and 64% would like more information concerning the way farm animals are treated in their home countries (European Commission 2016). Perceptions can be shifted once accurate information about agricultural practices is provided (Duncan and Broyles 2006). A research that involved surveying participants on their perceptions of the US agricultural industry before and after watching the docu-entertainment film Food Inc., revealed a significant shift in attitudes, and concluded that the medium of entertainment could be employed effectively to educate consumers about the agriculture industry (Holt and Cartmell 2013). However, this kind of 'before and after' testing of attitudes has not been conducted on a wider public scale.

Misleading or unclear labelling is reported as another key reason why consumers might be unwilling to pay for higher welfare goods. A major European study of knowledge levels of consumers of animal products found widespread misinterpretation of food labelling in relation to methods of production, even among consumers who felt they had a good or reasonable level of knowledge of farming practices, and that a majority of consumers were unable to correctly identify methods of production based on food packaging and labelling (Labelling Matters 2014). In Northern European countries such as Sweden and Norway, where there is more expectation that government will intervene to ensure high(er) standards of farm animal welfare, consumers place more trust in the livestock production systems and, as a result, routinely look for labelling that they understand to represent high(er) farm animal welfare standards (Clark et al. 2017, European Commission 2016).

It is also understood that consumer meat purchasing decisions are informed by quality assessments across four key areas: sensory characteristics, healthiness, convenience and aspects of production processes such as organic systems, and animal welfare (Grunert 2006). Because these characteristics are difficult to discern prior to, or in the process of, consumption, consumers rely on extrinsic or informational cues such as origin and place of purchase, e.g., butchers' shops being perceived as better sources of meat, to guide their behaviour (Grunert 2006). It is predicted that extrinsic cues will have an ongoing impact on consumer perceptions of quality, partly because consumers say they want more information, partly due to the long-established importance that origin of production and purchase play in consumer perceptions of quality, but also in recognition of the increasing role story-telling has to play in the marketing of consumer products (Grunert 2006).

Other findings that might explain behaviour, such as 47% of Europeans believing there is currently insufficient choice for consumers in retail locations (European Commission 2016) do not necessarily reflect the availability of high(er) welfare food items, but could reflect consumers' failure to recognize informational cues such as farm animal welfare labelling. This is supported by evidence of poor understanding of labelling of animal products (Evans and Miele 2007, Labelling Matters 2014). From an industry perspective, an increase in demand for high(er) welfare products and more extrinsic cues provide an opportunity for retailers and producers to differentiate their products and satisfy consumer in farm animal welfare via, for example, more transparent labelling systems.

Most studies on attitudes and beliefs about farm animal welfare have tended to focus on adults, despite the potential for children's education to positively affect demand (Jamieson et al. 2012). Research on the impact of knowledge on children's attitudes and behaviours in respect to farm animal welfare in particular species revealed a short-term increase in concern, but which diminished over a three-month period and did not generalize to other species (*ibid.*). This indicates that more research is needed into the effect of prolonged exposure to information about animal welfare, particularly in relation to buying behaviour as adults. Furthermore, the difference shown in the 2015 Eurobarometer survey between adults wanting information (64%) and those who believe campaigns are good sources of education for children (87%) could indicate that fewer adults feel they need the information a campaign could offer. These findings represent a nascent opportunity to look at specific forms of information and education among young people as a way of establishing accurate perceptions of farm animal welfare, and to remove barriers to shrinking the citizen-consumer gap, such as speciesism reinforced by inaccurate media portrayals, in the longer term. It is also recognized that self-declared ethical purchasers tend to be more informed (Vermeir and Verbeke 2006), however, there is a lack of research on how accurate ethical purchasers' knowledge levels are, and/or to what extent their decisions are driven by the aforementioned assumptions or misinterpretations about animal food products.

Livestock production and sustainability of food production

Animal production poses several challenges in relation to sustainability of food production, including its disproportionate contribution to the environmental cost of agriculture (Leip et al. 2010, Kastner et al. 2012, Garnett 2013), through resource use including water, soil, air and biodiversity (Steinfeld 2004), the environmental impacts of its waste (Tilman et al. 2002), and the increasing amounts of grain fed to animals (Bruinsma 2003). In addition, the efficiency gains through intensification may result in animal rights and welfare concerns (Austin et al. 2005, World Bank 2011), with Broom (2010) stating that "no system or procedure is sustainable if a substantial proportion of people find aspects of it now, or of its consequences in the future, morally unacceptable". Stakeholder's perceptions of animal production systems, including those of the public, therefore need to be taken into consideration, with systems as such may pose an unacceptable (ethical) cost to farm animal welfare (FAWC 2016).

Some will also argue that livestock production in particular, has positive and negative and direct and indirect effects on food security (Compassion in World Farming and World Society for the Protection of Animals 2012). Intensive animal production systems offer a means of rearing large volumes of animals for food production more efficiently than traditional extensive animal farming systems. The scale and control offered by the former systems contribute towards more stable food supplies and affordable food prices, while keeping pace with the growing global demand for animal products. Indeed, the consumption of animal products can improve diet and nutritional status as a result of higher protein and micronutrient intakes such as iron, zinc and vitamin A (World Health Organization 2009), with those in lower income countries benefiting from an increased in such consumption (Steinfeld and Gerber 2010). This contrast, however, with the situation in most developed countries where food intakes surpass nutritional needs and requirements (Sans and Combris 2015) and an increase in animal product consumption corresponded to a growth in the amount of saturated fat consumed and is associated with several non-communicable diseases (World Health Organization 2009), and some cancers (Bouvard et al. 2015). However, despite health concerns, animal product consumption in these regions is forecast to remain stable, or decrease only slightly (Vranken et al. 2014).

Although, globally, heterogeneity in dietary intakes exists (Kearney 2010), due mainly to cultural difference (de Boer et al. 2006), worldwide consumption of animal products is increasing (Alexandratos and Bruinsma 2012, Allievi et al. 2015). Developing nations such as India and China are demanding larger quantities of animal products following rise in their disposable incomes (Tilman et al. 2002), growing population numbers (Wathes 2013), urbanization, improved infrastructure (Kearney 2010) and changing tastes. Moreover, as the regions with increasing animal product consumption, such as Africa and Asia, are those where the largest population growth is predicted to occur (United Nations 2015), the nutritional transition and accompanying changes in food consumption patterns, would have implications for the global food supply, markets and trade (Kearney 2010), with traditional, more extensive farming approaches unable to keep up with demand.

As land limitations inhibit the expansion of extensive production systems in most regions (Steinfeld 2004) livestock production has shifted from ruminants, such as cattle, towards monogastric species such as pigs and chickens (Fraser 2005, Steinfeld and Gerber 2010). Due to their short reproduction cycles, high feed conversion ratios and grain-based diets (Steinfeld and Gerber 2010), the latter are easier and more straightforward to rear in confined, more intensive production systems (High Level Panel of Experts on Food Security and Nutrition 2016). Hence, their production has increased sharply over the last decades (Alexandratos and Bruinsma 2012), and it is now estimated that some 82% chicken and 38% pigs are kept in intensive production systems (High Level Panel of Experts on Food Security and Nutrition 2016). As livestock production and consumption continues to increase across the world, the ethical component is likely to become an inherently important component of sustainable intensive agriculture (Garnett 2013, Vinnari and Vinnari 2014, Allievi et al. 2015), and it is important to consider both the ethical and scientific challenges associated with this whilst ensuring the quality of lives experienced by the animals (Croney and Anthony 2014).

Conclusions

There is little doubt that animal welfare is a very complex issue scientifically and ethically. Research in animal welfare has flourished in recent years and animal welfare has been acknowledged as a science in itself. Furthermore, current research shows that public concern regarding farm animal welfare has continuously increased across different regions around the world and more people are prepared to pay more for animal welfare-friendly products. However, the literature shows that there is still a gap between people's ethical views (as citizens) and their buying behaviour (as consumers). There is also a knowledge gap around farming production systems and very little (if any) research that links the accuracy of knowledge of farm animal welfare and the provision of information, to attitudes or consumer behaviour *per se*. There is also no direct research available that unpacks the relationship between knowledge of farmed species and consumer behaviour. And this list is not exhaustive. Hence, by exploring gaps in knowledge levels and challenging the aforementioned barriers of assumptions, inferences and speciesism, it may allow to test if notions of accountability and willingness-to-pay are affected towards a potential reduction of the citizen-consumer gap. Moreover, all of the above raise another set of research questions, which remains to be addressed. Some are illustrated as follows.

How can those who are genuinely concerned be encouraged to purchase high(er) animal welfare products?

Included within this is the need for clearer and more consistent labelling that clearly indicates to consumers the welfare conditions under which animals were reared. Labels, however, can only be effective if trust exists between consumers and those offering the guarantees associated with a particular product attribute, with this credibility being essential for ensuring that displayed information is believed and used. This could be achieved through certification from an independent body, which may reduce risks associated with purchase. Labelling is indeed the preferred mechanism for conveying information about production, and would be aided but taking a clear, consistent approach. This is potentially important given the lack of knowledge consumers have in relation to farming. However, if the consumer was unable to ignore animal welfare through the presence of a clear labelling system and coherent assurance schemes coupled with education regarding farming practices, then demand for animal welfare friendly products could become more mainstream. This will support the 'virtuous circle' model of Napolitano et al. (2010) and endorse Harvey and Hubbard's (2013b) point regarding the importance of information at an early phase in the development of high(er) animal welfare standards in society.

How to encourage those who use disassociation strategies to eat meat more responsible?

Sustainability, including that of animal production, is currently subject to lively debate. Given that animal products are thought to contribute disproportionately and negatively to a number of environmental factors, such as greenhouse gases, encouraging people to eat less meat may actually be more ethical. Therefore, encouraging people to feel responsible and eat less meat may contribute in the long run to achieve sustainability. Using different approaches to understand the dissonance reducing activities, including more information on the disengagement mechanisms used to enable self-serving behaviours (Bandura 1999), such as meat consumption, will provide a greater understanding of individual decision making processes. This, subsequently, could be used to create more targeted communications (i.e., for different population segments) in relation to addressing concerns and perceived risks to reduce dissonance in the first instance, and also to chance behaviour, e.g., reduction of meat consumption for more sustainable diets. Reluctance to changing diets in relation to meat consumption has already been identified relative to environmental impacts (MacDiarmid et al. 2016). Hence, understanding behavioural intention in more detail concerning farm animal welfare may provide further insights into barriers and facilitators to changing behaviour, with a reduction in meat consumption thought to be an important part of a sustainable diet (FAO 2010).

How to create a better connection between the public and modern agriculture?

There is a clear lack of familiarity, and subsequent lack of knowledge, surrounding modern and intensive production systems. Given the potential for information to reduce the gap between citizen and consumer and increase concerns over high(er) welfare products, more should be done to inform the public about how livestock products are produced. Careful co-operation and co-ordination amongst the supply chain participants should be used, to ensure that clear and consistent messages are given to the general public. Consumer's perceptions regarding production systems are also thought to be influenced by the alternatives available at the time. For example, within Europe there are currently several alternatives to intensive animal production systems available to consumers, such as organic, free-range, and more friendly welfare labelled produce (e.g., Freedom Food in the UK), all of which have different implications for animal welfare, and which are becoming more widely publicised. Producers of these systems are generally more proactive in terms of communicating the benefits that these systems offer. Producers involved in more intensive production systems should therefore look to communicate their best practice and to highlight public good will. Thus, those involved in livestock production may need to focus their efforts on the role of information in moving society and industry towards a heightened state of consciousness, hence, an opportunity for actors across the entire supply chain to move the system in that direction by engaging with and educating consumers on farming processes.

As animal-based products play a central role in most Western diets, and are becoming increasingly important in diets across developing countries, research should be undertaken across a range of countries to understand similar or differentiating cultural factors and values in relation to animal welfare. This is especially important given the lack of research regarding consumer behaviour towards animal welfare in developing countries and regions where animal production is expected to increase the most, such as Asia, Africa and South America. Whether a more informed citizen would make significantly different consumer choices over time would be the ultimate acid test of the power of more and better information, though not within the capacity of this chapter.

References

Alexandratos, N. and J. Bruinsma. 2012. World agriculture towards 2030/2050: the 2012 revision. ESA Working paper. Rome: FAO.

Allievi, F., M. Vinnari and J. Luukkanen. 2015. Meat consumption and production—analysis of efficiency, sufficiency and consistency of global trends. J. Cleaner Prod. 92: 142–151.

Austin, E.J., I.J. Deary, G. Edwards-Jones and D. Arey. 2005. Attitudes to farm animal welfare: factor structure and personality correlates in farmers and agriculture students. J. Individ. Differ. 26: 107–120.

Baltzer, K. 2004. Consumers' willingness to pay for food quality–the case of eggs. Food Economics-Acta Agriculturae Scandinavica, Section C 1: 78–90.

Bauman, Z. 2009. Does Ethics Have a Chance in a World of Consumers? Cambridge: Harvard University Press.

Bellisle, F. 2006. The determinants of food choice. <http://www.eufic.org/en/healthy-living/article/the-determinants-of-food-choice> (accessed on 13 July 2017).

Bennett, R., A. Kehlbacher and K. Balcombe. 2012. A method for the economic valuation of animal welfare benefits using a single welfare score. Anim. Welfare 2: 125–130.

Boogaard, B.K., B.B. Bock, S.J. Oosting, J.S.C. Wiskerke and A.J. van der Zijpp. 2011. Social acceptance of dairy farming: The ambivalence between the two faces of modernity. J. Agr. Environ. Ethic. 24: 259–282.

Bouvard, V., D. Loomis, L.Z. Guyton, Y. Grosse, F. El Ghissassi, L. Benbrahim-Tallaa, N. Guha, H. Mattock and L. Straif. 2015. Carcinogenicity of consumption of red and processed meat. Lancet Oncol. 16: 1599.

Broom, D.M. 2010. Animal welfare: An aspect of care, sustainability, and food quality required by the public. J. Vet. Med. Educ. 37: 83–88.

Bruinsma, J. 2003. World Agriculture: Towards 2015/2030: An FAO Perspective. London: Earthscan Publications Ltd.

Buller, H and C. Morris. 2003. Farm animal welfare: a new repertoire of nature-society relations or modernism re-embedded. Sociol. Ruralis 43: 216–237.

Carrigan, M. and A. Attalla. 2001. The myth of the ethical consumer—do ethics matter in purchase behaviour? J. Consum. Mark. 18: 560–578.

Cicia, G. and F. Colantuoni. 2010. Willingness to pay for traceable meat attributes: A meta-analysis. Int. J. Food Syst. Dyn. 1: 252–263.

Clark, B., G.B. Stewart, L.A. Panzone, I. Kyriazakis and L.J. Frewer. 2016. A systematic review of public attitudes, perceptions and behaviours towards production diseases associated with farm animal welfare. J. Agr. Environ. Ethic. 29: 455–478.

Clark, B., G.B. Stewart, L.A. Panzone, I. Kyriazakis and L.J. Frewer. 2017. Citizens, consumers and farm animal welfare: A meta-analysis of willingness-to-pay studies. Food Policy 68: 112–127.

Compassion in World Farming and World Society for the Protection of Animals. 2012. Food security and farm animal welfare. <https://www.ciwf.org.uk/media/3758836/Food-security-and-farm-animal-welfare-report.pdf> (accessed on 21 March 2017).

Cornish, A., D. Raubenheimer and P. Mcgreevy. 2016. What we know about the public's level of concern for farm animal welfare in food production in developed countries. Animals 6: 74.

Croney, C. and R. Anthony. 2014. Food animal production, ethics, and quality assurance. Encyclopaedia of Food and Agricultural Ethics, pp. 845–853.

De Bakker, E. and H. Dagevos. 2012. Reducing meat consumption in today's consumer society: Questioning the citizen-consumer gap. J. Agr. Env. Ethics 25: 877–894.

de Boer, J., M. Helms and H. Aiking. 2006. Protein consumption and sustainability: Diet diversity in EU-15. Ecol. Econ. 59: 267–274.

DEFRA. 2015. Government response on the reform of animal welfare codes. <https://www.gov.uk/government/uploads/system/uploads/attachment_data/file/486162/welfare-code-reform-consult-gov-resp-final.pdf> (accessed on 2 June 2017).

Duncan, I.J.H. 2005. Science-based assessment of animal welfare: farm animals. Rev. Sci. Tech. Off. Int. Epiz. 24: 483–92.

Duncan, D. and T. Broyles. 2006. A comparison of student knowledge and perceptions toward agriculture before and after attending a governor's school for agriculture. NACTA J. 50: 16–21.

Duncan, I.J.H. and D. Fraser. 1997. Understanding animal welfare. pp. 19–31. In: Appleby, M.C. and B.O. Hughes (eds.). Animal Welfare, CAB International, Wallingford.

European Commission. 2016. Special Eurobarometer 442, Summary, Attitudes of Europeans towards Animal Welfare.

European Food Safety Authority (EFSA). 2015. The use of animal-based measures to assess animal welfare in EU-State of the art of 10 years of activities and analysis of gaps. Technical Report published 6 November 2015; EFSA Supporting publication 2015: EN-884; <https://www.efsa.europa.eu/sites/default/files/scientific_output/files/main_documents/884e.pdf > (Accessed on 5 October 2017).

Evans, A. and M. Miele. 2007. Consumers' views about farm animal welfare: Part I national reports based on focus group research. Welfare Quality Reports No. 4. Cardiff University.

Farm Animal Welfare Committee (FAWC). 2009. Farm animal welfare in Great Britain: Past, present and future, London. <https://www.gov.uk/government/uploads/system/uploads/ attachment_data/file/319292/Farm_Animal_Welfare_in_Great_Britain_-_Past__Present_and_Future.pdf> (Accessed 3 June 2017).

Farm Animal Welfare Committee (FAWC). 2016. Sustainable agriculture and farm animal welfare. <https://www.gov.uk/government/uploads/system/uploads/attachment_data/file/ 593479/Advice_about_sustainable_agriculture_and_farm_animal_welfare_-_final_2016.pdf> (Accessed on 21 March 2017).

Farming and Agriculture Organisation. 2017. Animal Health. <http://www.fao.org/animal-health/en/> (Accessed on 3 June 2017).

Fisher, M.W. 2009. Defining animal welfare—does consistency matter? NZ. Vet. J. 57: 71–73.

Fraser, D. 2005. Animal welfare and the intensification of animal production. An alternative interpretation. <http://www.fao.org/fileadmin/user_upload/animalwelfare/a0158e00.pdf> (Accessed on 22 October 2016).

Fraser, D. and D.M. Weary. 2004. Quality of life of farm animals: linking science, ethics and animal welfare. pp. 39–60. *In*: Benson, G.J. and B.R. Rollin (eds.). The Well-being of Farm Animals: Challenges and Solutions, Blackwell, Ames, IA.

Gallup. 2015. US support for animals having same rights as people. <http://www.gallup.com/ poll/183275/say-animals-rights-people.aspx> (accessed on 1 June 2017).

Garnett, T. 2013. Food sustainability: problems, perspectives and solutions. Proceedings of the Nutr. Soc. 72: 29–39.

Graça, J., M.M. Calheiros and A. Oliveira. 2016. Situating moral disengagement: Motivated reasoning in meat consumption and substitution. Pers. Indiv. Differ. 90: 353–364.

Grunert, K. 2006. Future trends and consumer lifestyles with regard to meat consumption. Meat Sci. 74: 149–160.

Grunert, K.G. 2005. Food quality and safety: Consumer perception and demand. Eur. Rev. Agr. Econ. 32: 369–391.

Harper, G. and S. Henson. 2001. Consumer concerns about animal welfare and the impact on food choice, EU FAIR CT98-3678, Centre for Food Economics Research, The University of Reading.

Hall, C. and V. Sandilands. 2007. Public attitudes to the welfare of broiler chickens. Anim. Welfare 16: 499–512.

Harvey, D. and C. Hubbard. 2013a. Reconsidering the political economy of farm animal welfare: An anatomy of market failure. Food Policy 38: 105–114.

Harvey, D. and C. Hubbard. 2013b. The supply chain's role in improving animal welfare. Animals 3: 767–785.

Heng, Y., H.H. Peterson and X. Li. 2013. Consumer attitudes toward farm-animal welfare: The case of laying hens. J. Agr. Resour. Econ. 38: 418–434.

Hewson, C. 2003. What is animal welfare? Common definitions and their practical consequences. Can. Vet. J. 44: 496–499.

High Level Panel of Experts on Food Security and Nutrition. 2016. Sustainable agricultural development for food security and nutrition: what roles for livestock? A report by the High Level Panel of Experts on Food Security and Nutrition of the Committee on World Food Security. <www.fao.org/cfs/cfs-hlpe> (Accessed on 21 March 2017).

Holt, J. and D. Cartmell. 2013. Consumer perceptions of the U.S. agriculture industry before and after watching the film food. Inc. J. Appl. Commun. 97: Iss.3.

Hubbard, C., M. Bourlakis and G. Garrod. 2007. Pig in the middle: farmers and the delivery of farm animal welfare standards. Brit. Food J. 109: 919–931.

Jamieson, J., M. Reiss, D. Allen, L. Asher, C. Wathes and S. Abeyesinghe. 2012. Measuring the success of a farm animal welfare education event. Anim. Welfare 21: 65–75.

Johansson-Stenman, O. 2006. Should Animal Welfare Count? Working paper in Economics 197. Göteborg, Sweden: Göteborg University: Department of Economics, School of Business, Economics and Law.

Kastner, T., M.J.I. Rivas, W. Koch and S. Nonhebel. 2012. Global changes in diets and the consequences for land requirements for food. Proc. Natl. Acad. Sci. 109: 6868–6872.

Kearney, J. 2010. Food consumption trends and drivers. Philosophical Transactions of the Royal Society B: Biological Sciences 365(1554): 2793–2807.

Kjaernes, U. and R. Lavik. 2007. Farm animal welfare and food consumption practices: Results from surveys in seven countries. Cardiff: School of City and Regional Planning, Cardiff University.

Kjaernes, U. 2012. Ethics and action: A relational perspective on consumer choice in European politics of food. J. Agr. Environ. Ethics 25: 145–162.

Knox, B. 2000. Consumer perception and understanding of risk from food. Brit. Med. Bull. 56: 97–109.

Labelling Matters. 2014. European Product Labels Research. <http://www.labellingmatters.org/ images/ Labelling_Matters_European_Product_Labels_report_V4f.pdf> (accessed on 20 May 2017).

Leip, A., F. Weiss, T. Wassenaar, I. Perez, T. Fellmann, P. Loudjani, F. Tubiello, D. Grandgirard, S. Monni and K. Biala. 2010. Evaluation of the livestock sector's contribution to the EU greenhouse gas emissions (GGELS)–final report. <http://ec.europa.eu/agriculture/analysis/ external/livestock-gas/ full_text_en.pdf> (Accessed on 23 March 2015).

Loughnan, S., B. Bastian and N. Haslam. 2014. The psychology of eating animals. Curr. Dir. Psychol. Sci. 23: 104–108.

Lusk, J.L. 2014. Are you smart enough to know what to eat? A critique of behavioural economics as justification for regulation. Eur. Rev. Agr. Econ. 41: 355–373.

Lusk, J.L. and F.B. Norwood. 2011. Animal welfare economics. Appl. Econ. Perspect. Policy 33: 463–483.

Lund, V. and H. Röcklinsberg. 2001. Outlining a conception of animal welfare for organic farming systems. J. Agr. Environ. Ethics 14: 391–424.

Mellor, D.J. 2016. Updating animal welfare thinking: Moving beyond the "Five Freedoms" towards "A Life Worth Living". Animals 6(3): 21; doi:10.3390/ani6030021.

MORI. 2005. Asian Nations Share British Concern for Animals. <https://www.ipsos.com/ipsos-mori/ en-uk/asian-nations-share-british-concern-animals-0?language_content_entity=en-uk> (accessed on 1 June 2017).

Napolitano, F., C. Pacelli, A. Girolami and A. Braghieri. 2008. Effect of information about animal welfare on consumer willingness to pay for yogurt. J. Dairy Sci. 91: 910–917.

Napolitano, F., A. Girolami and A. Braghieri. 2010. Consumer liking and willingness to pay for high welfare animal-based products. Trends. Food Sci. Tech. 21: 537–543.

Ong, A.S.J., L.J. Frewer and M.Y. Chan. 2017. Cognitive dissonance in food and nutrition—A conceptual framework. Trends. Food Sci. Tech. 59: 60–69.

Parente, S. and H. van de Weerd. 2012. Food security and farm animal welfare, Briefing paper, World Society for the Protection of Animals and Compassion in World Farming. <https://www.ciwf.org.uk/ media/3758836/Food-security-and-farm-animal-welfare-report.pdf> (accessed on 24 August 2017).

Park, M. and P. Singer. 2012. The globalization of animal welfare: More food does not require more suffering. Foreign Aff. 91: 122–133.

Pinar, J.P. 2006. Animal welfare in livestock operations. <http://www.ifc.org/wps/wcm/connect/ 7ce6d2804885589a80bcd26a6515bb18/AnimalWelfare_GPN.pdf?MOD=AJPERES> (accessed on 3 September 2016).

Rumble, J. and E. Buck. 2013. Narrowing the farm-to-plate knowledge gap through semiotics and the study of consumer responses regarding livestock images. J. Appl. Commun. 97: Iss.3.

Sans, P. and P. Combris. 2015. World meat consumption patterns: An overview of the last fifty years (1961–2011). Meat Sci. 109: 106–111.

Scott, M.E., A.M. Nolan and J.L. Fitzpatrick. 2001. Conceptual and methodological issues related to welfare assessment: A framework for measurement. Acta Agriculturae Scandinavica, Section A: Anim. Sci. 30: 5–10 (Supplement).

Shepherd, R. and P. Sparks. 1994. Modelling food choice. pp. 202–226. *In*: Thomson, D.M.H. and H.J.H. MacFie (eds.). Measurement of Food Preferences. New York: Springer.

Simonsen, H.B. 1996. Assessment of animal welfare by a holistic approach: behaviour, health and measured opinion. Acta Agriculturae Scandinavica 27, Sect. A, Anim. Sci. Suppl. 91–96.

Spooner, J.M., C.A. Schuppli and D. Fraser. 2014. Attitudes of Canadian citizens toward farm animal welfare: A qualitative study. Livest. Sci. 163: 150–158.

Steinfeld, H. 2004. The livestock revolution—a global veterinary mission. Vet. Parasitol. 125: 19–41.

Steinfeld, H. and P. Gerber. 2010. Livestock production and the global environment: Consume less or produce better? Proc. Natl. Acad. Sci. 107: 18237–18238.

Tilman, D., K.G. Cassman, P.A. Matson, R. Naylor and S. Polasky. 2002. Agricultural sustainability and intensive production practices. Nature 418(6898): 671–677.

Toma, L., A. McVittie, C. Hubbard and A.W. Stott. 2011. A structural equation model of the factors influencing British consumers' behaviour toward animal welfare. J. Food Prod. Market. 17: 261–278.

Tuyttens, F.A.M., F. Vanhonacker, E. Van Poucke and W. Verbeke. 2010. Quantitative verification of the correspondence between the Welfare Quality® operational definition of farm animal welfare and the opinion of Flemish farmers, citizens and vegetarians. Livest. Sci. 131: 108–114.

United Nations. 2015. World Population Prospects. Key Findings and Advanced Tables. <https://esa.un.org/unpd/wpp/Publications/Files/Key_Findings_WPP_2015.pdf> (accessed on 5 March 2017).

Vanhonacker, F., W. Verbeke, E. Van Poucke and F. Tuyttens. 2007. Segmentation based on consumers' perceived importance and attitude toward farm animal welfare. Int. J. Sociol. Agr. Food 15: 91–107.

Vermeir, I. and W. Verbeke. 2006. Sustainable food consumption: exploring the consumer 'attitude—behavioral intention' gap. J. Agr. Environ. Ethics 19: 169–194.

Veissier, I. and M. Miele. 2014. Animal welfare: towards transdisciplinarity—the European experience. Animal Production Science 54: 1119–1129.

Veissier, I., H. Spoolder, J. Rushen and L. Mounier. 2016. Editorial: Pigs crying, silent fish and other stories about animal welfare assessment. Animal 10(2): 292–293, © The Animal Consortium 2016, doi: 10.1017/S1751731115002682.

Vinnari, M. and E. Vinnari. 2014. A framework for sustainability transition: the case of plant-based diets. J. Agr. Environ. Ethics 27: 369–396.

Vranken, L., T. Avermaete, D. Petalios and E. Mathijs. 2014. Curbing global meat consumption: emerging evidence of a second nutrition transition. Environ. Sci. and Policy 39: 95–106.

Wathes, C.M., H. Buller, H. Maggs and M.L. Campbell. 2013. Livestock production in the UK in the 21st century: A perfect storm averted? Animals 3: 574–583.

Webster, A.J.F. 2001. Farm animal welfare: the five freedoms and the free market. Vet. J. 161: 229–37.

World Bank. 2011. Creating business opportunity through improved animal welfare. <http://www.ifc.org/wps/wcm/connect/633e46004885558fb714f76a6515bb18/Animal%2BWelfare%2BQN.pdf?MOD=AJPERES> (accessed on 3 October 2014).

World Health Organization. 2009. Global and regional food consumption patterns and trends. WHO Technical Report Series 916: 1–17.

13

Interpersonal and Institutional Trust Effects on Country of Origin Preference

Kar Ho Lim,[1,*] *Wuyang Hu,*[2] *Leigh J. Maynard*[2] and *Ellen Goddard*[3]

Introduction

Origin labeling is not new. Since ancient times, the Egyptians used it to indicate the resistance of bricks that built the Pyramids, the Greeks used it to indicate the quality of wine (Grote 2009). In the present day context of WTO and globalization, it is embroiled in controversy. Skeptics claim that it is merely a protectionist tool in disguise; its proponent maintains Country of Origin Labeling (COOL) preserve consumers' right and protection, as the label conveys food safety and quality information. And indeed, COOL affects consumers' perception, justifiably or not (Berry et al. 2015). Investigations into psychological factors have turned out promising insights, enriching understanding and sustaining a more informed policy debate. While ethnocentrism

---

[1] Department of Agricultural and Environmental Sciences, Tennessee State University, Nashville, TN.
[2] Department of Agricultural Economics, University of Kentucky, 313 C.E. Barnhart Bldg, Lexington, KY 40546 USA.
E-mail: wuyang.hu@uky.edu
E-mail: leigh.maynard@uky.edu
[3] Resource Economics and Environmental Sociology, University of Alberta, 547 General Services, Edmonton, AB Canada T6G 2H1.
E-mail: ellen.goddard@ualberta.ca
* Corresponding author: klim@tnstate.edu

and perceived food safety provide an intuitive explanation into consumers' aversion towards imported food, whether it is simply down to—I don't trust you—is not known.

Here, we investigate the role of interpersonal and institutional trust on consumers' preference for imported food with a choice experiment. The trust measurements strategically enter a random utility model interacted with products of imported origins, allowing marginal utility reflecting the interaction between the innate psychological factor and product's attribute to be parsed out. We found the skeptics are less likely to choose imports. But trust on government has a positive impact, highlighting the conflicted position of government in promoting the domestic product and maintaining consumers' preference for imported product.

Country of origin and consumer preference: striking a familiar tone

COOL's impact on consumer preference of food products has been extensively studied. Given the choice, consumers prefer products attached with origin information, compare to unspecified-origin product (Loureiro and Umberger 2005). And in general, consumers across different nations prefer domestic food products over imports. Example of the domestic preference is widespread and robust, for example: Americans are willing to pay significantly less for imported beef than domestic beef; the preference is much more pronounced on male, older, less educated and poorer consumers (Lim et al. 2013). Further, Americans prefer domestic-raised value-added tilapia and shrimp over similar products that are imported from China and Thailand (Ortega et al. 2014).

Canadians are observed to prefer and willing to pay a higher price for beef identified as the product of Canada than the product of U.S. (Lim and Hu 2016). So do Norwegians on the choice between Irish, Norwegian, and U.S. produced beef, where domestic Norwegian beef won out as the most preferred (Alfnes and Rickertsen 2003).

In East Asia, seafood-loving Japanese consumers are found to be willing to pay 27% of premium for Hokkaido salmon over similar Chilean product; the price premium of the Hokkaido product persists over the products of Alaska and products of Norway (Uchida et al. 2014). Likewise, South Koreans prefer Korean Hanwoo beef above all other choices; they are willing to pay $14/lb to $15/lb over imported beef from the U.S. and other origins (Chung et al. 2009). Even where exposure to imported products is high, Singaporean consumers rank bread and coffee produced domestically high in taste, prestige, quality, and likelihood of purchase (Ahmed et al. 2004). The findings of significant WTP for domestic food fuel the narrative that COOL as an effective product differentiation tool, and further cement the place of COOL in nations' food and agriculture trade policies (Verbeke and Roosen 2009). However, much remain unknown about the motivation of the domestic preference.

Underlying factors of domestic preference

Studies explore the underlying motivation of COOL via multiple angles. Two angles, food safety perception and ethnocentrism, have been frequently spotlighted. Focusing on these two areas are logical. Given that COOL has been argued by proponents to be a food safety measurement, and by opponents a technical barrier to trade that fuels by consumer ethnocentrism.

In particular, Berry et al. (2015) investigated the effect of COOL disclosure to purchase intention and perception of the product. They found that products labeled as "Born, Raised, and Slaughtered" in the U.S. receive a higher score in food safety, taste, freshness, and purchase intention against similar products labeled as "Born, Raised, and Slaughtered" in Mexico and unlabeled product. However, the effect is mitigated by disclosing information that meat processing system in Mexico is equivalent of those in the U.S., which lends further credence to the role of biased food safety perception in the choice process.

Lim et al. (2014), which uses the same choice experiment dataset as this study, investigated the role of food safety perception on consumers' willingness to pay for imported products from Australia and Canada. We found that Australian and Canadian beef received lower food safety score than American beef. In particular, a larger segment of the consumers stated they do not know about the food safety level of the imported products. And the lower food safety perception of the imports are translated into a statistically significant lower WTP for the imported products.

The linkage between perceived product quality and ethnocentrism has been investigated in Chryssochoidis et al. (2007). They divide Greek consumers into the ethnocentric and non-ethnocentric consumers segment. For the ethnocentric consumers, COO induce favorable views of domestic food products against imports.

Ethnocentrism—notably with CETSCALE, a psychometric method that measures consumers nationalistic behavior (Shimp and Sharma 1987)—maintains that domestic preference is partly due to consumers' innate nationalistic beliefs and the need to support domestic farmers (Balabanis and Diamantopoulos 2004, Lewis and Grebitus 2016, Ehmke et al. 2008). These results establish psychological factors' role in forming the domestic preference. Nevertheless, how *trust*, in particular, factors into the preference of food products' country of origin is scarcely explored.

The linkage between ethnocentrism and trust has been investigated. In a wider societal context, anti-immigrant attitudes are found to be correlated with regional and national interpersonal trust (Rustenbach 2010). Consumers' ethnocentrism is mediated by their level of trust associated with the country of origin where firms are based (Jiménez and San Martín 2010). Ethnocentrism may be traced further back into biochemical reaction in humans' brain, where experiments found that oxytocin induces ethnocentric behavior that facilitates within-group trust (De Dreu et al. 2011).

Trust's impact on food choice is widespread. For instance, Lobb et al. (2006) found that trust in food safety information provided by media and food supply chain influence consumers' intention to purchase food after a hypothetical food scare. Trust—of certifier and on provided information—plays a role in consumers' acceptance and willingness to pay for food labels (Wessells et al. 1999, Samant and Seo 2016, Chan et al. 2005, Mazis and Raymond 1997, Grunert et al. 2013). Siegrist et al. (2007) explore the acceptance of nanotechnology on food, where they suggest social trust influence willingness to buy food through heuristic and perceptions. Poortinga and Pidgeon (2005) explore the role of trust on the acceptability of GM food, where they similarly noted the correlation between trust and acceptability. Further, food supply gatekeepers state that trust in the integrity of production, certification and regulatory systems of exporting countries influence their country-of-origin preferences when sourcing food imports (Knight et al. 2007).

Trust is found to guide wide-ranging consumer behavior. Distrust of the Department of Energy forms the basis of public opposition to the building of a proposed nuclear waste facility (Pijawka and Mushkatel 1991). An intense research effort has investigated the role of trust in e-commerce (Teo and Liu 2007, Pavlou 2003, Gefen and Straub 2004, McKnight and Chervany 2001). Nevertheless, we do not have a good understanding on the role of either interpersonal trust or trust on the food chain of consumers on their preference of COOL.

While ethnocentrism and trust-related concepts, the implications towards managing consumers' behavior regarding COOL is arguably different. If ethnocentrism is the sole main driver of domestic preference, the effect might be mitigated by advertising that promotes the benefits of the imported products and education effort that illustrate the positive impact of trade. Mild manifestations of ethnocentrism perhaps can be persuaded with such corrections. However, if trust is the main driver of the behavior, the efforts outlined above might not be as effective, as those do not deal directly with trust. And if interpersonal trust is the main driver, economic tools available for policy intervention are substantially limited.

In this paper, we investigate the role of interpersonal and institutional trust to consumer preference of COOL. Our central contribution to the literature is that we point out a novel component of COOL's underlying motivation. We show that trust has a direct effect on consumer preference of domestic products. Because trust is seen as an underlying factor of ethnocentrism, this calls for a fuller investigation into the true magnitude of trust and ethnocentrism to the preference of COOL.

Method

We conduct an online survey that measures interpersonal trust, trust on the supply chain, and consumers' preference of beef steak from domestic and imported countries of origin. The survey consists of two parts. The first part deals with attitude, trust, and demographics. The second part features a choice experiment, which gauges consumer preference of beef products marked with the selected country of origins and value-added attributes.

The data is collected in May 2010, using the representative panel of TNS Global. The sample consists of 1079 American consumers, where it is comparable to the U.S. census data in gender, education, and household income (Table 1). Further, our sample possesses a desirable characteristic in that almost all of the respondents' participated household shopping. Our sample recorded higher age, where the mean age is 56.62 in our sample compare the median age of 36.8 in U.S. census data. However, part of the discrepancy could be due to the exclusion of children from our sample, whereas the U.S. census median age encompasses the whole population.

Trust measurements

Previous research furnished the trust measurements used in this study. We used the trust of food safety information by Frewer et al. (1996). These questionnaires measure how much consumers trust the food safety information provided by several institutions. They are the 5-points Likert scale that range from "no trust at all" to "complete trust".

Table 1. Sample description statistics.

Variable	Group	Percent	Sample Mean/ Median	2010 US Census Data
Age			56.62	36.8[a]
Gender	Male	47.54%		49.20%
	Female	52.46%		50.80%
Education	< High School	1.11%	14[a]	12[a]
	High School	23.08%		
	Some College	39.39%		
	4 year Degree	24.28%		
	Graduate	12.14%		
Household Income ($)	< 25k	24.10%	52.37k	51.42k
	25k–40k	23.54%		
	40k–65k	23.82%		
	65k–80k	9.55%		
	80k–100k	7.32%		
	100k–120k	6.12%		
	> 120k	5.56%		
Freq. shopping grocery	Never	1.85%		
	Sometimes	14.74%		
	Frequently	83.42%		

[a] Median values.

Further, we used the generalized interpersonal trust questionnaire that is found in the American General Social Survey, the European Values studies, and a wide range of research in psychological researches (Miller and Mitamura 2003). This is a questionnaire that elicits response in interpersonal trust in three categories: namely, "Can't be too careful in dealing with people", "People can be trusted", and an ambiguous "Don't know" option.

And lastly, we added two questionnaires that measure consumers' perception about the general food safety level and how much they are worried about food safety. Specifically, the food safety perception is a 5-point disagree-agree Likert scale item that asks "Generally, food products are safe", and for the worry questionnaire "I worry about the safety of food".

Over 70% of the respondents stated that food products are safe in general (Table 2). However, about a quarter stated worry about food safety, compared to 35% that stated they do not worry about food safety. These can be interpreted that Americans are in general, optimistic about food safety.

There has been a general decline in trust towards other institutions (Twenge et al. 2014, Robinson and Jackson 2001, Rotter 1980). This somewhat explains the lower trust measurements here. The spread appears similar across the sources on the trust of food safety information. As much as 15% of the sample noted that they do

Table 2. Tabulation of trust variables.

	Strongly Disagree				Strongly Agree		
	1	2	3	4	5	Mean	SD
Generally, food products are safe	1.5%	4.9%	22.2%	61.9%	9.5%	3.7	0.8
I worry about the safety of food	9.3%	34.9%	28.6%	22.5%	4.7%	2.8	1.0
	No Trust at all				Complete Trust		
To what extent do you trust information about the safety of food provided by …?	1	2	3	4	5	Mean	SD
Farmers	3.2%	22.7%	40.8%	27.9%	5.5%	3.1	0.9
The Government	15.8%	30.0%	31.3%	19.3%	3.6%	2.6	1.1
Manufacturers of food	8.8%	32.9%	37.4%	18.0%	3.0%	2.7	1.0
Retailers	5.5%	32.6%	41.4%	17.6%	2.9%	2.8	0.9
	People can be Trusted		Can't be too careful in dealing with people		Don't know		
Generally speaking, would you say that most people can be trusted?	41.1%				3.5%		

N = 1079

not trust food safety information from the government at all; this perhaps reflects the fractious political climate in the U.S. For illustration, since our survey used 1–5 Likert scale to measure trust, we combined the answers of 4 and 5 as *trusted*, and answers of 1 and 2 as *not trusted*; Farmers are the most trusted group when it comes to food related information, where about a third of the surveyed consumers falls in the trusted category, compared to about a quarter who distrust farmers. About one-fifth of the respondents trusted the other groups (government, manufacturers, and retailers), compared to slightly more than two-fifth who indicated distrust in them. The low percentage of complete trust in any of the entities is perhaps influenced by the level of interpersonal trust exhibited by the respondents. The responses are dichotomous in regard to how trusting they are towards other people in general. Over half of them (55.4%) reflect skepticism towards others, while about 40% stated that people can be trusted in general, compared to 3.5% who are uncertain. Slightly higher percentage of the respondents are skeptical towards others.

Choice experiment

The choice experiment features strip-loin beef steak as the representative products. Its examined attributes include countries of origin, food safety enhancements, tenderness assurance, the label "natural", and prices as the main attributes. The multiple attributes used in this study minimize concerns of single-cue bias, which manifest in single cue COO studies where the preferences are magnified compared to studies that include non-COO attributes (Bilkey and Nes 1982, Peterson and Jolibert 1995). Appendix 1 describes the instructions given to the respondents before the choice experiment tasks.

The countries of origin are the products of the U.S.A., Canada, and Australia. The U.S.A. is used as a base category to measure the marginal utility of switching to an imported product. Canada and Australia are chosen for its export volume to the U.S. market. While Mexico and other countries merit representation in research, they were not included there, to lessen the number of required choice sets.

The food safety enhancement attributes include the baseline level of No enhancement, BSE Testing, Traceability, and both BSE Tested and Traceable. Here, BSE Testing refers to beef derived from cattle that have undergone lab testing for the so-called "mad cow disease" inducing prions. Traceable refers to products that can be traced back from the point-of-sale to the origin farm. Further, the label "natural" refers to beef produced without the use of artificial hormones and antibiotics.

Tenderness assurance refers to beef that is guaranteed tender. Lastly, four levels of price, ranging from \$5.50/lb to \$16.00/lb, to encompass the price range of conventional to value-added beef steak were represented in this experiment.

We applied full-factorial design to generate the choice sets, thus the D-Efficiency score, which is a measure to show how precisely the effects of interest can be estimable, was 100%. A hundred and ninety-two choice profiles were distributed to 14 versions of experiments. The respondents each completed 10–14 choice sets—a level of task allowing efficient data collection while limiting data quality compromising respondents fatigue (Savage and Waldman 2008).

Theoretical and econometric model

We used a conditional logit model to analyze (McFadden 1974). We chose this simpler model for our interests in the marginal effect of trust on country of origin. A Random Utility Model describes consumers' utility:

$$U_{ijt} = \beta' x_{ijt} + \delta' z_{ijt} + \varepsilon_{ijt} \tag{1}$$

The left-hand side represents the utility of consumer i facing the choice set j and alternative t. The focal point of this study is the vector z, where it represents the interaction terms between the trust measurements and the imported products. Thus δ provides the marginal utility associated with the imported products when the trust measurements change, holding other factors constant. These paint a picture on how trust affects consumers' preference of the imported products. Vector x represents the main attributes in dummy variable format, and β represents utility associated with the main attributes. The error term, ε, represents all factors unobservable in the experiment; it is assumed to follow a Type I (Largest) Extreme Value distribution, which gives rise to the logit probability (Train 2003).

Further, we estimated an alternative conditional logic model that does not account for the interaction terms. With a likelihood ratio test, this model provides a mean to compare the additional variance explained by the interaction terms; failure to reject the null hypothesis would imply that the trust and food safety perception measurements do not significantly affect consumer preference of the imported beef. Further, this bare bone model would also provide a convenient means to estimate consumers' mean WTP associated with imported beef, which provides the premium American beef command over imported beef in consumers' willingness to pay, specifically,

$$WTP_{country} = -\frac{\beta_{country}}{\beta_{price}} \tag{2}$$

The econometric framework furnishes the testable hypotheses. Of main interest is whether interpersonal trust affects the preference for imported products. Specifically,

$H_o^1: \delta_{australia*trust\ people} = 0$

$H_o^2: \delta_{australia*don't\ know} = 0$

$H_o^3: \delta_{canada*trust\ people} = 0$

$H_o^4: \delta_{canada*don't\ know} = 0$

The omitted category is respondents who stated: "Can't be too careful in dealing with people". The coefficients thus reflect consumers' preference of the imported beef varied by their interpersonal trust towards others. Our expectation is that the coefficients associated with these variables to carry the positive sign, which would imply people who tend to trust people more are less averse towards consumption of imported products. Conversely, a joint test that supports the null hypotheses would suggest that interpersonal trust is not a factor in the preference.

The next set of hypotheses to be tested involves the interaction terms between the imported beef and institutional trust.

$H_o^5: \delta_{australia*institutions\ trust} = 0$

$H_o^6: \delta_{canada*institutions\ trust} = 0$

Specifically, the institutions are farmers, manufacturers, retailers, and governments, and the trust variables measure how much the respondents trust food safety information from these institutions. We did not have a prior expectation on the sign of the estimated coefficients.

Further, a set of hypotheses tests respondents' perception about the level of food safety.

$H_o^7: \delta_{australia*safe} = 0$

$H_o^8: \delta_{canada*safe} = 0$

The coefficients associated with the interaction terms denote how perception affects respondents' preference for imported beef. We expect positive signs on these coefficients, as it logically implies that the less safe consumers think food is, the more they tend to doubt the food safety level of the import.

Similarly, the last set of hypotheses reveals how worrying about food safety affect the preference for imported food.

H_o^9: $\delta_{australia*worry} = 0$

H_o^{10}: $\delta_{canada*worry} = 0$

We expect the coefficients to be negative, as it would imply that people who are worried about food safety would tend to consume the more familiar domestic products.

Results and Discussions

Table 3 presents the estimates from the conditional logit models, namely a main-effect-only model, and a full model that includes the trust interaction terms. Coefficients associated with price are negative and statistically significant across both models, resonating the theory of demand that higher price decreases utility. The full model reports a higher McFadden R^2 (0.157 compared to 0.153 of the main-effect-only model). The equivalency between the main-effect-only model and the full model is rejected at 99% confidence interval with a log-likelihood ratio test [χ^2 (16) = 147.43], which suggest the explanatory power of the interaction terms with trust variables.

The main-effect-only model illustrates the magnitude of the attributes. As expected, consumers prefer domestic beef over imported beef, as illustrated by the negative coefficients associated with Australia and Canada. The preference amounts to an average willingness to pay of –\$6.71/lb for Australian beef, and –\$5.22/lb for Canadian beef (Table 4). Other value-added attributes hold positive and significant coefficients as expected, except for the "natural" attribute, which is not statistically significant. The results suggest consumers rank the country of origin more importantly than tenderness assurance. Whereas the COO effect is similar in magnitude to the value created by BSE-tested and traceability individually; but smaller than when BSE-tested and traceability combined. Nevertheless, the COO effect appears larger than the statistically insignificant WTP for "natural" beef.

Note that the coefficients associated with the main level coefficient associated with products of Australia and products of Canada are different between the main-effect-only model and the full model. The differences are attributed to the inclusion of interaction terms between the countries and the trust measurements. Therefore, the COO main-effect coefficients cannot be interpreted independently of the interaction terms.

Interpersonal trust and preference of imported products

With the product of U.S.A. as the base category, the estimates $\delta_{australia*trust}$ and $\delta_{canada*trust}$ are both positive and significant at 99% level. The null hypothesis that these interpersonal trust coefficients are jointly insignificant is rejected [$\chi^2(4) = 24.61$]. These suggest that interpersonal trust is positively correlated to the preference of the imported products. The import aversion may be a reflection of ethnocentric behavior. Rustenbach (2010) found that natives who are less trusting towards others are more likely to hold anti-immigrant attitude. The logic extends that the anti-immigrant sentiment spills

Table 3. Conditional logit models.

RHS = Choice	Main Effect Only			Full Model		
	Coef.		**t-stat**	**Coef.**		**t-stat**
Price	−0.16	***	(41.80)	−0.16	***	(41.88)
Opt-out	−0.81	***	(14.03)	−0.81	***	(14.03)
Tenderness assured	0.68	***	(23.79)	0.68	***	(23.99)
BSE tested	0.90	***	(21.08)	0.91	***	(21.21)
Traceable	0.92	***	(21.57)	0.93	***	(21.69)
BSE tested & traceable	1.35	***	(31.78)	1.35	***	(31.83)
Natural	0.02		(0.84)	0.03		(0.90)
Australia	−1.08	***	(30.91)	−1.56	***	(7.34)
Canada	−0.84	***	(25.15)	−1.78	***	(8.67)
Interpersonal Trust						
Australia*Trust People				0.25	***	(4.18)
Australia*Don't Know				0.27	*	(1.77)
Canada*Trust People				0.15	***	(2.61)
Canada*Don't Know				−0.11		(0.71)
Trust of Source						
Australia*Farmer				0.02		(0.39)
Canada*Farmer				0.07	*	(1.89)
Australia*Manufacturer				0.01		(0.21)
Canada*Manufacturer				-0.02		(0.55)
Australia*Retailer				0.05		(1.17)
Canada*Retailer				0.04		(.89)
Australia*Government				0.12	***	(3.81)
Canada*Government				0.18	***	(5.94)
General Food Safety Attitude						
Australia*Worry				−0.03		(0.86)
Canada*Worry				0.03		(1.04)
Australia*Safety				−0.03		(0.81)
Canada*Safety				0.01		(0.18)
pseudo R^2	0.153			0.157		
log-likelihood	−13705.9			−13632.2		

* p < 0.1, ** p < 0.05, *** p < 0.01

Table 4. Willingness-to-pay from the main-effect only model.

Attributes	$/lb		Std. Err.	z	P > z
Canada	−5.22	***	0.24	−22.02	0.00
Australia	−6.71	***	0.26	−26.16	0.00
Tenderness assured	4.18	***	0.20	21.41	0.00
Traceable	5.72	***	0.29	19.62	0.00
BSE-tested	5.59	***	0.29	19.01	0.00
BSE-tested & traceable	8.33	***	0.31	26.81	0.00
Natural	0.15		0.18	0.84	0.40

* p < 0.1, ** p < 0.05, *** p < 0.01
Standard Errors are calculated with the Delta method

over to respondents' product choice, which is largely consistent with the literature on ethnocentrism. With Americans' declining interpersonal trust as a trend, we expect that the anti-imports sentiment to increase, holding other factors constant. It is not clear, however, whether the connection between interpersonal trust and ethnocentrism is monotonic. A nonlinear relation would change the implication above.

Translating the results into WTP measurements, which adds context to the results, reveals a significant difference in WTP guided by the interpersonal trust measurement (see Table 5). On an average, consumers who stated they can trust people are willing to pay $1.54/lb more for Australian beef, and $0.93/lb more for Canadian beef, than consumers who are in the reference category (of distrust).

Nevertheless, we cannot rule out that the interpersonal trust, by itself, has a significant impact on the preference. This alternative explanation assumes that there are sufficiently non-overlap between interpersonal trust and ethnocentric behavior, which seems a mild assumption. Using the estimates from the main-effect only model as a guideline, this would then mean 20% of consumers' aversion to imported beef is due to interpersonal trust.

Table 5. Willingness to pay inferred by trust and food safety attitudes variables.

WTP	$/lb		Std. Err.	z	P > |z|
Interpersonal Trust					
Australia*Trust People	1.54	***	0.37	4.17	0.00
Australia*Don't Know	1.65	*	0.93	1.77	0.08
Canada*Trust People	0.93	**	0.35	2.61	0.01
Canada*Don't Know	−0.67		0.94	−0.71	0.48
Institutional Trust					
Australia*Farmer	0.10		0.25	0.39	0.70
Canada*Farmer	0.44	*	0.23	1.89	0.06
Australia*Manufacturer	0.06		0.27	0.21	0.83
Canada*Manufacturer	−0.14		0.26	−0.55	0.58
Australia*Retailer	0.33		0.29	1.17	0.24
Canada*Retailer	0.25		0.28	0.89	0.38
Australia*Government	0.74	***	0.20	3.80	0.00
Canada*Government	1.13	***	0.19	5.89	0.00
General Food Safety Attitude					
Australia*Worry	−0.16		0.18	−0.86	0.39
Canada*Worry	0.18		0.17	1.04	0.30
Australia*Safety	−0.20		0.25	−0.81	0.42
Canada*Safety	0.05		0.25	0.18	0.85

* $p < 0.1$, ** $p < 0.05$, *** $p < 0.01$
Standard Errors are calculated with the Delta Method

Trust on food safety information provided by organizations in the food chain and preference of imported product

We interacted with respondents' trust towards the food safety information furnished by various institutions in the food chain with countries of origin in the conditional logit model. The estimated coefficients report how trust may affect the preference towards the imported product.

The estimated coefficients $\delta_{australia*government}$ and $\delta_{canada*government}$ are both positive and significant. The joint test is significant at 99% confidence internal with $\chi^2(2) = 44.47$. Trust on food safety information provided by the government has a positive impact on consumers' preference of imported product from Australia and Canada. This would suggest that consumers recognize government's role as a gatekeeper of unsafe imported products. The more they have confidence in the government, the more likely they are to choose imported food.

Translating into WTP suggests that a unit increases in the Likert scale measuring the trust towards government provided food safety information increase consumers' WTP by $0.74/lb for Australian beef, and $1.13/lb for Canadian beef (Table 5). These estimates suggest a wide gulf in WTP between low-trust and high-trust consumers in this category. This paints a rather unintuitive suggestion that building up consumers trust towards the importing country's government would benefit the exporting countries' product.

By contrast, the trust of food safety information provided by farmers, manufacturers and retailers are not significant factors. The joint test has a statistics of $\chi^2(6) = 9.76$, which is not statistically significant at 90% confidence interval. The insignificance on farmers' trust is counter-intuitive, as we expect confidence in farmers would translate into preference for domestic products. However, we may note that the variable does not make a distinction between domestic and foreign farmers similarly as on domestic and foreign food manufacturers. Such distinction could be unveiled with additional information, which can be a future research direction.

Summarizing these analyses suggests that domestic government plays an important role in consumers' acceptance of imported food given the positive role consumer trust in the government may play in their imported food purchases. This would suggest a stronger collaboration between the exporting and domestic governments, to increase consumers' confidence in the domestic government's ability in handling food safety, could ultimately translate into a higher preference for imported foods.

Perception and concern about food safety impacts on preference for imported product

All four of the estimates on perception and concerns about food safety are statistically insignificant. The joint test of the four coefficients [$\chi^2(4) = 2.44$] is not significant either. These suggest that a weak linkage between consumer preference for food and consumers' perception about the general level of food safety, or how much they are concerned with food safety. We note, however, that the origin-specific perception—i.e., how consumers perceived food safety of Canada, or Australia—is positively correlated with the choice (Lim et al. 2014). Nevertheless, based on the findings from this study, we draw an implication that the general perceptions are not a good predictor for the choice of imported products. This adds to the literature a logical distinction that general perceptions on food safety has less impact on preference of imported food than imported-origin-specific perceptions.

Conclusions

COOL serves the role to differentiate product quality and inform consumers. Studies have revealed that ethnocentrism, perception, and other psychological factors come into play. This study expands the literature by investigating the role of interpersonal and institutional trust. Using data from a choice experiment and trust measurements, we found that consumers who tend to distrust other people are more likely to avoid imported products. In addition, consumers who trust food safety information provided by government are more likely to choose imported products. The results suggest that trust plays an important role in shaping the preference on the products' origin.

Nevertheless, the limitation of our data prevents the clarification on whether the interpersonal trust merely mirrors ethnocentrism. Future research can explore the interconnectedness of interpersonal trust and ethnocentrism in food choice. Further, the trust measures used in this study is general and unspecific. Future research may explore trust variables specific to various countries for more insight.

Our results suggest that research into consumer aversion to imported foods should take in the level of interpersonal trust attitude of consumers. As for policy implication, exporting countries must realize the role the receiving countries' governments play as a food safety gatekeeper. Collaboration to raise consumers trust in the food safety information provided by the host government may also raise consumers' acceptance of the imported products.

Appendix 1: Instructions to the Respondents of the Choice Experiment

In this final section of this survey you are provided with 14 different pairs of alternative strip loin beef steaks (also known as Kansas City strip and New York steak) that could be available for purchase in the retail grocery store or butcher where you typically shop that possess differing attributes. Steak prices vary from $5.50/lb. to $16.00/lb. For each pair of steaks, please select the steak that you would purchase, or neither, if you would not purchase either steak. It is important that you make your selection like you would if you were actually facing these choices in your retail purchase decisions.

For your information in interpreting alternative steaks:

Country of Origin refers to the country in which the cow/animal was raised and includes USA, Canada, and Australia

Production Practice is the method used to produce the cow/animal where:

Approved Standards means the cow/animal was raised using scientifically determined safe and government-approved use of synthetic growth hormones and antibiotics (typical of cattle production methods used in Canada and the USA)

Natural is the same as typical except the cow/animal was raised without the use of synthetic growth hormones or antibiotics

Tenderness refers to how tender the steak is to eat and includes

Assured Tender means the steak is guaranteed tender by testing the steak using a tenderness measuring instrument

Uncertain means there are no guarantees on tenderness level of the steak and the chances of being tender are the same as typical steaks you have purchased in the past

Food Safety Assurance refers to the level of food safety assurance with the steak

None food safety means the steak meets current minimum government standards for food safety

Traceable means the product is fully traceable back to the farm of origin from your point of purchase

Animal Tested means that all animals are tested for BSE prior to meat being sold at your point of purchase

Appendix 2: A Sample Choice Set.

CHOICE SET 1

Steak Attribute	A	B	C
Price ($/lb.)	$16.00	$12.50	I would not purchase any of these products
Country of Origin	Canada	Canada	
Production Practice	Approved Standards	Approved Standards	
Tenderness	Assured Tender	Assured Tender	
Food Safety Assurance	Traceable	Traceable	
I would choose . . .	o	o	o

References

Ahmed, James P. Johnson, Xia Yang, Chen Kheng Fatt, Han Sack Teng and Lim Chee Boon. 2004. Does country of origin matter for low-involvement products? Int. Mark. Rev. 21: 102–120.

Alfnes, F. and K. Rickertsen. 2003. European consumers' willingness to pay for U.S. beef in experimental auction markets. Am. J. Agric. Econ. 85: 396–405.

Balabanis, G. and A. Diamantopoulos. 2004. Domestic country bias, country-of-origin effects, and consumer ethnocentrism: a multidimensional unfolding approach. J. Acad. Mark. Sci. 32: 80–95.

Berry, C., A. Mukherjee, S. Burton and E. Howlett. 2015. A COOL effect: The direct and indirect impact of country-of-origin disclosures on purchase intentions for retail food products. J. Retail. 91: 533–542.

Bilkey, W.J. and E. Nes. 1982. Country-of-origin effects on product evaluations. J. Int. Bus. Stud. 89–99.

Chryssochoidis, G., A. Krystallis and P. Perreas. 2007. Ethnocentric beliefs and country-of-origin (COO) effect: Impact of country, product and product attributes on Greek consumers' evaluation of food products. Eur. J. Mark. 41: 1518–1544.

Chan, C., C. Patch and P. Williams. 2005. Australian consumers are sceptical about but influenced by claims about fat on food labels. Eur. J. Clin. Nutr. 59: 148–151.

Chung, C., T. Boyer and S. Han. 2009. Valuing quality attributes and country of origin in the Korean beef market. J. Agric. Econ. 60: 682–698.

De Dreu, C.K., L.L. Greer, G.A. Van Kleef, S. Shalvi and M.J. Handgraaf. 2011. Oxytocin promotes human ethnocentrism. Proc. Natl. Acad. Sci. 108: 1262–1266.

Ehmke, M.D., J.L. Lusk and W. Tyner. 2008. Measuring the relative importance of preferences for country of origin in China, France, Niger, and the United States. Agric. Econ. 38, 277–285.

Frewer, L.J., C. Howard, D. Hedderley and R. Shepherd. 1996. What determines trust in information about food-related risks? Underlying psychological constructs. Risk Anal. 16: 473–486.

Gefen, D. and D.W. Straub. 2004. Consumer trust in B2C e-Commerce and the importance of social presence: experiments in e-Products and e-Services. Omega 32: 407–424.

Grunert, K.G., S. Hieke and J. Wills. 2013. Sustainability labels on food products: Consumer motivation, understanding and use. Food Policy 44: 177–189.

Jiménez, N.H. and S. San Martín. 2010. The role of country-of-origin, ethnocentrism and animosity in promoting consumer trust. The moderating role of familiarity. Int. Bus. Rev. 19: 34–45.

Knight, J.G., D.K. Holdsworth and D.W. Mather. 2007. Country-of-origin and choice of food imports: an in-depth study of European distribution channel gatekeepers. J. Int. Bus. Stud. 38, 107–125.

Lewis, K.E. and C. Grebitus. 2016. Why US consumers support country of origin labeling: Examining the impact of ethnocentrism and food safety. J. Int. Food Agribus. Mark. 28: 254–270.

Lim, K.H., W. Hu, L.J. Maynard and E. Goddard. 2013. U.S. consumers' preference and willingness to pay for country-of-origin-labeled beef steak and food safety enhancements. Can. J. Agric. Econ. 61: 93–118.

Lim, K.H., W. Hu, L.J. Maynard and E. Goddard. 2014. A taste for safer beef? How much does consumers' perceived risk influence willingness to pay for country-of-origin labeled beef. Agribus. Int. J. 30: 17–30.

Lim, K.H. and W. Hu. 2016. How local is local? A reflection on Canadian local food labeling policy from consumer preference. Can. J. Agric. Econ. 64: 71–78.

Lobb, A.E., M. Mazzocchi and W.B. Traill. 2006. Modelling risk perception and trust in food safety information within the theory of planned behaviour. Food Qual. Prefer. 18, 384–395.

Mazis, M.B. and M.A. Raymond. 1997. Consumer perceptions of health claims in advertisements and on food labels. J. Consum. Aff. 31: 10–26.

McFadden, D. 1974. Conditional logit analysis of qualitative choice behavior. Front. Econ. 8: 105–142.

McKnight, D.H. and N.L. Chervany. 2001. What trust means in e-commerce customer relationships: An interdisciplinary conceptual typology. Int. J. Electron. Commer. 6: 35–59.

Miller, A.S. and T. Mitamura. 2003. Are surveys on trust trustworthy? Soc. Psychol. Q. 62–70.

Ortega, D.L., H.H. Wang, O. Widmar and J. Nicole. 2014. Aquaculture imports from Asia: an analysis of US consumer demand for select food quality attributes. Agric. Econ. 45: 625–634.

Pavlou, P.A. 2003. Consumer acceptance of electronic commerce: Integrating trust and risk with the technology acceptance model. Int. J. Electron. Commer 7: 101–134.

Peterson, R.A. and A.J. Jolibert. 1995. A meta-analysis of country-of-origin effects. J. Int. Bus. Stud. 26: 883–900.

Pijawka, K.D. and A.H. Mushkatel. 1991. Public opposition to the siting of the high-level nuclear waste repository: The Importance Of Trust. Rev. Policy Res. 10: 180–194.

Poortinga, W. and N.F. Pidgeon. 2005. Trust in risk regulation: cause or consequence of the acceptability of GM food? Risk Anal. 25: 199–209.

Robinson, R.V. and E.F. Jackson. 2001. Is trust in others declining in America? An age–period–cohort analysis. Soc. Sci. Res. 30: 117–145.

Rotter, J.B. 1980. Interpersonal trust, trustworthiness, and gullibility. Am. Psychol. 35: 1–7.

Rustenbach, E. 2010. Sources of negative attitudes toward immigrants in Europe: A multi-level analysis. Int. Migr. Rev. 44: 53–77.

Siegrist, M., M.-E. Cousin, H. Kastenholz and A. Wiek. 2007. Public acceptance of nanotechnology foods and food packaging: The influence of affect and trust. Appetite 49: 459–466.

Teo, T.S. and J. Liu. 2007. Consumer trust in e-commerce in the United States, Singapore and China. Omega 35: 22–38.

Train, K. 2003. Discrete choice methods with simulation. Cambridge Univ. Press, New York.

Twenge, J.M., W.K. Campbell and N.T. Carter. 2014. Declines in trust in others and confidence in institutions among American adults and late adolescents, 1972–2012. Psychol. Sci. 25: 1914–1923.

Uchida, H., Y. Onozaka, T. Morita and S. Managi. 2014. Demand for ecolabeled seafood in the Japanese market: A conjoint analysis of the impact of information and interaction with other labels. Food Policy 44: 68–76.

Verbeke, W. and J. Roosen. 2009. Market differentiation potential of country-of-origin, quality and traceability labeling. Estey Cent. J. Int. Law Trade Policy 10: 20–35.

Wessells, C.R., R.J. Johnston and H. Donath. 1999. Assessing consumer preferences for ecolabeled seafood: the influence of species, certifier, and household attributes. Am. J. Agric. Econ. 81: 1084–1089.

14

Do Consumers and Producers Benefit from Labels of Regional Origin? The Case of the Czech Republic

Iveta Bošková[1,*] and *Tomáš Ratinger*[2]

Introduction

The maturity and saturation of the EU food market have forced the food industry to accelerate product differentiation to gain or increase market shares. In accordance with the information asymmetry theory, producers generally know the product quality much better than buyers do (Luhmann 1979, Darby and Karni 1973). Further, due to the ongoing advancements in food systems, consumer knowledge of and engagement with the production of food has declined (Meyer et al. 2012). However, it does not mean they are not attentive in their food choices; in fact, they show a growing interest in product characteristics such as food composition, methods of production, origin, and quality certification (Teuber 2011). Therefore, it is imperative that consumers should receive information so they can make "informed food choices". Certification labeling schemes have become a popular means for suppliers to deliver information about food quality and other product characteristics (Janssen and Hamm 2012). As Kneasfey (2013) argues, they are very important when consumers do not buy directly

---

[1] Institute of Agricultural Economics and Information (IAEI), Senior researcher, Mánesova 75, 12000 Prague 2.
[2] Technology Centre of the Academy of Sciences (TC ASCR), Senior researcher, Ve Struhách 27, 16000 Prague 6.
E-mail: ratinger@tc.cz
* Corresponding author: boskova.iveta@uzei.cz

from the producers, and are of less importance when purchase takes place through direct contact. According to Tonkin et al. (2016) in modern food systems, food labeling plays a primary role in facilitating information transfer between the food system and consumers. Therefore, developing an understanding of how food labelling influences consumer choice is essential.

In recent years, regional and local food has gained significant market interest. Scholars confirm that regional affiliation has become an important consumer criterion (Lorenz et al. 2015, Verbeke and Roosen 2009) and expectations of related product attributes may influence consumer preferences (van Ittersum et al. 2002). This may happen either as an act of resistance to globalization; disapproval of industrial, unknown, and unseasoned food (Benedek and Balász 2014); health- and environment-related reasons (Capt and Wawresky 2014); or others. Several reasons can be advanced why regional-origin labels may address consumer concerns and why the other agri-food chain participants might benefit from the regional-origin labeling.

First, due to food industrialization and globalization, production has shifted to specific states or regions, and long-distance shipment of commodities has increased. Jones (2014) referring to the proponents of "local" foods argues that product quality is inversely related to transport distance. Second, separating production from consumption pushes consumers to rely on an overall institutional arrangement for their food (Giddens 1990, Meyer et al. 2012). Third, technological progress makes food technology less understandable for consumers (Bruce et al. 2014). Moreover, a range of new actors have emerged in the chain who produce, handle, transport, and retail food so that product information is transmitted to the consumer via a series of mediators (Busch and Bain 2004, Burch and Lawrence 2009, Hendrickson and Heffernan 2002).

Some scholars place importance on geographical proximity and direct interaction, such as short food supply chains, as a mechanism for generating consumer trust and securing credible local foods (Kirwan 2006, Sage 2003, Jarosz 2008, Renting et al. 2003). On the other hand, some authors argue that either modern technologies, such as the Internet, or product leaflets or information materials may generate trust (Renting et al. 2003, Lamine 2005, Thorsøe and Kjeldsen 2015), which may be as much a predisposition as an outcome (Mount 2012). Renting et al. (2003) draw attention to extended food systems, where products are offered to consumers outside the production region who may have no personal experience of that locality. Hence, product packing contains the information enabling consumers to make a connection to places, values, and people involved in the production. These systems use a complex of institutional arrangements for securing the authenticity of the products, and they rely on systemic rather than on personal trust. Bildtgård (2008), distinguishing the three theoretical bases for trust in food, namely emotional, habitual, and reflexive, emphasizes on habitual trust, and shows four different bases for it: community, rational organization, policy, and systems of knowledge. In accordance with the author's confidence that nation relates to the trust through community it can be derived that region can take the role of community too. However, the question is whether regions (either administrative or referring to landscape protection areas) [can] constitute food communities providing habitual trust in food in the post-modern European/Czech society. Sassatelli and Scott (2001) and Kjaernes et al. (2007) argue that market authorities have largely replaced community and personal relations as the nexus for the creation of trust in food. Appadurai (1996) and Virilio (1991) highlight that the creation of communities, traditionally restricted by distance and traversability, has been eased by the arrival of the electronic media, which provide an opportunity for formation of non-geographic communities in which membership is dictated by interest rather than place.

According to the United Nations' definition, the quality label is a symbol or mark on a product's packing, indicating that the product or its production process complies with given standards and has been certified. Certification is defined as a procedure by which a third party gives written assurances that a product or a process is in conformity with a corresponding standard (United Nations 2007). Bildtgård (2008) argues that products sold under a label use scientific knowledge to legitimize their production methods and create trust through an institutionalized control system. Thus, it can be deduced that certification fosters trust by promoting certain community-based norms and values in food production. However, this might not always be successful. Eden et al. (2008) refers to empirical findings showing that consumers' reconnection to regional foods does not necessarily occur because assurance information is not always received positively by consumers.

In the Czech Republic, quite a few quality labels on the food products have appeared. They refer to various food attributes, such as region of origin, health aspects, environmental approach, absence of non-desirable additives, unique taste, recipe, tradition, solo or in combination, and others (Horáček 2014a). Their various types of labeling schemes and number of providers make consumer understanding difficult (Horáček 2014b). The labeling systems in the country have been developed by a wide range of certifiers including the EU and national authorities, producers themselves, producers' associations, retailers, marketing companies, fair organizers, independent groups of experts, and other agents. Velcovska and Chiappa (2015) confirm that Czech consumers are confused, and they do not know what these labels mean. According to the authors, the main problem is the lack of information about some labels. Numerous agri-food agents concur that consumers are not able to fully use the labels in their purchase decisions, and some of the labels are not perceived as credible (Velčovská and Del Chiappa 2015). Thus, credible communication of information via the labeling schemes seems to be a challenge in the country.

Within a range of quality labels identified in the Czech Republic, the main attention has been paid to two regional schemes. Both were initiated with the government's financial support. The first one, Regional Brand, runs as a bottom-up scheme, connecting 30 mostly touristic regions each of which has developed its own label while the second one, Regional Foods, is a top-down scheme based on the administrative regions using a uniform label for all of them. In the backdrop of the formation of these two labeling schemes, including the policy support, the meaning of the regional logos to consumers has been investigated and whether the meaning is in line with the labeling schemes' intentions.

The paper is organized into six parts. After the methods and data are specified in the next section (2), the features of the certification schemes including the policy incentives are briefly specified (3). This is followed by the core section: the analysis of adoption and benefits of the regional-origin labeling by Czech producers and consumers based on survey results (4). The discussion (5) is followed by the final part (6) where the findings are summed up and conclusions including policy recommendations are drawn.

Methods and Data

The analytical part begins with a brief overview of the recent policy measures supporting production and marketing of regional products as well as the introduction of two regional-origin labeling schemes from the official documents and interviews with the representatives of the implementation bodies. Another desk research was carried out to review web pages of the producers of certified regional products, the retailers of food and the certification bodies (organizers of the regional-origin labeling schemes).

The empirical analysis builds on the online surveys of experience and opinions of consumers (households). Since the list of quality and origin labels used on food products was not readily available, five colleagues were asked to collect/note labels by their shopping for one month. Together they identified 32 quality labels (schemes). To them 43 logos of the Regional Brand scheme were added. Through an online survey, the panel of 500 consumers across the regions, locality sizes, age, and education were addressed. In total 12 (11 closed and 1 open) questions were put forth to consumers to get information on their use of different information channels concerning food, their motives to buy or decline regional products, and their expectations.

In the discussion, results of the consumer survey are supplemented by findings of the Czech funding agency's study on the Regional Food label carrying a survey of 1430 consumers (SZIF 2016). These two results are contrasted with some results of the authors' previous research based on the survey of opinions of the producers participating in the Regional Brand scheme (Boskova et al. 2016). This survey included 160 respondents (38% of all food producers included in the Regional Brand scheme).

Descriptive statistics was used to evaluate the survey data and simple statistical analysis (cross tables with *Chi Square test*) was performed to assess the relationships between consumer characteristics, expectations, and practices. Probit regression (e.g., Gujarati 1988) was applied to investigate the characteristics of respondents that influenced their interest in the regional origin of food, knowledge of regional-origin labels, and their role in the decision to buy regional products. In this model, the dependent variable takes only two values 0 and 1 (e.g., 1 for interested in regional origin and 0 for not interested). The probit regression estimates the probability that an observation with a particular characteristic will fall into one of the categories (0 or 1). The probability distribution is assumed to be normal. Maximum likelihood technique is used for the estimation. Probit routine of STATA 14 was used in the research.

The Policy Framework of Regional Schemes

The recent EU and national policies favor regional products by a series of incentives to encourage their production and marketing. While already the EU 2007–13 policy contained the agenda of regional product support (European Commission 2006), the EU countries were emboldened by the "CAP towards 2020" strategy (European Commission 2010) to fund programmes focused on the development of short supply chains and marketing the locally produced foods which was later embedded in the Regulation on Support for Rural Development within the EU 2014–20 policy, particularly in Article 35 (European Commission 2013). Implementation at the national level was ensured by several measures in the National Rural Development Plans for 2007–13 and 2014–20, such as the support to the establishment and development of micro enterprises, the support of local action groups (LAGs) within the Leader axis, and the support to the horizontal and vertical collaboration among the short supply chains (SSC). Moreover, various government platforms have been created to popularize the quality of the products and to attract consumers' attention to the farmers' products. These take the form of certifications—such as "Klasa" (high quality food), "The Czech product, guaranteed by the Federation of the Food and Drink Industries", "Only the best from our country", "The regional product", and so on—or information services—such as, "Find your producer", "Farm festivals" websites, and similar others.

Regional food scheme

Encouraged by the policy, two main national schemes of labeling regional products merged. The "Regional Food" scheme, organized by the Ministry of Agriculture (MoA) and its State Agricultural Intervention Fund (SAIF), consists of annual awards for the best quality foods produced in each of the 14 regions using a high proportion of regional agricultural products. The aim of the scheme, according to the MoA, is to support the domestic producers of local foods and to motivate consumers to look for them in shops, farm markets or buying them directly from producers' outlets/farms. The uniqueness of the award-winning products in each region would rest in the quality, local inputs, traditional recipe, and an outstanding taste. Owing to the presumed short distribution chain, the products are expected to be fresh and have exceptional sensory properties (e.g., taste) (Regional Foods 2017). The regions follow the official administrative structure of the country at the NUTS 3 level defined in 1997 (Parliament of the Czech Republic 1997). Area of the 14 regions ranges between 4000 sq. km and 11,000 sq. km with population from 0.3 million to 1.3 million inhabitants. Producers who enter the Regional Foods competition should comply with several general rules: at least 70% of the material inputs have to come from the region or from within the country if necessary; products have to be made within the region; and they must show some extraordinary quality features. Besides these common rules, each region is allowed to specify its own additional criteria. Within nine product categories, a limited number of best products are selected by regional committees consisting of representatives of the MoA, other government authorities, the respective regional government, and the regional offices of the Czech Agrarian Chamber and the Federation of the Food and Drink Industries. The competition is open for small- and medium-sized enterprises (SME) and the winning products are designated to use the award label for the next four years. The list of all certified regional food products counted 475 in 2016. The programme, running since 2007 and sourcing around 2.7 million Euros annually in recent years, is accompanied by the government information campaign with occasional direct supports to winning producers; for example by subsidizing their participation in exhibitions and other promotional events. The logo/label, showing a schematic picture of the Czech countryside and complemented by the words "Regional Foods" is uniform for all the 14 regions.

Regional brand scheme

The initial idea of the other leading platform, coordinated by the "Association of Regional Brands" (ARB), is historically related to the government programmes of the early 2000s that aimed at enhancing sustainability of several well-established protected landscape areas by attracting visitors' attention to local products, in addition to those introduced with the intention to provide additional jobs opportunities for local people. The local action groups (LAGs), under ARB coordination, played a key role in widening the platform, by initiating and organizing local stakeholders to start and then to maintain production and marketing of regional products. The ARB and the LAGs have gradually achieved the consensus of local stakeholders and have managed to define new regions, usually covering a small geographic area with a specific feature, carrying the name inspired by a distinctive element of the geographical typography or historical connotations. The sizes of such regions differ from very small ones to ones even as big as the administrative regions (NUTS 3). In fact, only Vysocina region overlaps fully with the administrative region; others are smaller and have their own identity. The initial governmental scheme was taken over by local organizations, usually LAGs of LEADER. It makes the sub-schemes autonomous and self-financed (of course partly funded from the RDP budget). The promotion of the individual label (regional brand) depends on the activity of the organizing body while the ARB runs promotions for the whole scheme. In 2016, the ARB coordinated 27 individual regions with 450 certified food products. Besides food, handicraft products and some services are certified too. It allows for synergies in promoting the label and the region. Tourism has remained the leading marketing idea. The individual regional logos follow similar design concept; however, each of the regions has its own variant indicating a dominant regional characteristic and carrying its name. Producers who want their products certified should fulfill the criteria of origin and quality defined by the ARB and pay a membership fee. Expansion of the scheme, however, has led to some weakly defined certification criteria as follows; reference to the region (can be a name), final output produced within the region, some share of regional material input, established/active in the region for some time (at least 5 years), originality, unique quality, handmade, and environment friendliness. The applicants collect points for fulfillment of the criteria, and crossing the threshold constitutes the minimum requirement for the award of the label. It means that producers need not satisfy every criterion; being strong in any one criteria can compensate even lacking in others. The label is valid for three years and can be renewed. The regional committees supervised by the ARB comprises mostly representatives of local authorities and local LAG, environment and/or ethnographic experts or other local stakeholders. The ARB promotions are funded mostly from its own budget being occasionally supplemented by grants from structural and development funds.

Results

Consumer attitude to regional-origin labels

The panel of 500 consumers roughly followed the socio-economic distribution of the population of the country, namely rural and urban structure, education and age distributions, and regional (NUTS 3) division. About one-fourth (24%) of respondents came from rural municipalities with less than 2000 inhabitants, 24% lived in large cities with over 100,000 inhabitants, and the rest was spread in small and midsized towns. Out of all respondents, 11% had elementary education, 73% reached secondary education and professional training, and 14% had completed university degree. Because the survey focused on the main food purchasers in the households, the sample was gender-imbalanced in favour of women (63%).

The consumers were offered a range of options on how they used to find, and if ever, the information about the foods, and they could choose more than one options. Forty percent of respondents affirmed that they did not-search purposely for any information as they shopped habitually, while 60% checked at least the consume-by date.[1] For those who cared about some additional information, the most important sources were the text on the package, the logos, and the published quality test results. In total 37% of all respondents confirmed reading the text on the package while 17% of the respondents focused on the logos and quality tests. Some respondents (12%) were interested in the experience of others through Internet discussions, while options such as visiting producers' websites (7%), direct communication with the shop assistant (4%), or use of the QR code (1.6%) received negligible attention.

Consumers were shown 33 different logos to identify the ones they knew and would be encouraged to buy. The logos referred to the quality (six of them), exceptional taste (six), Czech origin (five), regional or other geographical origin (five), the absence of non-desirable ingredients (five), organic or environment-friendly production (three) and other relating to specific aspects (such as tradition, health, and fair trade; three). The results showed that two logos referring to Czech origin and one referring to quality (but originally started as the Czech-origin logo as well) were well known to 54–79% of respondents. These were followed by a group of eight logos known to 20–36% of consumers. The remaining 23 logos, including three European certifications, showed weaker awareness of less than 20%, out of which a one-half was known by less than 5% of consumers. On average, consumers indicated knowledge of five logos.

The three best-known logos were also identified as the top three ones for stimulating the purchase, indicated by 44–29% of respondents. About 5–17% of consumers took about one-third of logos into account when buying products, and more than one-half of logos had minimum effect on the purchase (for less than 5% of consumers). The "Regional Food" logo was known by 36% of all consumers while the "Regional Brand" only by 11% of respondents. Knowledge of either of geographical origin labels (KNO_reg) was shown by 47% of respondents (Fig. 1).

Three questions followed the respondents' attitude to regional foods and their expectations of regional food label, their affection to particular regions, and its importance in purchasing regional products. Slightly more than one-half of the respondents-consumers (54.6%) expressed their interest in regional products; they indicated "wishing to support the regional development" as the most frequent reason for buying regional products (Table 1); almost every third consumer expressed this motive. On the other hand, almost one-half of non-interested respondents justified their negative position by distrusting the quality of the product based on its regional

---

[1] Best before date or expiry date.

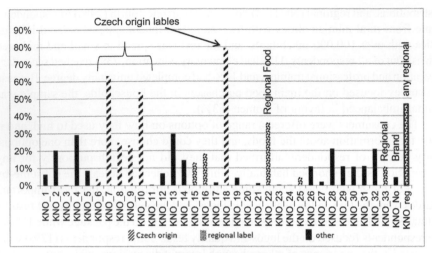

Fig. 1. Knowledge of quality labels (logos).
Note: KNO_x means knowledge frequency of logo "x".
Source: own overview based on the survey among consumers.

Table 1. Consumers attitude to regional foods: the reason for purchase.

Interested in regional food	#	% of all	Not interested in regional food	#	% of all
Supporting regional development	143	28.6	No effect on the quality	112	22.4
Belief in health regions	53	10.6	Not trusting labels nor sellers	60	12.0
Regional food specificity	46	9.2	Too expensive	20	4.0
Personal rel.	29	5.8			
Other	2	0.4	Other	35	7.0
Total	273	54.6	Total	227	45.4

Source: Own investigation.

origin. Only a very low share of non-interested consumers (4%) considered regional foods to be too expensive.

Concerning price premium, about 71% of respondents expressed their willingness to pay some premium for certified regional products (in comparison to standard products). One would expect that respondents' interest in regional-origin labels, their logo knowledge, and their use in shopping would affect willingness to pay price premium for the guarantee of the product origin. In fact, it is confirmed using cross tables with Chi-square test for all the three interest/knowledge variables at the confidence level $\alpha = 0.05$ (or even smaller). Table 2 summarizes the respective Chi square tests. The cross tables are included in the Appendix. Looking at the table for "interest in regional products" (Table A1) it can be seen that 85% of the interested respondents are willing to pay some premium for certified geographical origin and almost one-half would pay a premium higher than 5%. In contrast, 80% of non-interested respondents would consider a premium below 5% or none at all. Similar

Table 2. Test of the effect of interest in regional-origin labels on willingness to pay a price premium.

	Interest	Knowledge	Importance
Chi square statistics	65	22.77	21.86
Degree of freedom	3	3	3
P value (sig. level)	< 0.01	< 0.01	< 0.01

Source: Own calculations based on the survey among consumers.

distributions hold for the other two cross tables ("knowledge of regional labels" and "use of regional-origin labels when shopping").

An interesting question is if consumers prefer some of the 14 NUTS 3 regions. In the survey, respondents were asked to choose up to two most favourable NUTS 3 region particularly in respect of the attractiveness of food products. The two most favoured regions preferred substantially over the other 12 were CZ064 Southern Moravia for its fresh and canned fruits and vegetables and alcoholic drinks and CZ031 Southern Bohemia for its milk and milk products.

Because regional brands target "tourism" and "tourist experience", the authors investigated whether consumers paid more attention to the labeled product when they visited the region or when they were outside, say in their home region. Of the 182 respondents who paid attention to regional brands, one-third did so when visiting the region while the rest admitted them outside the region. As it is likely that all survey respondents would have visited some of the 43 "Regional Brand" regions, one-third of them (i.e., 182) were influenced by these visits. However, it is important to stress that these answers are not fully consistent with the answers on the knowledge of the regional-origin labels—only 11% showed knowledge of "Regional Brands" (logos).

Supporting the development of a region was the most important reason for those who paid attention in the region (62%) as well outside or away from it (74%). This is in line with the question ranking the motivations for the first purchase of a regional product. Out of the respondents who knew about regional-origin labels, almost one-third (31%) gave "personal relationship to a region" as the most important reason and 40% as within the first two most important reasons. The answer "purchased by chance" received top rank in 16% of cases. The answer "recommendation of friends" ranked as the second most important reason. Table A2 in the Appendix ranks all the reasons.

To explore what characteristics of respondents influence their interest in the regional origin of food, three probit models were used deploying the same independent variables: age of the respondent, level of education, size of the site lived in, and a dummy indicating whether the respondent lives in one of the four most favored regions resulting from the question discussed above. However, the models differed in its dependent variables: interest in the regional origin of food (INT_reg), knowledge of regional-origin labels (KNO_reg), and use of the regional-origin labels in the decision to buy regional products (IMP_reg). In fact, only the model of the knowledge of regional-origin labels appears to be statistically valid (Table 3). Age, education, and size of the site are significant determinants of the respondents' knowledge of regional-origin labels; age and size of the place where the respondent lives decrease the probability that he/she knows a regional-origin label while higher education level increases this probability. "Coming from a popular region" has no significant effect on

Table 3. The effect of respondents' characteristics on the knowledge of regional-origin labels; results of the probit analysis.

	Coefficient	Standard Error	Probability
Age	–0.012	0.005	0.009
Size of the site (inhabitants)	–0.058	0.025	0.020
Education (4 cater)	0.171	0.068	0.012
Region of the resp. (dummy)	–0.070	0.115	0.543
Constant	0.272	0.324	0.401
Number of observations	500		
Degree of freedom (parameter)	5		
Degree of freedom (model)	4		
Pseudo R^2	0.032		
Wald Chi square	20.873		
P value (sig. level)	< 0.01		

Note: Figures in bold indicate significance at $\alpha = 0.05$.
Source: Own calculations based on the survey among consumers.

the knowledge of regional-origin labels (on the probability). However, it is important to stress that the model explains only very little of the variation (pseudo-$R^2 = 3.6\%$).

Answers to the multiple-choice question on the expected characteristics of certified regional product suggest that households are mostly concerned about the regional origin of the raw material and that the product is Czech (291, 281 answers respectively). Characteristics like "traditional production", "better taste", and "healthy ingredients" received much attention—around 200 responses. The option "limited issue", that is only a small quantity produced, stood at the bottom of the list of choices (see Fig. 2).

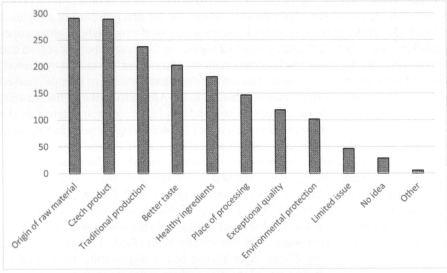

Fig. 2. Expected properties of regional products.
Source: Own overview based on the survey among consumers.

Discussion

"Regional" is not embedded in the purchasing practice/routine

Starting with the first question, most of the consumers declared rather routine shopping without seeking additional information except for the expiration date. Jastran et al. (2009) argue that routines are beneficial as they simplify daily activities and tasks. Obviously, supermarkets are typical places of routine shopping and, in general, the supermarkets are the most frequent shopping place for foods in the Czech Republic according to shopping monitors of GfK; for example, 90% of dairy products, and 73% of meat and meat products are bought there[2] (GfK 2016). The dominance of routine in shopping makes delivery of additional information, particularly, of geographic origin, to customers difficult. Further, Ilmonen (2001) states that supermarket food shopping has the potential to be either a planned and critical enactment of agency or an unreflective and reactive set of habitual behaviors. Then it can be derived that "regional" products must be included in shopping when it is planned and displayed adequately on the shelves to attract response of routine shoppers.

In line with Bildtgard (2008), the challenge for regional food schemes and related logos is to build consumers' trust in (i) specific product characteristics and (ii) in the regional uniqueness. This might be achieved by personal experience (e.g., Kirwan 2006, Sage 2003) or by institutional arrangements (Renting 2003).

Mismatch between consumers' expectations and producers' perception of regional logos

Great mismatch exists between the expectations of consumers (Fig. 2) and the perception of producers (Fig. 3). Producers' perceptions were obtained from the survey among producers participating in the Regional Brand scheme conducted by the

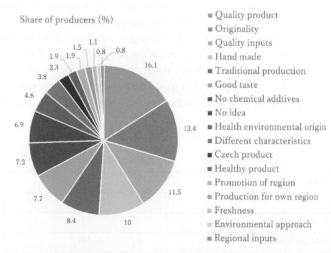

Fig. 3. Producers perception of regional food attributes.
Source: Own survey among producers participating in the Regional Brand scheme.

---

[2] Counts for purchases in hypermarkets, supermarkets, and discount stores.

authors in the other research.[3] Comparing the two charts (Figs. 2 and 3), the critical disparity is between the top interest in the regional origin of inputs by consumers and the bottom position of this attribute in the producers' perception of regional food. In general, consumers claim regional (or national) aspects of food labeled as important (quite naturally) leaving qualitative aspects and uniqueness aside while producers, in contrast, understand regional-origin labels as quality guarantee in terms of low level of contaminants, high content of quality inputs (like high share of meat, desired animal breeds or crop varieties, etc.) undervaluing the "geographical" attributes (ranking them low). According to Gruenert (2005), due to this inconsistency, producers can hardly build their strategy on the product quality as, according to the author, only when food producers can translate consumer wishes into physical product characteristics, and only when consumers can then infer desired qualities, will quality be a competitive parameter for producers.

The State Agricultural Intervention Fund (funding agency) carried out an extensive survey of consumers about their perception of the governmental "Regional Food" label in 2015 (SZIF 2016). The study uses two samples: the sample of 1500 women between 25 and 55 years, and a smaller, general population sample of about 550 respondents of both genders and aged between 15 and 55 years. Three results of this study are highly relevant to the research presented here.

i)   Only one-fourth of the target group respondents (women 25–55 years) and 15% of the general group (15 to 55 years including also men) were aware about the "regional food" label.
ii) Only two-thirds of the consumers aware of quality labels trust them.
iii) Most people who have not experienced regional products before expressed that the label associates them with the landscape. The four most frequent associations are presented in Table 4.

Regarding the first point, the survey presented in this research showed greater knowledge of the Regional Food label than in the SZIF study, likely as an effect of the continuous governmental promotional campaigns. However, the effect of adopting the label as a criterion for the purchase decision is rather negligible; only 13% of respondents indicated so. It might be partly due to the second finding of the SZIF study (insufficient trust in the certification) but also due to the reason discussed above that producers (and retailers) as well as the schemes themselves fail to present the aspects

Table 4. Consumer associations regarding "regional food" label.

Imagination (association with)	The share of respondents (%)
Landscape	72
Know place	45
Czech product	41
Clean environment	39

Source: SZIF 2016.

---

[3] For more details, see Boskova et al. 2016.

which are expected by consumers. The third finding of the SZIF study suggests that it is the landscape and some "intimacy" of location which attracts consumers while they are only weakly presented in the two selected schemes.

For example, the Regional Foods logo is the same for all regions and it is promoted nationally; thus, no outstanding regional characteristics are transmitted to consumers. In fact, it is a deep-seated problem of the governmental scheme; the presented research showed that consumers have preference for two administrative regions. If the governmental scheme and promotion accepted it, it would break the equal treatment/opportunity principle of the national policy (unless there was an argument of "lagging behind", which was not definitely the case of these two regions). The Regional Brand seems to be better embedded in the region at least in terms of self-governance of the scheme and the landscape aspect being an integral part of the regional-origin label. However, neither the scheme nor producers build sufficiently on consumers interests in regional input or regional specificity—neither are critical conditions for certification.

Henseleit et al. (2007) and Hensche et al. (2007) found two main groups of characteristics dominating consumers' perception of regional foods: (i) product freshness, good taste, and nutritional quality and (ii) environmentally and socially sustainable production. In both the Czech surveys (the one presented in this study and the one of SZIF, neither freshness nor environmentally friendly production were frequent answers. Landscape or intimacy of the location are more difficult to include as guarantee in certification and marketing/selling process. On the other hand, origin of raw material is an easy attribute which unfortunately is not popular among producers.

Knowledge implies willingness to pay premium, and there are no obvious socio-economic determinants of consumer interest in regional origin

There are two other results to discuss. These are as follows: (a) knowledge of labels implies willingness to pay premium for regional products and (b) neither this nor SZIF study delivered clear socio-economic determinants of consumer interest in regional origin. The former is a positive message for producers. The current study showed that aware consumers are willing to pay more than 10% price premium for certified regional origin of food. It only emphasizes the failure of communicating the origin of food to consumers in the pervasive disappointment with the effectiveness of the schemes. The latter message is more complex. First, there were not enough characteristics gathered in the survey, and it implies the need for further research on this aspect. Second, if it was true in general, it would complicate the marketing strategy of producers and scheme organizers. The finding that consumers pay more attention to regional-origin labels outside the region than in the region is surprising; probably, regional food is not sufficiently visible in the region. In this case the producers and label organizers will need to build closer relationship between the region and the consumers. We can regard the information that consumers want to support regional development by buying regional products as useful (and perhaps guiding) for designing the strategy for the schemes in the future.

Finally, filling up the questionnaire, can be perceived as a learning process. In the initial questions, the respondents exhibited rather neutral or cold relationship to geographical information and regional schemes. But in the later phases, it appeared that they had (or formed) their positions and preferences toward regional food. It emphasizes what has already been said that the label can only be successful if the right information is provided and in the way that consumers can understand and "digest" it and perhaps gradually transmit it in the shopping habit.

Conclusions

The objective of the present investigation was to show whether the certification schemes and related logos do attract consumers' attention by meeting their notion of regional food and whether the logos/labels provide sufficient guidance for consumers' shopping decisions.

It is an important question as regional food is a subject of public policy. The findings of this research can be summarized as follows:

- Regional logos/labels are known to households but rarely used in the decision on food shopping.
- The meaning of the most known label "Regional Food" overlaps with the meaning of "Czech" food. The scheme lacks regional differentiation and hence can hardly build and establish links between consumers and specific regions for certain foods. It misses the opportunity that consumers prefer some regions for certain foods.
- Households (consumers) have their own idea about attributes of regional food; first of all, it is the origin of raw material, but it does not meet with the rules of the "Regional Brand" certification, and it is not recognized by producers.
- In households, the category "regional" is associated with landscape and known place rather than anything else (freshness, environmental sustainability, etc.) as it follows particularly from the survey of SZIF. Familiarity with the region motivates consumers to support the development in the region by buying food from there which is stronger than "regional food specificity" in the survey conducted in the presented study.
- Households are in general willing to pay premium for regional food, of course, such food which complies with their notion about "regional". This willingness is substantially higher among those who know some of regional-origin labels.

The implications for producers and labeling scheme organizers including public policymakers can be summarized as follows:

i) There is some potential for encouraging production and labeling of regional food.
ii) The current schemes are not much successful in attracting the attention of households (consumers).

iii) Improvement of the performance can only be achieved if the schemes build on the knowledge of households (consumers) expectations.

iv) Promotion nationally might only help to recognize the scheme, it should be regionally differentiated to communicate specific reasons why consumers should buy a particular product from a particular region.

v) More should be done on the recognition of other regions outside home regions.

Following Bildtgard (2008), the scheme organizers should promote forthwith the regional food characteristics through institutional arrangements or promote such practices which enable to create trust in consumers based on personal experience.

Acknowledgement

The authors thank Dr. Katerina Cadilova, the president of the Association of Regional Brands for her provision of information on the labelling scheme. The research was conducted in cooperation of the Project "Science and Technology in the Society (VATES)" of the Ministry of Education, Youth and Sports of the Czech Republic, Task 4 Sustainable Consumption, and the Internal Research Project of the Institute of Agricultural Economics and Information. "Information Transfer Concerning Producers and Products across the Agri-food Chain".

Appendix

Table A1. The effect of interest in regional-origin labels on willingness to pay the price premium.

	#	Yes
No price premium	101	42
0 < price premium ≤ 5%	79	97
5 < price premium ≤ 10%	37	96
Price premium > 10%	10	38
Total	227	273
Test:		
Chi square statistics	65	
Degree of freedom	3	
P value (sig. level)	< 0.01	

Source: Own calculations based on the survey among consumers.

Table A2. The reasons for the first purchase of a regional (labelled) product.

Rank	Origin label	Personal relationship to the reg.	Information of the seller	Producer's story	Text on the packaging	Recommendation of friends	Leaflet	Information campain in media	Other sources	Purchased by chance	Other reason
1	23%	31%	2%	4%	5%	10%	4%	1%	1%	16%	4%
2	8%	10%	3%	4%	9%	13%	5%	2%	0%	3%	1%
3	4%	2%	1%	4%	7%	9%	4%	1%	0%	3%	1%
4	1%	0%	1%	2%	4%	3%	2%	2%	0%	0%	0%
5	1%	1%	0%	3%	0%	2%	3%	1%	0%	0%	0%
6	1%	2%	2%	0%	1%	0%	1%	1%	1%	1%	0%
7	1%	0%	1%	0%	1%	0%	1%	1%	0%	0%	0%
8	0%	0%	2%	1%	0%	1%	0%	2%	0%	0%	0%
9	0%	0%	0%	0%	0%	0%	0%	1%	0%	2%	0%
10	0%	0%	0%	0%	0%	0%	0%	0%	1%	0%	0%
11	0%	0%	0%	0%	0%	0%	0%	0%	0%	0%	1%
1+2	31%	40%	5%	8%	14%	23%	10%	3%	1%	19%	5%

Source: Own survey among consumers (households).

References

Appadurai, A. 1996. Modernity at Large: Cultural Dimensions of Globalization. Minneapolis: University of Minnesota Press.

Benedek, Z. and B. Balász. 2014. Regional differences in Hungary: the current stage of local food production at the county level. Paper prepared for presentation for 142nd Seminar Growing Success? Agriculture and Rural Development in an Enlarged EU, May 29–30, Budapest.

Bildtgård, T. 2008. Trust in food in modern and late-modern societies. Soc. Sci. Inform. 47: 99–128.

Boskova, I., T. Ratinger and K. Kličková. 2016. Do the agri-food stakeholders benefit the regional foods certification? The case of the Czech Republic. The Proceedings of the 149th EAAE Seminar Structural Change in Agri-food Chains: New Relations Between Farm Sector, Food Industry and Retail Sector, Rennes, France.

Bruce, A.S., J.L. Lusk, J.M. Crespi, J.B.C. Cherry, J.M. Bruce, B.R. McFadden, C.R. Savage, W.M. Brooks and L.E. Martin. 2014. Consumers' neural and behavioral responses to food technologies and price. J. Neurosci. Psychol. and Econ. 7: 164–173.

Burch, D. and G. Lawrence. 2009. Towards a third food regime: Behind the transformation. Agr. and Hum. Val. 26: 267–279.

Busch, L. and C. Bain. 2004. New! improved? The transformation of the global agrifood system. Rural Sociol. 69: 321–346.

Capt, D. and P. Wavresky. 2014. Determinants of direct-to-consumer sales on French farms. Rev. Agr. Env. Stud. 95: 351–377.

Darby, M. and E. Karni. 1973. Free competition and the optimal amount of fraud. J. of Law and Econ. 16: 67–88.

Eden, S., C. Bear and G. Walker. 2008. Understanding and (dis)trusting food assurance schemes: Consumer confidence and the 'knowledge fix'. J. of Rural Stud. 24: 1–14.

European Commission. 2006. Regulation (EU) No. 1974/2006 of 15 December 2006 laying down detailed rules for the application of Council Regulation (EC) No. 1698/2005 on support for rural development by the European Agricultural Fund for Rural Development (EAFRD). <http://eur-lex.europa.eu/legal-content/EN/ALL/?uri=celex:32006R1974> (accessed on 18 June 2017).

European Commission. 2013. Regulation (EU) No. 1305/2013 of the European Parliament and of the Council on support for rural development by the European Agricultural Fund for Rural Development (EAFRD). <http://eur-lex.europa.eu/legal-content/EN/TXT/?uri=celex%3A32013R1305> (accessed on 5 June 2017).

GfK Czech. 2016. Shopping Monitor 2016. GfK Czech, s.r.o. Prague. (In Czech).

Giddens, A. 1990. The Consequences of Modernity. Cambridge: Polity Press.

Gruenert, K.G. 2005. Food quality and safety: consumer perception and demand. Eur. Rev. Agr. Econ. 32: 369–391.

Gujarati, D.N. 1988. Basic Econometrics. McGraw-Hill, New York, Second edition. 705pp.

Hendrickson, M.K. and W.D. Heffernan. 2002. Opening spaces through relocalization: Locating potential resistance in the weaknesses of the global food system. Sociologia Ruralis 42: 347–369.

Henseleit, M, S. Kubitzki, D. Schotz and R. Teuber. 2007. Consumer preferences for locally produced foods—A representative analysis of the influencing factors. Berichte uber Landwirtschaft 85: 214–237.

Horáček, F. 2014a. Přehled: Nejpoužívanější značky kvality potravin v Česku. I-Dnes.cz: Ekonomika. (The survey: the most used quality labels in Czechia.) (In Czech) Online (cited 3 April 2017). <http://ekonomika.idnes.cz/loga-na-potravinach-04c-/ekonomika.aspx?c=A140107_114302_ekonomika_fih> (accessed on 20 July 2017).

Horáček, F. 2014b. Chaos za stovky milionů. Značek kvality je na českém trhu příliš. I-Dnes.cz: Ekonomika. (The chaos at hundred millions: there is too much quality labels on the Czech market.) (In Czech) Online (cited 3 April 2017). <http://ekonomika.idnes.cz/chaos-v-ceskych-znackach-kvality-do4-/ekonomika.aspx?c=A140120_133534_ekonomika_fih> (accessed on 20 July 2017).

Ilmonen, K. 2001. Sociology, consumption and routine. pp. 9–24. *In*: Gronow, J. and A. Warde (eds.). Ordinary Consumption. Routledge, London.

Janssen, M. and U. Hamm. 2012. Product labelling in the market for organic food: consumer preferences and willingness-to-pay for different organic certification logos. Food Qual. Preference 25: 22–29.

Jarosz, L. 2008. The city in the country: Growing alternative food networks in metropolitan areas. J. Rural Stud. 24: 231–244.

Jastran, M.M., C.A. Bisogni, J. Sobal, C. Blake and C.M. Devine. 2009. Routine and ritual elements in family mealtimes: contexts for child well-being and family identity. Appetite 52: 127–136.

Jones, E. 2014. An Economic Analysis of "Local" Production: Is it Efficient or Inefficient. Paper prepared for Presentation at the Agricultural & Applied Economics Association's 2014 AAEA Annual Meeting, Minneapolis, MN, July 27–29, 2014.

Kirwan, J. 2006. The interpersonal world of direct marketing: Examining conventions of quality at UK farmers' markets. J. Rural Stud. 22: 301–312.

Kjearnes, U., M. Harvey and A. Warde. 2007. Trust in food: A comparative and institutional analysis. Hampshire, UK, New York: Palgrave Macmillan.

Kneafsey et al. 2013. Short Food Supply Chains and Local Food Systems in the EU. A State of Play of their Socio-Economic Characteristics. European Commission Joint Research Centre, Institute for Prospective Technological Studies. <http://ipts.jrc.ec.europa.eu/publications/pub.cfm?id=6279> (accessed on 25 May 2017).

Lamine, C. 2005. Settling shared uncertainties: Local partnerships between producers and consumers. Sociologia Ruralis 45: 324–345.

Lorenz, B.A.S., M. Hartmann and J. Simons. 2015. Impacts from region-of-origin labeling on consumer product perception and purchasing intention—Causal relationships in a TPB based model. Food Qual. Preference 45: 149–157.

Luhmann, N. 1979. Trust and Power. Wiley, Chichester, Toronto.

Meyer, S.B., J. Coveney, J. Henderson, P.R. Ward and A.W. Taylor. 2012. Reconnecting Australian consumers and producers: Identifying problems of distrust. Food Policy 37: 634–640.

Mount, P. 2012. Growing local food: Scale and local food systems governance. Agr. Hum. Val. 29: 107–121.

Parliament of the Czech Republic. 1997. Ústavní zákon č. 347/1997 Sb., o vytvoření vyšších územních samosprávných celků a o změně ústavního zákona České národní rady č. 1/1993 Sb. (Constitutional Act No. 347/1997 Coll., On the Establishment of Higher Territorial Self-Governing Units and on the Amendment to the Constitutional Act of the Czech National Council No. 1/1993 Coll.).

Regional Foods. 2017. Regional Foods label awards the best products from each region. The regional Foods website. (In Czech) <http://www.regionalnipotravina.cz/o-projektu/> (accessed on 10 June).

Renting, H., T. Marsden and J. Banks. 2003. Understanding alternative food networks: Exploring the role of short food supply chains in rural development. Environ. Plan. A 35: 393–411.

Sage, C. 2003. Social embeddedness and relations of regard: Alternative 'good food' networks in southwest. Ireland. J. Rural Stud. 19: 47–60.

Sassatelli, R. and A. Scott. 2001. Novel food, new markets and trust regimes: reponses to the erosion of consumers' confidence in Austria, Italy and UK. Eur. Soc. 3: 213–44.

SZIF. 2016. The position of quality labels in the Czech food market. Study carried out by the Czech Agricultural University on the request of the State Agricultural Intervention Fund (SZIF). (In Czech) 121pp.

Teuber, R. 2011. Consumers' and producers' expectations towards geographical indications: Empirical evidence for a German case study. Br. Food J. 113: 900–918.

Thorsøe, M. and C. Kjeldsen. 2015. The constitution of trust: Function, configuration and generation of trust in alternative food networks. Sociologia Ruralis 56: 157–175.

Tonkin, E., J. Coveney, S.B. Meyer, A.M. Wilson and T. Webb. 2016. Managing uncertainty about food risks—Consumer use of food labelling. Appetite 107: 242–252.

United Nations. 2007. Safety and quality of fresh fruit and vegetables. A training manual for trainers. [online] New York and Geneva: United Nations. <http://unctad.org/en/docs/ditccom200616_en.pdf> (Accessed 7 May 2015).

Van Ittersum, K. 2002. The Role of Region of Origin in Consumer Decision-Making and Choice. Den Haag: LEI. <https://www.researchgate.net/publication/40141802_The_Role_of_Region_of_Origin_in_Consumer_Decision-Making_and_Choice> (accessed on 5 June 2017).

Velčovská, Š. and G. Del Chiappa. 2015. The food quality labels: Awareness and willingness to pay in the context of the Czech Republic. Acta Universitatis Agriculturae et Silviculturae Mendelianae Brunensis 63: 647–658.

Verbeke, W. and J. Roosen. 2009. Market differentiation potential of country-of-origin, quality, and traceability labeling. The Estey Centre J. Int. Law and Trade Policy 10: 20–35.

Virilio, P. 1991. The Lost Dimension (transl. D. Moshenberg). New York: Semiotext.

15

Organic and Local Foods

Substitutes or Complements?

Carolyn Dimitri[1,]* and *Samantha Levy*[2]

Introduction

Growth in consumer demand for organic and local foods has made these two niche segments critically important to the food system. In 2015, the dollar value of farm level sales marketed locally totaled $8.7 billion, while farm level sales of certified organic products amounted to $6.2 billion (US Department of Agriculture, NASS 2016b, 2016a). While local and organic farm level sales comprise a small portion of total farm sales, these market segments have benefits for farm income, business profits and consumer welfare. Organic farmers, for example, earn higher profits, largely due to their ability to command higher selling prices (Crowder and Reganold 2015, McBride et al. 2015). In contrast to organic products, the relative profitability of local marketing is less clear (Brown and Miller 2008). Consumer benefits include greater availability of organic and local foods, as retailers plan to meet growing consumer demand for organic and local foods (Packaged Facts 2017).

Long-time participants in the organic market consider organic and local foods as "two sides of the same coin" (Lipson 2008). In practice, organic farmers in the U.S. have relied on local markets as a primary outlet for their production, suggesting a complementarity between the two labels. In 2014, 70% of organic farms sold direct to consumers or institutions, although these transactions comprised less than 25% of

[1] Department of Nutrition and Food Studies, New York University, 411 Lafayette St, 5th Floor, New York, New York 10003.
[2] New York Policy Manager, American Farmland Trust, 112 Spring St, Suite 207, Saratoga Springs, NY 12866.
E-mail: slevy@farmland.org
* Corresponding author: carolyn.dimitri@nyu.edu

total farm level sales of organic food (US Department of Agriculture, NASS 2016c). The share of direct sales is remarkably similar to the share in 1997, when, according to the Organic Farming Research Foundation, organic farmers marketed 20% of their products directly to consumers or retailers (Walz 1999). It is unclear whether organic farmers relied on local markets due to a conscious decision to sell directly to consumers or whether it was the result of circumstance caused by early difficulties in accessing traditional market channels. Regardless of the motive, a historical relationship between local markets and organic producers is well established.

Organic agricultural products are raised on certified organic farmland around the world. In 2015, certified organic agricultural land reached nearly 126 million acres, up from 27 million acres in 1999 (Greene et al. 2017). In most countries, organic food is produced and handled according to standards that are set by government bodies. In 2015, 87 countries had organic regulations (Lernoud and Willer 2016). While the specific definition of organic production and handling varies across countries, the following key provisions are universal: (1) definition of organic, including substances that can be used or those that are prohibited, (2) certification bodies, to ensure that each farm and business has a plan consistent with the organic standards, and (3) enforcement, so there is a way to punish those who cheat. The organic standards and enforcement mechanisms are designed to maintain consumer confidence in the organic product, assuring consumers that organic food adheres to well-defined standards. Research has shown that organic farming systems are less damaging to the agro-ecosystem, on a per-acre basis, as they largely avoid use of synthetic chemicals for fertilization of plants and for pest and weed management (Tuomisto et al. 2012). Many organic farming practices enhance on-farm biodiversity (Hole et al. 2005).

Local food, in contrast, is grown near the consumer, and has no constraint placed on practices used on a farm. Consequently, there is no relationship between on-farm environmental benefits and food produced and marketed locally. Furthermore, a concrete definition of local is lacking. Instead, there are multiple definitions of local food, which include being produced within 100 miles of where it is consumed, within an eight-hour drive, or within the boundaries of a state. Thus, the organic and local labels describe discrete and distinct attributes of the food system. Some organic certifiers already certify producers and handlers to a number of other standards—including food safety and animal welfare standards—while international organic standards incorporate a social justice component (IFOAM 2009). Thus a product might easily carry both an organic label, denoting the ecologically based production and handling system used, and a locally grown logo, denoting low mileage between the farm and the consumer.

On the farm, raising organic food is likely to be more costly: yields tend to be lower when a farm is managed organically, labor needs are higher, and farmers incur certification costs (McBride et al. 2015). The primary benefit of the organic label is to assure consumers that a farm and handler used approved practices in the production and processing of food sold as organic. Local farms (that are not organic) are not constrained by federal regulations and incur no certification costs. Thus from an economic perspective, the local farm has an advantage in terms of production costs. Those producing for local markets furthermore have an incentive to claim the use of 'sustainable practices', which have no generally accepted meaning but often make consumers feel better about their purchases.

The Organic and Local Food Markets in the United States

Describing the organic and local food markets is challenging, as the available data do not fully describe the market channels. In the U.S., as in most developed countries, agricultural commodities raised on the farm are distributed to handlers, which are businesses that process, manufacture, pack and move products along the supply chain. Ultimately, the food products eaten by consumers are purchased from food retail venues, such as supermarkets, grocers, and specialty shops, or in institutions, such as cafeterias, airplanes, and schools, or directly from farmers (Dimitri and Oberholtzer 2009). The U.S. government statistics are consistently available for farm level production but not for distribution and consumption. Furthermore, the government did not collect data focusing on organic farmers until 2008. The final complicating factor is that, because of differences in distance, separate supply chains handle foods marketed locally and those marketed nationally. Both conventionally and organically produced local foods follow shorter supply chains, which include local foods bought directly from farmers, in restaurants, and in retail outlets (Low et al. 2015). Longer supply chains are commonly used for foods marketed nationally, and are designed for moving large quantities of agricultural commodities from large farms to large retail outlets, under the power of large-scale intermediaries (Dimitri and Oberholtzer 2009).

An accounting of the organic farm sector is a relatively new component of the *Agricultural Census*, and provides data for select years on the number of organic farms, certified organic acreage and farm level sales since 2008 (see Table 1). Between 2008 and 2014, the number of organic farms and acres of certified organic farmland declined. The 2014 (and earlier) and 2015 (and later) data are not directly comparable, as 2014 includes organic farms that are exempt from being certified due to their low level of sales (less than $5,000 a year) while 2015 excludes exempt farms. Between 2015 and 2016, the number of certified farms and acres increased. Farm level sales increased between 2008 and 2016, and within the eight-year period, farm level sales more than doubled to $7.6 billion. Note that even with this growth, certified organic farmland accounted for less than one percent of all U.S. cropland and pasture in 2015. Other characteristics of organic farms, in comparison to the rest of agriculture in the U.S., include a younger farmer population, more diversified crop production and higher level of direct-to-consumer sales and value-added marketing (Greene et al. 2017).

Table 1. United States organic farm sector: 2008, 2014–2016.

	2008	2014	2015	2016
Farms (number)	14,540	14,093	12,816	14,217
Acres (number)	4,077,337	3,670,560	4,361,849	5,019,496
Farm level sales (1,000 USD)	3,164,995	5,456,732	6,163,472	7,553,872
Farms in transition (number)	1,938	1,365	1,530	n/a
Acreage in transition	194,383	122,175	150,880	n/a

Notes: 2008 and 2014 includes certified and exempt farms and farmland; 2015 and 2016 include certified (but not exempt) farms and farmland. Exempt farms are those with annual sales of $5,000 or below. Land and farms in transition are those in the process of converting farmland to organic production.
Sources: (US Department of Agriculture, NASS 2010, 2016c, 2015b, 2017).

Estimates of the size of the organic market at the retail level vary. The organic industry trade association (OTA) estimates 2015 U.S. organic retail sales at $43.3 billion while IFOAM data puts this number at $39.7 billion (Organic Trade Association 2016, FiBl 2017). In 2015, the U.S. market consumed approximately half of total global organic sales (FiBl 2017). While organic sales continue to represent a relatively small share of total U.S. food sales, annual double-digit growth since 2000 shows the strength and growth potential of the organic market. The largest categories of sales in the organic industry are fruit and vegetables followed by dairy.

Millennial consumers are an important part of the organic market, which suggests that growth in the organic market is likely to continue as the Millenials age. More than half of the respondents to the 2014 Gallup food consumption survey between the ages of 18 and 29 indicated that they actively try to include organic products in their diet, compared with only one-third of respondents 65 or older (Greene et al. 2017). Industry research suggests that parents in the 18–35-year-old range are now the biggest group of organic buyers in America, driven by the belief that buying organic food makes them better parents (Organic Trade Association 2017b). Concerns about pesticide effects, hormones and antibiotics along with the desire to avoid highly processed foods were top reasons for buying organic. For these reasons, organic baby foods and food for children were rated as the most important foods to purchase, outranking fruits and vegetables for the first time ever (Organic Trade Association 2017b).

Quantifying the size of the local food market is more difficult, largely due to the relatively small size of the market and the lack of data covering farmer marketing practices. In the United States, local foods are either sold directly by farmers to consumers or through 'intermediated' channels. Direct sales are those made at farmers markets and farm stands, for example, while intermediated sales are those made to schools or other institutions or to regional distributors (Low et al. 2015). The U.S. Department of Agriculture has collected data on direct to consumer sales since 1978. Both the number of farms selling directly to consumers and the value of sales (thousands of dollars) increased between 1992 and 2012, although, as Fig. 1 shows, sales increased at a faster rate.

The extent of local sales made through intermediated market channels is more difficult to track over time. Comprehensive data, for all local market channels, are available for just one year, 2015, when the federal government conducted a survey of farmers regarding their local marketing practices (US Department of Agriculture, NASS 2016b). During 2015, the $8 billion of local market sales, measured at the farm level, were marketed through the following channels: $3.4 billion through regional distributors and institutions, $3.0 billion, direct to consumer, and $2.4 billion directly to retailers (US Department of Agriculture, NASS 2016b). Earlier work estimated the value of farm level sales sold through all local market channels as approximately $6.1 billion in 2012 (Low et al. 2015).

Evidence supports the perception of a close link between organic farms and local markets, in terms of the numbers of farms selling their organic products locally. In 2008, the first year such data are available, 77% of the approximately 11K certified organic farms sold at least some of their products directly to consumers or through the intermediated markets, yet these local market sales made up just 17% of farm level sales into local markets (US Department of Agriculture, NASS 2010). In 2014, 70%

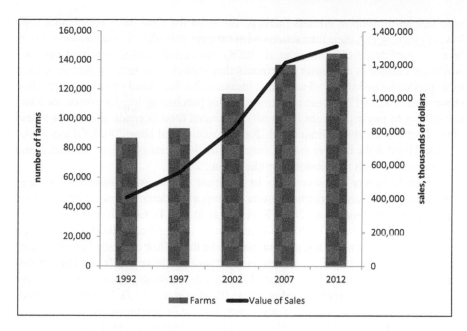

Fig. 1. Direct to consumer farm level sales and number of farms: 1992–2012.
Source: US Department of Agriculture, NASS 1999, 2009, 2014.

of the 14K certified organic farms accessed local markets, selling 22% of total farm level sales into local markets (US Department of Agriculture, NASS 2015a). Overall, the share of certified organic farms selling into the local market channels declined over the six years, but the share of sales increased, indicating that certified organic farmers still view the local market as a key outlet for their production.

The census data furthermore suggest that smaller farms are selling locally, with fewer, larger farms selling the bulk of organic products through the national supply chains. This finding reflects that of the entire agricultural sector, thus it is not surprising that organic farm products also flow through the large-scale marketing infrastructure. In the absence of data reflecting the years prior to 2008, it is not possible to precisely pinpoint the date that organic foods made a wholesale shift out of local markets into the national, large-scale food system. That said, the shift into the national markets appears to have resulted from widespread mergers between large food companies and small organic companies, in the mid to late 1990s, that made it possible for organic products to gain access to the mass market supermarket, placing the U.S. organic market on a new growth trajectory (Dimitri and Richman 2000).

Motivations for Consumer Purchases of Organic and Local Food

Local food consumers enjoy the food for its taste, freshness, and in some cases, the economic and psychological benefits of supporting local farms and businesses (DeWeerdt 2009). Attitudes towards cooking and shopping behavior are related to the probability of purchasing local food (Zepeda and Li 2006). For consumers who frequently shop at farmers markets, the need for third party certifications are low, suggesting that their interactions with farmers provides consumers with a high level of confidence (Thilmany et al. 2008). So-called 'collectivist motivations,' which are defined as consumer preferences that explicitly include a desire to support local producers and the local economy, increase the likelihood of buying local food (Baumann et al. 2017). Some consumers prefer purchasing locally grown food and are willing to pay a premium for locally produced food (Zepeda and Leviten-Reid 2004, Darby et al. 2008, Giraud et al. 2005, Loureiro and Hine 2002). Of those who buy local food at the farmers market, many enjoy the social aspects of the market and the interactions with the vendors (Smithers et al. 2008).

Multiple factors lead consumers to purchase organic food. Research indicates that environmental and altruistic reasons are important determinants of organic food consumption (Hughner et al. 2007, Kareklas et al. 2014). Organic food consumers are motivated by personal health (Akhondan et al. 2015, Mintel 2017). Organic food consumption (in Germany) is higher for consumers that follow a holistic lifestyle, which includes eating a healthy diet (Aschemann-Witzel and Zielke 2017). A meta-analysis finds that individual attitudes and social norms are important drivers underlying the decision to consume organic food products (Scalco et al. 2017). Similar work finds that older consumers were more likely to buy organic food if they perceived doing so was socially desirable (Hwang 2016). The research that points to the importance of personal health is consistent with work suggesting that while green consumers wish to improve the environment through their purchases, they also expect to receive personal benefits (Marchand and Walker 2008).

Changing Roles of Organic and Local Foods

Considering local food and organic food as substitute, rather than complementary, products is a recent trend. Some opposed to the influence and roles of larger food businesses in the organic sector have argued that the organic standards, as promulgated by the United States Department of Agriculture (USDA), made possible the 'corporate co-opting of organic'. The argument suggests that government regulation is captured by corporate influences, and thereby weakens standards in favor of corporations (Jaffee and Howard 2010). The watchdog organization, the Cornucopia Institute, has challenged USDA's implementation of the organic standard, pointing to lack of oversight of select dairy and egg farms, and imported organic products (Whoriskey 2015, 2017a, 2017b, Ross 2017).

Some suggest there is a growing skepticism about organic, in response to industry consolidation and the presence of large food companies in the organic sector (Fromartz 2007, Howard 2009, Hauter 2014). As organic foods continue to penetrate conventional and mass-market supermarkets, it is possible that (1) consumer valuation of organic will change and (2) organic food prices will decline (Aschemann-Witzel and Zielke 2017). In the United States, experimental data suggests that consumer trust in certified organic brands was higher for those sold Target, in comparison to those sold in Walmart (Ellison et al. 2016).

At the same time, questioning the integrity of the organic label in the U.S. has become more frequent (Steinmetz 2015, Behar 2017b). A recent audit by USDA's

Inspector General recommended that the National Organic Program strengthens control over imported organic products, including addressing fumigation of products with pest infestations and verification of authenticity of organic certifications at the border (US Department of Agriculture, Office of Inspector General 2017). Such questions about the ability of USDA to enforce organic standards further contribute to an erosion of confidence in the organic label in the U.S. While the European Union has tackled fraud head on starting in 2007, the U.S. has lagged in addressing enforcement and transparency in the organic supply chain (Behar 2017a). The task force, Global Organic Supply Chain Integrity Task Force, was formed in 2017, and is an industry led effort to combat fraud and maintain organic integrity (Organic Trade Association 2017a).

Industry research examining trends in grocery shopping reports that consumers want transparency and a closer connection with their food (Food Marketing Institute 2017). Contrast the two following scenarios describing organic and local foods. For organic food consumers, media attention questioning the integrity of the organic label in the U.S. has eroded consumer trust in the label (Behar 2017b). In addition to questions about integrity of the label, when a consumer purchases organic food, produced by a large corporation and sold in a large supermarket, there is a large gap between the consumer and her food. In contrast, consumers who buy directly from farmers, develop a relationship that builds trust in the local food farmer. Typically, this food is sold in a farm stand, at a farmers market, or in a supermarket where the food is accompanied by signage describing the farm or town where the food was grown. As described, the consumer buying local food scenario has an experience that is closer to what she wants; that is, one that helps her cultivate a relationship to her food. One consequence of these two scenarios is that some consumers prefer unregulated local foods purchased directly to regulated organic foods purchased in supermarkets or other retail venues.

Conclusion

Organic and local foods are not the dominant types of food purchased by consumers. Yet many consumers seek out these niche products for at least some of their food purchases. Such products satisfy consumer demand, contribute to the potential for greater farm profitability, and have fewer negative environmental impacts associated with on-farm production. At this juncture, the organic farming sector faces a real threat resulting from weak enforcement of the federal organic standard. The integrity issues compound the unease that some consumers have about the role of large corporations in the organic food industry, which are largely the result of the maturing organic industry.

Many have turned to local food to satisfy their desire for more transparency in the food supply chain. While food marketed locally has value to society in terms of supporting local communities, it is likely that at least a portion of the local food consumers perceive that the products are raised sustainably or are produced in a way that reduces the agro-environmental impact of agricultural production. The major issue may be that consumer perception of local farm practices are based on ideas consumers have about farming and sustainability, and how they are related. The concepts of farm sustainability are complex and subject to individual interpretation, in the absence of a uniformly accepted definition of sustainability. Furthermore, consumer perception of transparency in the local food supply chain may be misguided. The transparency relies on the reputation and integrity of the farmer producing and marketing the food, as well as consumer knowledge about farming practices and what questions to ask their farmers.

Current questions about integrity and transparency point out several challenges and contradictions prevailing in the U.S. organic and local food systems. Firms in the food sector, both farms and businesses, face incentives to increase profits by reducing costs. Consumers who consistently choose food items based on price, create another incentive for cost cutting. Thus, the evolution of the organic industry into one that captures the economies of scale and scope of the food system is a rational response to consumer demand. As a result, organic food is widely available, and nearly every food product has an organic version. The reaction to market success, by a small section of consumers calling for transparency and higher priced food, has created a market opportunity for an alternative to organic food—locally produced food, which is not subject to regulation. Yet the farms responding to this demand face similar pressures to lower costs, and one way to do so is through the choice of farm practices, which are unobservable to consumers. Note that this potential for misrepresentation of quality is what led the organic industry to seek federal standards, which were created to protect both consumers and organic producers from fraudulent products.

Over the course of the next few years, it will be interesting to watch the evolution of the organic and local food markets. We expect that farmers, businesses, and consumers will continue to grapple with issues of integrity and transparency, and struggle with what it means to farm sustainability. Local food markets appear to be on the same trajectory that organic food markets were in the 1980s and 1990s, which ultimately led to federal level organic policy. Yet at this point in time, US federal organic standards need better enforcement to prevent the market from unraveling. Leaving the fate of organic and local food product integrity to businesses suggests that the resulting solution will meet the needs of firms and farm businesses, and not consumers. Yes, these are all contradictions, leading one to a perception that the organic and local food markets are the modern, agricultural version of the Wild West.

References

Akhondan, H., K. Johnson-Carroll and N. Rabolt. 2015. Health consciousness and organic food consumption. J. Fam. Consum. Sci. 107: 27.

Aschemann-Witzel, J. and S. Zielke. 2017. Can't buy me green? A review of consumer perceptions of and behavior toward the price of organic food. J. Consum. Aff. 51: 211–51.

Baumann, S., A. Engman, E. Huddart-Kennedy and J. Johnston. 2017. Organic vs. local: comparing individualist and collectivist motivations for 'ethical' food consumption. Can. Food Stud. 4: 68–86.

Behar, H. 2017a. Protecting organic integrity: Too little, too late? *Organic Broadcaster*, 2017. <http://mosesorganic.org> (accessed on 19 October 19 2017).

Behar, H. 2017b. Should we achieve organic integrity through flexibility or consistency? *Organic Broadcaster*, 2017. <http://mosesorganic.org> (accessed on 19 October 2017).

Brown, C. and S. Miller. 2008. The impacts of local markets: A review of research on farmers markets and community supported agriculture (CSA). Am. J. Agr. Econ. 90: 1298–1302.

Crowder, D.W. and J.P. Reganold. 2015. Financial competitiveness of organic agriculture on a global scale. Proc. Natl. Acad. Sci. 112: 7611–16.

DeWeerdt, S. 2009. Is local food better? *Worldwatch Magazine*, 2009. <http://www.worldwatch.org/node/6064> (accessed on 19 October 2017).

Dimitri, C. and N. Richman. 2000. Organic food markets in transition. Greenbelt, MD, Henry A. Wallace Center for Agricultural & Environmental Policy.

Dimitri, C. and L. Oberholtzer. 2009. Marketing U.S. organic foods: Recent trends from farms to consumers. EIB-58. US Department of Agriculture, Economic Research Service. <https://www.ers.usda.gov/publications/pub-details/?pubid=44432> (accessed on 19 October 2017).

Ellison, B., B.R.L. Duff, Z. Wang and T.B. White. 2016. Putting the organic label in context: Examining the interactions between the organic label, product type, and retail outlet. Food Qual. Preference 49: 140–150.

FiBl. 2017. The World of Organic Agriculture 2017. <http://www.fibl.org/fileadmin/documents/ en/news/2017/mr-world-organic-agriculture-2017-english.pdf> (accessed on 19 October 2017).

Food Marketing Institute. 2017. How is the new grocery shopping marketplace evolving? <https://www.fmi.org/our-research/research-reports/u-s-grocery-shopper-trends> (accessed on 19 October 2017).

Fromartz, S. 2007. Organic, Inc.: Natural Foods and How They Grew. Orlando: Harcourt.

Greene, C., G. Ferreira, A. Carlson, B. Cooke and C. Hitaj. 2017. Growing organic demand provides high-value opportunities for many types of producers. Amber Waves. US Department of Agriculture, Economic Research Service. <https://www.ers.usda.gov/amber-waves/2017/januaryfebruary/growing-organic-demand-provides-high-value-opportunities-for-many-types-of-producers> (accessed on 19 October 2017).

Hauter, W. 2014. Walmart squeezes organics. Progressive 78: 18–20.

Hole, D.G., A.J. Perkins, J.D. Wilson, I.H. Alexander, P.V. Grice and A.D. Evans. 2005. Does organic farming benefit biodiversity? Biol. Conserv. 122: 113–30.

Howard, P.H. 2009. Consolidation in the north American organic food processing sector, 1997 to 2007. Int. J. Sociol. Agr. Food 16: 13–30.

Hughner, R.S., P. McDonagh, A. Prothero, C.J. Shultz II and J. Stanton. 2007. Who are organic food consumers? A compilation and review of why people purchase organic food. J. Consum. Behav. 6: 94–110.

Hwang, J. 2016. Organic food as self-presentation: The role of psychological motivation in older consumers' purchase intention of organic food. J. Retail. Consum. Serv. 28: 281–287.

Jaffee, D. and P.H. Howard. 2010. Corporate cooptation of organic and fair trade standards. Agr. Hum. Val. 27: 387–399.

Kareklas, I., J.R. Carlson and D.D. Muehling. 2014. I eat organic for my benefit and yours': Egoistic and altruistic considerations for purchasing organic food and their implications for advertising strategists. J. Advertising 43: 18–32.

Lernoud, J. and H. Willer. 2016. Organic Agriculture Worldwide 2014. Research Institute of Organic Agriculture. <http://www.fibl.org/en/media/media-archive/media-archive16/media-release15/article/bio-waechst-weiter-weltweit-437-millionen-hektar-bioflaeche.html> (accessed on 19 October 19).

Low, S.A., A. Adalja, E. Beaulieu, N. Key, S. Martinez, A. Melton, A. Perez et al. 2015. Trends in US Local and Regional Food Systems: Report to Congress. Administrative Publication AP-068. United States Department of Agriculture, Economic Research Service. <https://naldc.nal.usda.gov/download/60312/PDF> (accessed on 19 October 2017).

Marchand, A. and S. Walker. 2008. Product development and responsible consumption: designing alternatives for sustainable lifestyles. J. Cleaner Prod. 16: 1163–69.

McBride, W., C. Greene, L. Foreman and M. Ali. 2015. The profit potential of certified organic field crop production. ERR-188. US Department of Agriculture, Economic Research Service. <https://www.ers.usda.gov/publications/pub-details/?pubid=45383> (accessed on 19 October 2017).

Mintel. 2017. The Natural/Organic Food Shopper-US, July 2017.

Organic Trade Association. 2016. U.S. organic sales post new record of $43.3 billion in 2015. Press Release. <https://www.ota.com/news/press-releases/19031> (accessed on 19 October 2017).

Organic Trade Association. 2017a. Global organic supply chain integrity task force—News From OTA. 2017. <http://www.newsfromota.com/ota-members/government-affairs-policy/global-organic-supply-chain-integrity-task-force/> (accessed on 19 October 2017).

Organic Trade Association. 2017b. Today's millennial: tomorrow's organic parent. <https://ota.com/news/press-releases/19828> (accessed on 19 October 2017).

Packaged Facts. 2017. U.S. produce market: 6 key trends driving sales of fresh fruits & vegetables. Press release. Packaged Facts. 2017. <https://www.packagedfacts.com/about/release.asp?id =4232> (accessed on 19 October 2017).

Ross, A. 2017. Imports citizen petition, 2017. <https://www.cornucopia.org/wp-content/uploads/ 2017/07/ImportsCitizenPetition.pdf> (accessed on 19 October 2017).

Scalco, A., S. Noventa, R. Sartori and A. Ceschi. 2017. Predicting organic food consumption: a meta-analytic structural equation model based on the theory of planned behavior. Appetite 112: 235–48.

Smithers, J., J. Lamarche and A.E. Joseph. 2008. Unpacking the terms of engagement with local food at the farmers' market: insights from Ontario. J. Rural Stud. 24: 337–50.

Steinmetz, K. 2015. Why many consumers think 'organic' labels are hogwash. Time. 2015. <http://time.com/3857799/organic-label-standards-poll> (accessed on 19 October 2017).

Thilmany, D., C.A. Bond and J.K. Bond. 2008. Going local: Exploring consumer behavior and motivations for direct food purchases. Am. J. Agr. Econ. 90: 1303–9.

Tuomisto, H.L., I.D. Hodge, P. Riordan and D.W. Macdonald. 2012. Does organic farming reduce environmental impacts? A meta-analysis of European research. J. Environ. Manage. 112: 309–20.

US Department of Agriculture, NASS. 1999. 1997 Census of Agriculture. AC97-A-51. <http://agcensus.mannlib.cornell.edu/AgCensus/getVolumeOnePart.do?year=1997&part_id=949&number=51&title=United%20States> (accessed on 19 October 2017).

US Department of Agriculture, NASS. 2009. 2007 Census of Agriculture. <https://www.agcensus.usda.gov/Publications/2007/ Full_Report/Volume_1,_Chapter_1_US> (accessed on 19 October 2017).

US Department of Agriculture, NASS. 2010. 2008 Organic Production Survey (2007 Census of Agriculture Special Study). <https://www.agcensus.usda.gov/Publications/2007/Online_Highlights/Organics/> (accessed on 19 October 2017).

US Department of Agriculture, NASS. 2014. 2012 Census of Agriculture. AC-12-A-51. <https://www.agcensus.usda.gov/ Publications/2012/Full_Report/Volume_1,_Chapter_1_US/> (accessed on 19 October 2017).

US Department of Agriculture, NASS. 2015a. 2014 Organic Survey (2012 Census of Agriculture Special Study). <https://www.agcensus.usda.gov/Publications/2012/Online_Resources/Organics> (accessed on 19 October 2017).

US Department of Agriculture, NASS. 2015b. Table 1. Farms, land, and value of sales of organic agricultural products—Certified and exempt organic farms: 2014. In. 2012 Census of Agriculture: 2014 Organic Survey. <https://www.agcensus.usda.gov/Publications/2012/Online_Resources/Organics> (accessed on 19 October 2017).

US Department of Agriculture, NASS. 2016a. 2015 Certified Organic Survey. Farms, Land, and Sales Up. 2016–8. NASS Highlights. <https://www.nass.usda.gov/Publications/Highlights/2015_Certified_Organic _Survey_Highlights.pdf> (accessed on 19 October 2017).

US Department of Agriculture, NASS. 2016b. Direct Farm Sales of Food. Results from the 2015 Local Marketing Practices Survey. ACH12-35. 2012 Census of Agriculture Highlights. <https://www.agcensus.usda.gov/ Publications/2012/Online_Resources/Highlights/Local_Food/LocalFoodsMarketingPractices_Highlights.pdf> (accessed on 19 October 2017).

US Department of Agriculture, NASS. 2016c. Organic Survey 2014. Census. Census of Agriculture. <https://www.agcensus.usda.gov/Publications/2012/Online_Resources/Organics/index.php> (accessed on 19 October 2017).

US Department of Agriculture, NASS. 2017. 2016 Certified Organic Survey. <http://usda.mannlib.cornell.edu/usda/current/ OrganicProduction/OrganicProduction-09-20-2017_correction.pdf> (accessed on 19 October 2017).

US Department of Agriculture, Office of Inspector General. 2017. National Organic Program – International Trade Arrangements and Agreements. Audit Report 01601-0001-21. <https://www.usda.gov/oig/webdocs/01601-0001-21.pdf> (accessed on 19 October 2017).

Whoriskey, P. 2015. Millions of 'organic' eggs come from industrial scale chicken operations, group says. *Washington Post*, 2015. <https://www.washingtonpost.com/news/wonk/wp/2015/ 12/15/millions-of-organic-eggs-come-from-industrial-scale-chicken-operations-group-says> (accessed on 19 October 2017).

Whoriskey, P. 2017a. Why your 'organic' milk may not be organic. *Washington Post*, 2017. <https://www.washingtonpost.com/business/economy/why-your-organic-milk-may-not-be-organic/2017/05/01/708ce5bc-ed76-11e6-9662-6eedf1627882_story.html> (accessed on 19 October 2017).

Whoriskey, P. 2017b. Bogus 'organic' foods reach the U.S. because of lax enforcement at ports, inspectors say. *Washington Post*, September 18, 2017, sec. Wonkblog. <https://www.washingtonpost.com/news/wonk/wp/2017/09/18/lax-enforcement-at-ports-allows-bogus-organic-foods-to-reach-u-s-government-report-says> (accessed on 19 October 2017).

Zepeda, L. and J. Li. 2006. Who buys local food? J. Food Distrib. Res. 37: 1–11.

PART IV
Marketing and Regulation

16

Marketing and Regulation Associated With Food Attributes

Robert P. Hamlin

Introduction

The nature of food attributes

In the literature on food marketing, a distinction is made between a food product's intrinsic and extrinsic attributes (Espejel et al. 2007). An intrinsic attribute is an actual (tangible) feature of the product itself. Thus a meat patty will possess certain ratios of fat, protein and water. These primary intrinsic attributes combine with others to produce secondary intrinsic attributes, including flavour, mouth feel and satisfaction. These form the direct inputs for a consumer's pre- and post-consumption evaluation of a food product. The product will possess other intrinsic attributes that may contribute significantly to consumer pre-purchase evaluation of it. These include colour, size, and evidence of freshness, etc. (Enneking et al. 2007).

Up to a century ago consumers were almost entirely reliant upon these intrinsic attributes to guide their food purchasing behaviour. However since then the advent of packaging and self-service retail formats has increased both the quantity and importance of extrinsic attributes in determining consumer choice (Marsh and Bugusu 2007). An extrinsic attribute does not form part of the food product itself, but nevertheless may be used as an input to a consumer's evaluation of it. Predominantly these extrinsic attributes are communicated by 'cues' on the product's packaging. Such cues may include brand marks, colours and claims as to the food's nature and

Department of Marketing, University of Otago, P.O. Box 56, Dunedin, 9054, New Zealand.
E-mail: rob.hamlin@otago.ac.nz

price (Wells et al. 2007). The value of these package cues as an input to consumer evaluation and choice may be reinforced by communications that are delivered to them either at the point of sale (e.g., posters) or remote from it (e.g., advertising on T.V. billboards and various public relations communications exercises). The advance of extrinsic cues has now reached a conclusion in that a consumer who selects and purchases items in the fast growing Web market for food products is entirely reliant upon them (Canavan et al. 2007).

It is important to appreciate the differences between extrinsic cues and extrinsic attributes. The latter is inferred by the consumer from the former. Thus a consumer who is purchasing a meat patty via a website may be presented with the extrinsic cue that it is 'organic'. From this cue they may infer (probably correctly) the extrinsic attribute that it has been grown in accordance with organic principles. They may also infer (probably incorrectly) the extrinsic attribute that the patty is lower in fat and salt than its non-organic equivalent. It is also worth noting that different consumers may infer different extrinsic attributes from the same extrinsic cue, and that these extrinsic attributes may be close analogues of actual intrinsic attributes possessed by the product (See the patty fat/salt content example above). These consumers generated extrinsic attribute analogues may or may not be an accurate representation of the intrinsic reality of the product itself.

The distinction between *direct* and *indirect* food attributes must also be understood before any meaningful discussion relating to the regulation of their communication to consumers can be undertaken. Intrinsic attributes of food products such as the actual fat, salt and protein content of the patty noted above are all *direct* attributes. The possible future consumer outcomes that are related to these intrinsic attributes are *indirect* attributes. Thus the three intrinsic direct attributes of the meat patty above (protein/fat and salt) can also be indirectly related to specific future consumer outcomes (muscle bulk, obesity and heart disease). The first of these three would usually be viewed positively and is thus a positive extrinsic indirect attribute of the product. The second two would usually be viewed negatively and are thus negative extrinsic indirect attributes of the same product. However, if the consumer's inferred extrinsic indirect attributes are not equivalent to the actual intrinsic indirect attributes of the product, then the consumer's attribution of indirect attributes to the food product and their subsequent decisions will be potentially erroneous (Fig. 1).

Intrinsic, extrinsic, direct and indirect food attributes can interact with one another in the process of consumer choice. They can also be effectively managed by marketing managers to influence that choice (Hamlin 2010). Indirect food attributes communicated via extrinsic analogues of intrinsic food attributes that are communicated to consumers by extrinsic cues are already a significant feature of the food marketplace in developed countries, and they are growing fast. Thus the extrinsic brand cue ProActiv® is used to create the extrinsic attribute of 'lower cholesterol' for all the products that carry it, which is a direct analogue of an intrinsic attribute. This direct attribute is then used to create the indirect attribute that the ProActiv® branded product will reduce the risk of heart disease relative to its alternatives. The planned outcome of such a chain of reasoning is increased purchase of ProActiv® products on the basis of their perceived superiority with regard to a single indirect attribute—their capacity to reduce the risk of heart disease (Phillips et al. 2000).

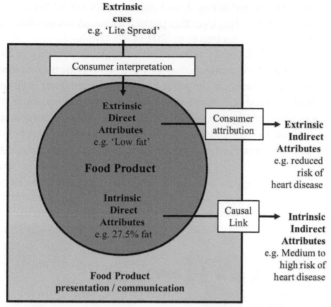

The reality of this product is represented by its intrinsic direct attributes (27.5% fat)   This is causally linked to the <u>intrinsic indirect attribute</u> of an *elevated risk of heart disease*. However, this reality may be bypassed as an input to product evaluation by a consumer interpretation of a related <u>extrinsic cue</u> *'Lite'* that creates the <u>extrinsic direct attribute</u> *low fat* within the consumer's mind.   Once created this <u>extrinsic direct attribute</u> may support the consumer attribution that in turn creates the <u>extrinsic indirect attribute</u> *reduced risk of hear disease*, which is not analogous to the actual <u>intrinsic direct product attribute</u> of *medium to high risk of heart disease*.

Fig. 1. The relationship between the four types of food attributes as part of a consumer food product evaluation.

Marketing managers will attempt to use these tools to influence food consumer choice to maximize their organization's sales and profit. That, after all, is their legal responsibility (McDonald and Wilson 2016). However, there is no reason why these objectives should coincide with maximizing the sales of food products with the best indirect attribute profiles, and thus the maximization of public good. As most developed economies have government agencies that are specifically charged with maximizing public good in this area, then from time to time the objectives of these public agencies and those of the private marketing agencies of the food industry will come into conflict (Josling et al. 2004). It is here that public regulation of marketing's communication of food attributes has a potential role to play in communal welfare.

The history of regulation of marketing's communication of food attributes

It is impossible to consider the history of the regulation of how food attributes are marketed without first considering the evolution of food attributes and attribute based consumer choice in developed economies over the last sixty years. These developments are summarized in Fig. 2, and they broadly reflect the advance of large format self-service food retailing over this period, although social changes have also had a significant impact (Spaargaren et al. 2013).

The overall situation faced by a consumer wishing to reliably infer the actual intrinsic attributes of the food products that they are buying has become progressively more difficult as both products and the purchase environment itself have become both more complex and inscrutable, making the indirect inferential and attributional route shown in the upper part of Fig. 1 either the dominant or the only route to an attribute based product evaluation. This situation has been aggravated by social developments that have also led to a broader but more diffuse and less interconnected body of

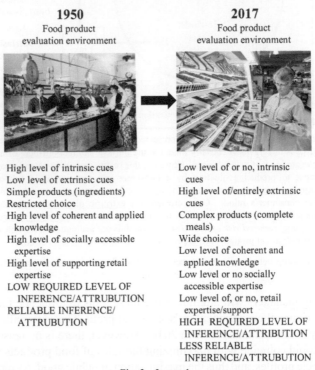

1950	**2017**
Food product evaluation environment	Food product evaluation environment

High level of intrinsic cues	Low level of or no, intrinsic cues
Low level of extrinsic cues	High level of/entirely extrinsic cues
Simple products (ingredients)	
Restricted choice	Complex products (complete meals)
High level of coherent and applied knowledge	Wide choice
High level of socially accessible expertise	Low level of coherent and applied knowledge
High level of supporting retail expertise	Low level or no socially accessible expertise
LOW REQUIRED LEVEL OF INFERENCE/ATTRUBUTION	Low level of, or no, retail expertise/support
RELIABLE INFERENCE/ ATTRUBUTION	HIGH REQUIRED LEVEL OF INFERENCE/ATTRIBUTION
	LESS RELIABLE INFERENCE/ATTRUBUTION

Fig. 2 – Image 1

Serving Counter, Shaw's Butchery, Katoomba, Notes: annotated on back—serving counter, Tom Porter, Bill Gilroy, George Shaw, Ern Howard, Eric Shaw, Jack Breen. Format: B&W photograph, Licensing: Attribution, share alike, creative commons. Repository: Blue Mountains City Library www.bmcc.nsw. gov.au/library/ Part of: Local Studies Collection PF475, Provenance: Miss M Fawcett. https://www.google.co.nz/search?as_st=y&hl=en&tbs=sur%3Afc&tbm=isch&sa=1&q=old+butcher+&o q=old+butcher+&gs_l=psy-ab.3..0l4.17084.23022.0.23614.37.19.0.0.0.0.398 .2374.2-5j3.8.0....0...1.1.64.psy-ab..29.7.2153...0i67k1.mqfTn8rcLVw#imgrc=FCGicxSqJRKG6M:

Fig. 2 – Image 2

A shopper examines a package of meat in a grocery store for freshness on Jan. 17, 2012 USDA photo by Stephen Ausmus. https://www.flickr.com/photos/usdagov/8411827143

By U.S. Food and Drug Administration—http://www.fda.gov/Food/GuidanceRegulation/ GuidanceDocumentsRegulatoryInformation/LabelingNutrition/ucm385663.htm, Public Domain, https:// commons.wikimedia.org/w/index.php?curid=31375590

Fig. 2. The development of the retail food product evaluation environment.

knowledge with regard to food products. In addition to these issues with regard to their own expertise, many consumers do not have access to either immediate or background expertise from third parties such as relatives and retail assistants while making these indirect inferences and attributions (Spaargaren et al. 2013).

All of these developments have made the indirect inferential attributional route to consumer food product evaluation simultaneously more important and less reliable. This increasing unreliability has in turn led to more opportunity for marketing communication to intervene effectively in the evaluation process, with outcomes that may not always be desirable for the public good. The opportunity has been aggressively taken up in some quarters (Lang and Heasman 2015).

The development of regulatory intervention with regard to the marketing of food attributes has reflected this degradation in consumer capability, and the consequential increase in the potential impact of marketing communications on consumer behaviour (Fig. 3). The development shown in this figure is not progressive, but is rather incremental from one level to the next, over time. As can be seen, regulation has become increasingly proactive, or aggressive, from certain perspectives, over time. Initially regulation assumed a target consumer that could make the right choice for themselves as long as they were provided with a sufficient level of accurate information.

As time has gone by the assumption of a fully sovereign consumer has gradually been displaced by the assumption that a consumer requires a level of third party evaluation and/or persuasion to come to the 'correct' decision. By definition the 'correctness' of any such decision has to be determined by the third party or agency that is doing the evaluation/persuasion. As regulation invariably involves a level of interference in the activities taking place within a market, it is likely that any definition of correctness that is used as the basis for regulation will be subject to formal challenge by one or more market participants. As a result the determination of what is a correct consumer decision by third party regulatory agencies usually involves a level of 'scientific' support that is either already publicly available, commissioned for the purpose or both. The role of empirical science in food marketing regulatory

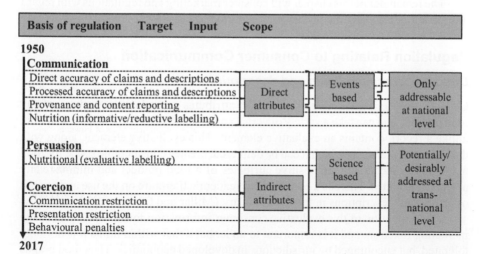

Fig. 3. The development of the scope of food marketing regulation.

policy has thus been increasing over time (European Commission 2007, Vaqué and Melchor 2016).

Recently, persuasion has been supplemented by coercion. This actual or proposed coercion can take several forms: Direct restriction on communications, taxes of one form or another, denial of publicly funded health services for individuals whose diet is deemed to be 'incorrect', and outright bans on the display of certain foodstuffs. Such regulation is based upon the assumption that the consumer does not possess any sovereignty whatsoever. They are thus incorrigible and cannot be brought to the 'correct' decision, as defined by the regulator, by means based upon either information or persuasion.

As regulation has progressed from informing the consumer to coercing them, so its objectives have shifted from the direct attributes of a food product (what it is), to the indirect attributes (what it may do to the individual who consumes it). 'Correctness' in purchasing behaviour has thus tended to focus on indirect attributes related to health outcomes that are linked to specific intrinsic attributes of the product to which either persuasive or coercive regulation can be attached. These direct intrinsic attributes are usually a specific component of the food—the most notable being saturated fat, sodium chloride and sucrose.

Regulations are now established or proposed at all levels, which creates an all-encompassing role for food regulation, best expressed by the relevant clause in European Food Law which can be fairly applied to almost any aspect of how food attributes are communicated by marketing:

> *Without prejudice to more specific provisions of food law, the labelling, advertising and presentation of food or feed, including their shape, appearance or packaging, the packaging materials used, the manner in which they are arranged and the setting in which they are displayed, and the information which is made available about them through whatever medium, shall not mislead consumers. (European Union 2002, p. 11)*

The remainder of this chapter will consider marketing and regulations with regard to food attributes using the framework presented in Fig. 3.

Regulation Relating to Consumer Communication

Direct accuracy of claims and descriptions

Claims and descriptions are extrinsic direct attributes carried on the package of any food product, and can take a variety of verbal and visual forms. A claim or description, unlike a report, contains an evaluative element. This evaluative element, along with the variety of means by which it can be delivered, creates considerable scope for food marketers to accentuate the positive attributes of a food product and minimize any negative attributes. There is a considerable academic literature on the use and misuse of claims and descriptions in the food industry (Millar and Cassady 2014).

The generation of this evaluative consumer information forms a core part of the marketing function, and it is important to understand that within limits it is not only tolerated, but encouraged by jurisdictions in developed economies. Thus food brands are some of the most valuable tradable entities within such economies. Their value is based upon two pillars: First, brands create incremental revenue over and above an equivalent unbranded food product, which can only occur if consumers are paying a premium based upon an evaluation that is not in their direct economic interests. Second, this situation can be captured, perpetuated and transferred to a new owner as the brand itself can be protected by laws developed by the jurisdiction specifically for the purpose. The encouragement of food brands and related marketing claims and descriptions is based upon the jurisdictions' position that as long as certain limits are not exceeded, the communications allow consumers to make more reliable choices. This position is supported by recent theoretical research in psychology (Ashton and Pressey 2011).

The key regulatory question is: Where do the limits of beneficial communication via claims and descriptions lie? It is of course possible to create claims and descriptions that are deceptive, and then apply them to food products. However, the evaluative nature of these claims and descriptions makes them highly context-specific, which makes statutory regulation of them exceedingly difficult, even within a single jurisdiction. The EU Unfair Commercial Practices Directive 2005 (EU 2005) illustrates the issues involved. This regulation required the individual member states of the EU to pass their own statutory versions—all of which vary to a greater or lesser extent. The U.K. version is the Consumer Protection from Unfair Trading Regulations 2008 (OPSI 2008), which reveals a regulatory framework so vague that establishing where the limits of desirable claims and descriptions lie on food products can only be achieved on a case to case basis, usually in a court of law. Nevertheless, the EU has made strong efforts to regulate claims on food packages by direct statute (EC No. 1924/2006, EC No. 1047/2012, EC No. 1048/2012) (European Union (EU) 2006, 2012a, 2012b). These regulations exercise, but do not suppress, the ingenuity of marketers in the field.

Thus case by case common law is the primary platform by which regulation of claims and descriptions affixed to food products develops within jurisdictions—a process that is slow, expensive and unpredictable. The additional complexity generated by cross-cultural impacts upon where the limits of beneficial consumer communication lie makes the development of effective trans-jurisdictional regulatory statutes relating to them a very challenging task.

Processed accuracy of claims and descriptions

The consumer processes the extrinsic direct attributes of claims and descriptions into extrinsic indirect attributes for the product. This additional processing adds a layer of complexity to the situation described in the paragraph above, which is acknowledged in the EU regulations' (European Union (EU) 2006) specification of three types of consumer (average/targeted/vulnerable) to which individual legal tests may be applied.

Even accurate descriptions of food products may be processed by consumers into inaccurate extrinsic indirect attributes—and food marketers along with academic researchers are well aware of this possibility (Stancu et al. 2017). For example, most baked beans outside of North America do not contain meat. In such an area a manufacturer may choose to put their new baked beans product into a green can, charge 50% more for them and label them 'vegetarian'. These three extrinsic direct attributes are not in and of themselves inaccurate, or deceptive, but together they may cause a consumer to conclude (incorrectly) that the other baked beans at the point of sale are not vegetarian, and then to apply this erroneous indirect extrinsic attribute to them and to then react accordingly by paying a premium for the distinctive 'vegetarian' product.

It is almost impossible to regulate such behaviours and consumer reactions to them by direct statute, and insofar as they are able to be dealt with, they are dealt with by the within-jurisdiction statutes and regulations that have been developed to deal with direct accuracy of claims and descriptions.

Provenance and content reporting

The direct accuracy of reporting of the actual constituents and origins (direct intrinsic attributes) of food products on their packaging is regulated by well-developed statutes in most developed economies and supports a significant research literature (Fortin 2016). Unlike the claims and descriptions in the preceding paragraphs, this reporting is not a normally matter of opinion or interpretation. For example, the European Food Information to Consumers Regulation No. 1169/2011 (FIC) (European Union (EU) 2014) regulates what information must be placed on packaged and other foods within the jurisdiction of the EU, with enforcement of these regulations left up to the existing food agencies of the individual EU member states. These arrangements are closely reflected in other major developed jurisdictions, with differences largely being in the detail of what is required, and how the rules are administered and enforced. Nevertheless, 'food fraud' which involves the deliberate misreporting of direct intrinsic attributes has been estimated to run at 10% in developed economies, and thus remains a major issue (Spink 2012).

The development of regulation in this area may be driven by both events and science within individual jurisdictions. Issues with the accuracy of direct intrinsic attribute reporting at the point of sale can lead to major scandals, the contamination of the food supply chain in the UK by horsemeat being an example of this (Sentandreu and Sentandreu 2014). Science is also a major driver of regulation in this area as it makes the detection of misreporting easier. While the basic nutritional analysis of any food product is a matter of fairly straightforward chemistry, establishing the type, nature, amount and origins of the individual ingredients used in a product is a far more complex process. DNA analysis is offering some promise in this area (Armani et al. 2015). Some aspects of regulation enforcement, such as the organic status of any ingredient, can still often defy after the fact of chemical analysis, and policing thus must continue to rely upon records and audit.

Provenance of food products and their ingredients has historically been a matter for documentation and audit, but recent advances in isotopic analysis has now made this issue more tractable with regards to enforcement (Camin et al. 2017). Nevertheless provenance fraud is still a major problem, one extreme case being manuka honey, with 1,700 tons being produced each year in New Zealand, but 10,000 tons plus being sold around the world—indicating a misreporting/fraud rate of 83% (Fairfax Media 2016a). At the present time the international focus on regulation in the food provenance and content area is on cooperative policing rather than harmonized regulation (Basu 2014).

Regulation relating to consumer nutrition information

The provision of nutrition information on food packages has been mandated by regulation in many developed countries for a considerable period. For most of this time the required presentation format has been in the form of a 'nutrition facts' (NF) or 'nutritional information panel' (NIP). The information varies only slightly from jurisdiction to jurisdiction and the American NF panel shown in Fig. 4 is typical of the information that is presented to the consumer. All label panel formats require a statement of nutrient content per 100 g, and/or per defined serving weight in grams. Some labels also include the amounts as percentage of a daily dietary intake (Malike et al. 2016).

Any NF/NIP panel regulation/compulsion regime relies upon two assumptions: firstly that the consumer will be motivated to locate the panel, read it, process it and use it in their evaluations, and secondly that they will understand the information that is presented, and possess the knowledge to correctly appreciate its implications. Research has cast doubt on the validity of both of these assumptions (Hamlin 2015).

One obvious drawback of the NF/NIP label has been that it is not usually carried on the prime facing of the food package that is visible to the consumer at the point of sale. This has been addressed by the development of an even more tightly reduced nutrition information format that is placed on the prime facing, usually referred to as the 'front of pack' (FoP). This Percentage Dietary/Daily Intake (PDI)/Recommended/ Reference Dietary/Daily Intake (RDI) label repeats a complete or partial summary of the information on the NIP (Fig. 4). The PDI/RDI information may be additional to the information contained within the NIP. However, the shaky assumption as to whether the consumer can effectively process the information still remains, as it is a fully reductive label that presents no evaluation of the nutritional information or its implications to the consumer (Temple and Fraser 2014).

Fig. 4. Nutritional Fact Panel and five item PDI/RDI label.

At present no jurisdiction mandates the use of PDI/RDI labels on the prime facings of packages, although there are calls for it in some jurisdictions that favour this label format (Emrich et al. 2017). Regulations that are in force as to what information is presented via these PDI/RDI labels, or how it is presented (format/content/size) remain fragmentary (e.g., Food and Drink Federation 2012 vs Australian Food and Grocery Council 2016). Thus a bewildering array of complete or partial PDI formats may be encountered by the consumer in a variety of informative, uninformative or misinformative colour schemes. It is not uncommon to see PDI labels on the prime face that consist of one nutritional fact of the seven or eight that make up the entire NIP suite of information, or that use a colour (e.g., green) as an element of them that may be misinterpreted by the consumer as an evaluative traffic light cue.

It would be reasonable to observe that a necessary precursor to any regulations that requires mandatory use of this label format would be an enforced consistency of its content and presentation, in which trans-national cooperation and regulation may have a role to play. A consistent name and acronym for it, in place of the dozens that currently exist, would be a logical and helpful start point.

Regulation Relating to Consumer Persuasion

The previous section dealt with regulations relating to consumer information. This section will deal with regulations that relate to consumer persuasion. As the majority of purchases are made without picking up any of the food products on offer all of this activity occurs on the front of the retail food package, and is thus a front of pack (FoP) affair (Hamlin 2010). While FoP labels are often lumped together there is a massive paradigm shift between FoP labels that purely inform (reductive), and those that seek to persuade (evaluative). Persuasion requires that some third party evaluation is presented to the consumer, which indicates a presumption on the part of those who manage and regulate these labels that the consumer is simply not able to act as a sovereign entity while evaluating products as part of a purchase decision.

The addition of evaluative content to an FoP nutritional label introduces a number of significant issues as the regulator advances onto ground that has historically been occupied by the commercial food marketer. Evaluative FoP labels are probably the most actively researched area of food marketing regulation at the present time, and support a massive literature (Hamlin 2015). While few are as yet mandatory, there are calls for them to become so on a wider basis (European Food Information Council 2016). The area is characterized by a marked lack of consensus as to how the food consumer might best be persuaded by these labels, and a variety of formats are in a late stage of development or are now fully deployed in individual jurisdictions (Fig. 5). This shows the fully reductive label types discussed in the previous section at the extreme left hand end of the reductive/evaluative continuum.

At the other end of the reductive/evaluative continuum lie the fully evaluative FoP nutrition label formats. These are the closest to brands in the way that they are designed and communicate to consumers. They are also quite rare. Two major formats are currently deployed—the Dutch 'Tick' programme, and the Scandinavian Keyhole. The Tick is privately owned, and has run into some issues with how it is perceived

Entirely Factual (Reductive)	A combination of fact and opinion (Hybrid)	Entirely Recommendation (Evaluative)

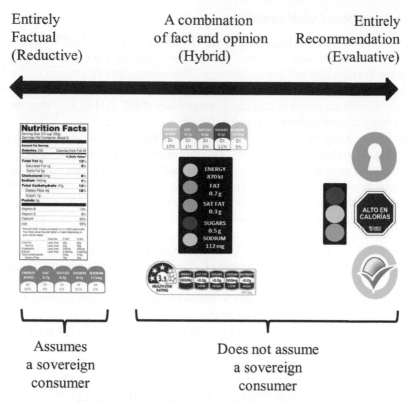

Assumes a sovereign consumer	Does not assume a sovereign consumer

Fig. 5. A typology of Front of Pack (FoP) nutritional labels.

and used by consumers, largely as a result of its privately held status and the two confusingly similar versions of it that have been deployed (Michail 2016).

The publicly controlled Scandinavian Keyhole has been deployed for nearly thirty years and is the closest of any FoP label to a standard food brand in its function. The Keyhole contains no factual information and is a purely evaluative brand that communicates nutritional status by the presence or otherwise of a green keyhole symbol on a product (Nordic Council of Ministers 2010). Uniquely, it is also calibrated at a category level; the healthiest products in each category can qualify for the keyhole mark, whether that category is full fat cheese or pre-prepared salads. This calibration policy corresponds with the commercial food industry's adoption of the category as the primary strategic unit of marketing (Chimhundu and Hamlin 2007). Neither of these label formats is compulsory, and there appears little likelihood of their becoming so in the foreseeable future.

The recently introduced Chilean food labelling system is also a fully evaluative system that uses a binary system of communication. The black labels are negative rather than positive reinforcers. It is also compulsory (USDA 2015). In this system octagonal black labels are affixed to products that have more than a specific amount of energy, sugar, salt and fat. As there is one label for each of these nutritional inputs, it is possible to have a product that has several of these substantial black labels attached to its prime facing. Its effectiveness has not yet been established.

Hybrid nutritional label systems

Research on FoP labels is beginning to focus very heavily upon 'traffic light' based systems that communicate a product's nutritional status by colour: green for good, amber for average and red for undesirable. This system has its origins in the UK, which is also the jurisdiction within which its deployment is most developed, with moves now well advanced to making it compulsory (BBC 2016). It is thus the FoP system that is closest to being a regulated entity.

In its pure form the traffic light system is an entirely evaluative label with three states (Fig. 5, right), but it is hardly ever encountered in this form. Instead it is encountered as an evaluative/reductive hybrid 'battery' of traffic light type indicators for various nutritional information items drawn from the NIP, or it occurs as a 'coloured in' version of the PDI/RDI label formats discussed in the previous section (Fig. 5, centre). In both cases the numerical data from the PDI/RDI format is incorporated into the hybrid label. One major issue with the traffic light system in this hybrid format is its potential for complexity. The 'pure' traffic light system shown on the right of Fig. 5 has only three states. The number of states that a hybrid label can express is 3 raised to the power of the number of traffic lights in the battery. So a six traffic light battery (not an uncommon FoP label to encounter in the UK) can express $3^6 = 729$ individual nutritional states to the consumer.

Figure 6 illustrates this situation for a simpler four traffic light battery with only 81 states. The situation is not improved by the fact that nearly all traffic light systems are calibrated across the entire food product range found in the supermarket, rather than within individual categories as is the case with the Scandinavian Keyhole. Thus a consumer selecting product within a category may only encounter a relatively narrow

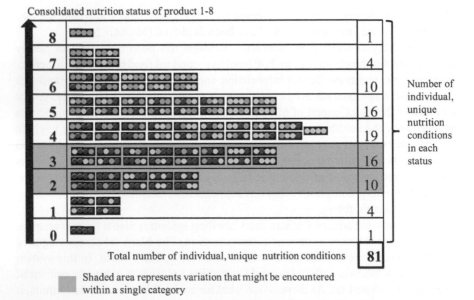

Consolidated nutrition status of product 1-8

Number of individual, unique nutrition conditions in each status

Total number of individual, unique nutrition conditions   **81**

Shaded area represents variation that might be encountered within a single category

Fig. 6. An expansion of all the consolidated nutrition statuses and unique individual nutrition states that may be expressed by a four-battery, hybrid traffic light FoP label.

range of states (indicated by the shaded area in Fig. 6). This is likely to compound their difficulties in interpreting the label. Perhaps for this reason, research that directly measures the impact of these systems on consumer food product choice are rare in the literature, relative to those that seek to develop/promote them. Research results which clearly indicate that they have a significant impact upon consumer choice in retail situations are rarer yet (Hamlin 2015). The recently introduced Health Star Rating (HSR) system in Australasia (Fig. 5 centre) is a hybrid system that significantly reduces this level of complexity using a star system that has only five states (1–5 stars). However, research indicates that this system too may have limitations in its effectiveness (Hamlin and McNeill 2016).

The HSR has also run into trouble with its calibration in terms of how the star score rating for any particular product is arrived at, with criticism aimed at the fact that products that are very high in refined sugar (c. 30%) can still score four and a half out of five health stars (Edmunds 2017). This merely illustrates the issues are inherent within any system that seeks to evaluate food products in an environment where both scientific and public opinion as to what is desirable and what is not are subject to steady evolution and occasional abrupt change (e.g., Broughton 2017).

The hybrid FoP label is thus an area that is probably experiencing the highest level of activity in creating calibrated instruments that might at some future point form a basis for regulation of food marketing communication. At the present time it can only be described as chaotic. Figure 5 does not begin to illustrate the enormous variety of reductive, hybrid and evaluative FoP nutrition label formats that are deployed in the food marketplace. Even the HSR system, which is the most regulated system at present, allows an extraordinarily wide range of variation within it (Department of Health (Aus.) 2016). Most of these systems are backed up by little or no testing that might indicate whether they are effective in influencing consumer choice. Nevertheless moves towards regulation on the basis of them are clearly discernable (e.g., Hagen 2015).

Deploying an ineffective system by regulation may come at considerable economic and social cost. The economic costs to the general population of deploying an ineffective system will be significant, and may impact disproportionately upon smaller participants in the food industry. To this may be added a significant social opportunity cost if regulation based upon an ineffective system eliminates or blocks a system that is more effective. The retirement of the established and successful Heart Foundation Tick FoP label in New Zealand as an outcome of the deployment of the untested HSR may be an example of exactly this type of undesirable development (Fairfax Media 2016b). The larger players in the food industry may welcome the regulated imposition of an ineffective FoP system for these reasons. The fact that high-sugar children's cereals may be encountered with various versions of these favoured FoP labels voluntarily and prominently placed on their prime facings may be indicative of the outcomes of the industry's own proprietary research on the effectiveness of these label formats.

Regulation Relating to Industry and Consumer Coercion

The previous section examined regulation with regard to consumer persuasion; this section looks at regulation that seeks to coerce either industry or consumers. While the effectiveness of regulation to inform or persuade consumers may be questioned, the tobacco experience leaves little doubt that coercion is effective (e.g., Shang et al. 2016). The primary obstacle to coercive regulation of food markets is political rather than technical. Food is a very sensitive economic and social area, and even totalitarian dictatorships approach the regulation of the supply, purchase and consumption of food with considerable caution if the population might consider its effects to be undesirable (Bellemare 2015). Needless to say the leaders of developed economies that answer to their populations via the ballot box view the matter with even more trepidation (Caraher and Cowburn 2015). Coercive regulation of food purchase and consumption is thus advancing at a slow pace. Similar regulation of the food industry in the area of marketing communication, presentation and children's food is slightly more advanced.

Communication restriction

Regulations restricting food marketing communications away from the point of sale are now in force in several jurisdictions, and are being considered in many more. The target in all cases is food products with high sugar, fat and sodium content targeted at children, collectively referred to as 'junk foods' (Boyland and Whalen 2015). The restrictions take a variety of forms, including bans on the use of cartoon characters (Ogle et al. 2017), restrictions on advertising in programming targeted specifically at children in both traditional and virtual media (Clarke 2017), and a ban on advertising for junk foods prior to a certain 'curfew' time in the evening. Initiatives of this type date back to Quebec's introduction of restrictions in 1980 (Dhar and Bayliss 2011), and there is a considerable body of literature devoted to the subject, in the areas of historical reporting, consumer research and consumer behaviour mapping.

There is also a rather smaller, but possibly highly relevant stream of literature on the food industry's reactions to such initiatives in the form of food reformulations to avoid the restrictions. The capacity for this reaction to reduce intakes without requiring action from the consumer has attracted significant interest (e.g., Magnusson and Reeve 2015).

Over the last two years the deployment of regulation in the area of remote advertising of junk foods to children appears to have acquired a strong momentum, and it seems likely that these restrictions will become the norm across the developed world. At present, there appear to be no active plans to extend these restrictions on junk food advertising to the general population, although many of these restrictions aimed at children such as curfew and sponsorship also significantly impact upon adults' exposure. This lack of activity appears to be due to a lack of political will, rather than a shortage of pressure from various groups that wish to impose general restrictions on the advertising of junk foods in addition to a variety of other foods (e.g., meat) that they disagree with (Carrington 2016).

Presentation restriction

In many ways regulation of food marketing has followed that of tobacco. In recent years the focus of tobacco marketing regulation has switched to point of sale, firstly by eliminating the display of tobacco packages at the point of sale (Robertson et al. 2014), and then by eliminating distinctive brand images on the packages (Waa et al.

2017). Research reports suggests that these moves have had a significant impact upon tobacco sales, and there are indications that a similar path in may be pursued food marketing regulation (Fairfax Media 2017).

Behavioural penalties

The only direct penalty law currently in force is the Japanese 'Metabo' law that requires employers to measure the waistlines of men and women between 40 and 75. The personal penalties for those who do not 'measure up' are mild—including counselling. However, employers who have more than the prescribed percentage of overweight employees can face significant fines (Curry 2008).

Penalising consumers by taxing them has a rather wider basis of experience, with a number of regulatory regimes in place in the developed economies (Khan et al. 2015). High sugar soda is by far the most common target for such taxes, with considerable experience built up in the United States where local authorities are empowered to impose these taxes leading to multiple examples. While there has been reasonable acceptance of soda taxes where they are imposed, it appears that the acceptance of a tax is highly dependent upon what the proceeds of such taxes are used for. If any indication is given that revenue generation is a motive for levying the tax, then consumer resistance to such taxes increases markedly, and in the most recent and ambitious case to a politically dangerous level (CNBC 2017).

There have been few attempts to levy taxes on foods other than soda, largely as an outcome the technical difficulties in doing so, and a strong and generalized consumer resistance to tax on food due to its potentially regressive nature (hitting poorer people harder) (Khan et al. 2015). The most comprehensive attempt was made by Denmark starting in 2011, when a tax was introduced on all food products containing above a certain level of fats, sugar and sodium. The tax collapsed a year later amid a welter of consumer resistance and consumer trips to neighbouring countries where tax was not levied (Bødker et al. 2015). There have been no major initiatives in this area since.

Conclusions

This chapter has given a far broader overview of marketing and regulation than is usually the case. The regulation of food marketing cannot be usefully considered without an understanding of the consumer decision processes that marketing is trying to influence. Over the last fifty years retail food consumers have become increasingly reliant upon extrinsic cues to guide their choice. Over the same time period, the 'hard' knowledge and social support resources and access to intrinsic cues that used to inform these choices has declined. The result is a consumer decision process that is far more pliable in its response to marketing messages delivered via extrinsic cues. These marketing messages may not be inconsistent with good dietary choices.

Over the same time period, governments around the world have responded to this trend by progressively introducing proactive regulations that relate to increasingly aggressive and sophisticated commercial marketing activity. These regulations initially sought to inform, then to persuade, and now finally to coerce consumers into good dietary habits. At the present time, the majority of research and regulatory activity is centered upon consumer persuasion, with a wide variety of reductive, evaluative or hybrid front of pack nutrition labels at various stages of development and deployment.

When looking at the regulation of food marketing within economies where consumer food purchases are made on a basis of discretion rather than urgent necessity, the following broad observations can be made with regard to the current situation:

1) Regulations that seek to inform the consumer of the nutritional status of a product in some way or another via extrinsic cues carried somewhere on the package are almost universally in place. While there is no doubt that the consumers in most developed economies now have access to the information that they require in order to make an informed choice, there is no consistent evidence that they actually use it.

2) Regulations that aim to persuade the consumer to buy healthy food products via extrinsic nutritional cues that are carried on the front of the package (FoP) are in a chaotic state. Multiple systems exist, many of which do not conform to known systems of consumer behaviour or to established and proven market communication techniques, and none of which have been adequately tested to establish if they significantly affect consumer choice. The two systems that most closely conform to consumer behaviour and industry practice are the Scandinavian Keyhole and the Chilean black label system. Neither of these two systems are 'mainstream'. The dominant format is the hybrid 'Traffic Light System' which is critically short of empirical support with regard to its effectiveness in influencing consumer choice. Nevertheless, some of these systems are moving towards regulated/mandatory status on the basis of opinion and inertia. The outcome of this may be unfortunate. The introduction of a mandatory but ineffective FoP nutritional labelling system would incur significant economic and social opportunity costs.

3) A variety of coercive regulations relating to food product and market communications have been introduced in recent years; and the pace and scope of their introduction appears to be accelerating rapidly. While coercive regulation is undoubtedly effective; it should be remembered that coercion requires the consent of the coerced within any democracy. Food is a particularly sensitive social and political area, and care needs to be taken with regard to maintaining this democratic consent. There are signs in some areas that the limits of public tolerance may have been reached. Overreach in coercive food regulation risks not only the regulation in question, but maybe others too if the consequent consumer backlash against the measure and those who are perceived to have perpetrated it is significant.

Governments of developed economies could contribute positively to this situation by specifically and urgently promoting the systematic consumer testing of these systems in order to establish which (if any) work and which do not. If a superior system can be identified, then considerable benefit might be gained from cooperative government action to achieve a global standardisation on that label type. In addition governments should carefully consider imposing domestic coercive regulation in this area if such coercion runs appears to be running significantly ahead of majority consumer sentiment, as a significant anti-regulatory kick back may be the outcome.

References

Armani, A., L. Guardone, R. La Castellana, D. Gianfaldoni, A. Guidi and L. Castigliego. 2015. DNA barcoding reveals commercial and health issues in ethnic seafood sold on the Italian market. Food Control. 55: 206–214.

Ashton, J.K. and A.D. Pressey. 2011. The regulatory challenge to branding: An interpretation of UK competition authority investigations 1950–2007. Journal of Marketing Management 27(9-10): 1027–1058.

Australian Food and Grocery Council. 2016. Best Practice Guide 2016: Daily Intake Guide Style Guide. <https://www.afgc.org.au/wp-content/uploads/AFGC-Best-Practice-Guide-DIG-Style-Guide-June-2016-1.pdf> (accessed on 26 August 2017).

Basu, G. 2014. Combating illicit trade and transnational smuggling: key challenges for customs and border control agencies. World Customs J. 8: 15–26.

Bellemare, M.F. 2015. Rising food prices, food price volatility, and social unrest. Am. J. Agr. Econ. 97: 1–21.

Bødker, M., C. Pisinger, U. Toft and T. Jørgensen. 2015. The rise and fall of the world's first fat tax. Health Policy 119: 737–742.

Boyland, E.J. and R. Whalen. 2015. Food advertising to children and its effects on diet: review of recent prevalence and impact data. Pediatric Diabetes 16: 331–337.

British Broadcasting Corporation (BBC). 2016. Food 'traffic light' labelling should be mandatory, councils say. <http://www.bbc.com/news/health-37389804> (accessed on 24 August 2017).

Broughton, K. 2017. Consumer trust in food Health Star Ratings dropping, New Zealand Herald June 28 2017. <https://www.stuff.co.nz/national/health/94125290/consumer-trust-in-food-health-star-ratings-dropping> (accessed on 31 August 2017).

Camin, F., M. Boner, L. Bontempo, C. Fauhl-Hassek, S. Kelly, J. Riedl and A. Rossmann. 2017. Stable isotope techniques for verifying the declared geographical origin of food in legal cases. Trends Food Sci. Technol. 61: 176–187.

Canavan, O., M. Henchion and S. O'Reilly. 2007. The use of the internet as a marketing channel for Irish speciality food. Int. J. Retail & Distrib. Manage. 35: 178–195.

Caraher, M. and G. Cowburn. 2015. Guest commentary: Fat and other taxes, lessons for the implementation of preventive policies. Preventive Med. 77: 204–206.

Carrington, D. 2016. Tax meat and dairy to cut emissions and save lives, study urges, The Guardian, London. UK. <https://www.theguardian.com/environment/2016/nov/07/tax-meat-and-dairy-to-cut-emissions-and-save-lives-study-urges> (accessed on 1 September 2017).

Chimhundu, R. and R.P. Hamlin. 2007. Future of the brand management structure in FMCG. J. Brand Manage. 14: 232–239.

Clarke, J. 2017. Ban on junk food aimed at children extended to online and social media. The Independent, London. UK, 8 December 2016. <http://www.independent.co.uk/life-style/ health-and-families/ health-news/junk-food-ads-children-advertising-banned-ban-online-social-media-ofcom-a7462521. html> (accessed on 29 August 2017).

CNBC (USA). 2017. Chicago-area soda tax may carry political price for backers, 26 August 2017. <https:// www.cnbc.com/2017/08/26/chicago-soda-tax-causes-political-problems.html> (accessed on 31 August 2017).

Curry, J. 2008. Japanese firms face penalties for overweight staff, The Guardian, London. UK. <https:// www.theguardian.com/world/2008/mar/19/japan> (accessed on 28 August 2017).

Department of Health (Aus.). 2016. Health Star Rating System Style Guide. <http://healthstarrating.gov. au/internet/healthstarrating/publishing.nsf/Content/651EEFA223A6A659CA257DA500196046/$Fi le/HSR%20Style%20Guide%20v4.pdf> (Accessed on 31 August 2017).

Dhar, T. and K. Baylis. 2011. Fast-food consumption and the ban on advertising targeting children: the Quebec experience. J. Mark. Res. 48: 799–813.

Edmunds, S. 2017. Health Star rating system 'may mislead shoppers', New Zealand Herald, Auckland. NZ, 27 April 2017. <http://www.stuff.co.nz/business/91971947/health-star-rating-system-may-mislead-shoppers> (accessed on 25 August 2017).

Emrich, T.E., Y. Qi, W.Y. Lou and M.R. L'Abbe. 2017. Traffic-light labels could reduce population intakes of calories, total fat, saturated fat, and sodium. PloS One 12: e0171188.

Enneking, U., C. Neumann and S. Henneberg. 2007. How important intrinsic and extrinsic product attributes affect purchase decision. Food Qual. Preference 18: 133–138.

Espejel, J., C. Fandos and C. Flavián. 2007. The role of intrinsic and extrinsic quality attributes on consumer behaviour for traditional food products. Managing Serv. Qual.: Int. J. 17(6): 681–701.

European Commission, Directorate General for Research. 2007. Food consumer science: Lessons learnt from FP projects in the field of food and consumer science. Directorate General for Research. <http:// cordis.europa.eu/pub/food/docs/booklet-consummer.pdf> (accessed on 28 August 2017).

European Food Information Council. 2016. Global Update on Nutrition Labelling, Brussels, Executive Summary. <http://www.eufic.org/images/uploads/ files/Executive Summary.pdf> (accessed on 24 August 2017).

European Union (EU). 2002. Regulation (EC) No. 178/2002 of the European Parliament and of the Council of 28 January 2002 laying down the general principles and requirements of food law, establishing the European Food Safety Authority and laying down procedures in matters of food safety. <http:// eur-lex.europa.eu/legal-content/EN/TXT/PDF/?uri=CELEX: 32002R0178&from=EN> (Accessed on 1 September 2017).

European Union. 2005 Directive 2005/29/EC of the European Parliament and of the Council. <http:// eur-lex.europa.eu/LexUriServ/LexUriServ.do?uri=OJ:L:2005:149:0022:0039:EN: PDF> (accessed on 26 August 2017).

European Union. 2006. Regulation No. 1924/2006 of the European parliament and of the Council of 20 December 2006 on nutrition and health claims made on foods. Available at: http://eur-lex.europa. eu/legal-content/EN/TXT/PDF/?uri=CELEX:02006R1924-20121129 &from=EN. Accessed 25 August 2017.

European Union. 2012a. Commission Regulation (EU) No. 1047/2012 of 8 November 2012 amending Regulation (EC) No. 1924/2006 with regard to the list of nutrition claims. <http://eur lex.europa.eu/ LexUriServ/LexUriServ.do?uri=OJ:L:2012:310:0036:0037:EN:PDF> (accessed on 28August 2017).

European Union. 2012b. Commission Regulation (EU) No. 1048/2012 of 8 November 2012 on the authorisation of a health claim made on foods and referring to the reduction of disease risk. <http:// eur-lex.europa.eu/LexUriServ/LexUriServ.do?uri=OJ:L:2012:310:038:0040:EN: PDF> (accessed on 28 August 2017).

Fairfax Media. 2016a. Riddle of how 1,700 tons of manuka honey are made... but 10,000 are sold. New Zealand Herald, Auckland. NZ. 23 August 2016, p21. <http://www.nzherald.co.nz/business/news/ article.cfm?c_id=3&objectid=11699412> (accessed on 29 August 2017).

Fairfax Media. 2016b. Why is the Heart Foundation ditching the Tick programme? New Zealand Herald, Auckland. NZ. 18 October 2016. <http://www.nzherald.co.nz/lifestyle/news/ article.cfm?c_ id=6&objectid=11731045> (accessed on 30 August 2017).

Fairfax Media. 2017. Half of food marketed at kids is junk, obesity researchers find. New Zealand Herald, Auckland. NZ. 27 June 2017. <http://www.nzherald.co.nz/nz/news/article.cfm?c_id =1&objectid=11882456> (accessed on 27 August 2017).

Food and Drink Federation (FDF) (UK). 2012. GDA Labelling Industry Style Guide. <https://www.fdf. org.uk/publicgeneral/GDAStyleGuide.pdf> (accessed on 4 September 2017).

Fortin, N.D. 2016. Food Regulation: Law, Science, Policy, and Practice. John Wiley and Sons, Hoboken. NJ.

Hagen, K. 2015. Health star rating system for packaged food should be compulsory: poll. Sydney Morning Herald, Sydney Australia. 20 January 2015. <http://www.smh.com.au/national/ health/health-star-rating-system-for-packaged-food-should-be-compulsory-poll-20150119-12thnn.html> (Accessed on 29 August 2017).

Hamlin, R.P. 2010. Cue-based decision making. A new framework for understanding the uninvolved food consumer. Appetite 55: 89–98.

Hamlin, R.P. 2015. Front of pack nutrition labelling, nutrition, quality and consumer choices. Current Nutr. Rep. 4: 323–329.

Hamlin, R.P. and L. McNeill. 2016. Does the Australasian "health star rating" front of pack nutritional label system work? Nutrients 8(6): 327.

Josling, T., D. Roberts and D. Orden. 2004 (August). Food regulation and trade: toward a safe and open global system-an overview and synopsis. In: American Agricultural Economics Association Annual Meeting.

Khan, R., K. Misra and V. Singh. 2015. Will a fat tax work? Market. Sci. 35: 10–26.

Lang, T. and M. Heasman. 2015. Food wars: The global battle for mouths, minds and markets. Routledge, Abingdon. UK.

Magnusson, R. and B. Reeve. 2015. Food reformulation, responsive regulation, and "Regulatory scaffolding": strengthening performance of salt reduction programs in Australia and the United Kingdom. Nutrients 7: 5281–5308.

Malike, V.S., W.C. Willett and F.B. Hu. 2016. The revised nutrition facts label: a step forward and more room for improvement. Jama 316: 583–584.

Marsh, K. and B. Bugusu. 2007. Food packaging—roles, materials, and environmental issues. J. Food Sci. 72: 39–55.

McDonald, M. and H. Wilson. 2016. Marketing Plans: How to Prepare Them, How to Profit from Them. John Wiley and Sons, Hoboken. NJ.

Michail, N. 2016. Dutch Industry backed label ticked by mounting criticism, Food Navigator.com, 26 October 2016. <http://www.foodnavigator.com/Policy/Dutch-industry-backed-nutrition-label-ticked-off-with-mounting-criticism> (accessed on 4 September 2017).

Millar, L.M.S. and D.L. Cassady. 2015. The effects of nutrition knowledge on food label use. A review of the literature. Appetite 92: 207–216.

Nordic Council of Ministers. 2010. The Keyhole: Healthy Choices Made Easy. <http://norden.diva-portal.org/smash/get/diva2:700822/FULLTEXT01.pdf> (accessed on 1 September 2017).

Office of Public Sector Information (OPSI) UK. 2008. Consumer Protection from Unfair Trading Regulations. <http://www.legislation.gov.uk/uksi/2008/1277/pdfs/uksi_20081277_en.pdf> (accessed on 28 August 2017).

Ogle, A.D., D.J. Graham, R.G. Lucas-Thompson and C.A. Roberto. 2017. Influence of cartoon media characters on children's attention to and preference for food and beverage products. J. Acad. Nutr. Diet. 117: 265–270.

Phillips, C., J. Belsey and J. Shindler. 2000. Flora pro. activ: a clinical and financial impact analysis. J. Drug Assess. 3: 179–194.

Robertson, L., R. McGee, L. Marsh and J. Hoek. 2014. A systematic review on the impact of point-of-sale tobacco promotion on smoking. Nicotine & Tobacco Res. 17: 2–17.

Sentandreu, M.Á. and E. Sentandreu. 2014. Authenticity of meat products: Tools against fraud. Food Res. Int. 60: 19–29.

Shang, C., J. Huang, K.W. Cheng, Q. Li and F.J. Chaloupka. 2016. Global evidence on the association between POS advertising bans and youth smoking participation. Int. J. Environ. Res. Pub. Health. 13: 306.

Spaargaren, G., P. Oosterveer and A. Loeber (eds.). 2013. Food practices in transition: changing food consumption, retail and production in the age of reflexive modernity. Routledge, Abingdon, UK.

Spink, J. 2012. Defining food fraud and the chemistry of crime. pp. 195–216. In: Ellefson, W., L. Zak and D. Sullivan (eds.). Improving Import Food Safety. John Wiley and Sons, Ames, Iowa.

Stancu, V., K.G. Grunert and L. Lähteenmäki. 2017. Consumer inferences from different versions of a beta-glucans health claim. Food Qual. Preference 60: 81–95.

Temple, N.J. and J. Fraser. 2014. Food labels: a critical assessment. Nutrition 30: 257–260.

USDA Foreign Agricultural Service. 2015. Chile's new nutritional labelling law. GAIN Report CI1513. <https://gain.fas.usda.gov/Recent%20GAIN%20Publications/Chile's%20New%20Nutritional%20Labeling%20Law_Santiago_Chile_6-26-2015.pdf> (accessed on 25 August 2017).

Vaqué, R.G. and S.R. Melchor. 2016. A yankee in King Arthur's Court: a lawyer's perspective on EFSA. In: Alemanno, A. and S. Gabbi (eds.). 2016. Foundations of EU food law and policy: Ten Years of the European Food Safety Authority. Routledge, Abingdon. UK.

Waa, A.M., J. Hoek, R. Edwards and J. Maclaurin. 2017. Analysis of the logic and framing of a tobacco industry campaign opposing standardised packaging legislation in New Zealand. Tobacco Control 26: 629–633.

Wells, L.E., H. Farley and G.A. Armstrong. 2007. The importance of packaging design for own-label food brands. Int. J. Retail & Distrib. Manage. 35: 677–690.

17

Consumer Views of Health-Related Food Labelling

From Front-of-Pack Logos to Warning Labels

Gun Roos

Introduction

Concerns related to food and nutrition have from the last century shifted from food shortage to abundance and personal health. Overweight, obesity and diet-related diseases are today considered to be key issues in public health worldwide, and these increasing health problems have often been associated with easy access and increased consumption of cheap, unhealthy, energy-dense and processed food products. Global health and nutrition policy responses have likewise focused on individual responsibility, empowerment and consumer information (WHO 2004). It seems like people today are expected to be responsible for their own health, and new policy measures are mainly designed to increase knowledge and make it easier for consumers to put together healthy and nutritious diets. Consumers often describe that health is an important motivator for food choice. However, health and nutrition are food credence attributes that are difficult for consumers to evaluate when choosing and purchasing food. Consequently, food labelling, including front-of-pack nutrition labelling, has been put forward as an important strategy and source of nutritional information for consumers who wish to improve their health. Consumers today buy most of their food

Consumption Research Norway - SIFO, Oslo and Akershus University College of Applied Sciences, P.O. Box 4 St. Olavs plass, N-0130 Oslo Norway.
E-mail: Gun.Roos@sifo.hioa.no

in supermarkets where the majority of foods and beverages are pre-packaged and the packages have various product and health-related information and labelling, including a variety of logos, symbols and images. For food manufacturers the use of logos, text and images on packages are also a source of marketing and food packages are designed to attract attention and influence consumers' purchasing decisions.

Health-related food labelling consists of both mandatory and voluntary information that may be relevant for health, and has been defined to incorporate lists of ingredients, nutrient declarations, supplementary nutrition information, and nutrition and health claims (Rayner et al. 2013). Health-related food labelling is shaped by both national and international guidelines and regulations. International standards, such as the Codex Alimentarius guidelines, give recommendations for food labelling, and national food labelling legislation usually mandates what information has to be included on the food packages (Perez and Edge 2014). For example, current mandatory labelling in Europe (EC Regulation No. 1169/2011) covers food safety related issues (e.g., it is mandatory to highlight allergens in the ingredients list), nutritional composition (energy, fat, saturated fat, carbohydrates, sugars, protein and salt), ingredients and identification. However, in addition to the mandatory information, contemporary food labelling also includes a range of voluntary nutrition logos, rating systems and other types of information used for marketing and product differentiation. In Europe, front-of-pack is voluntary information (EC Regulation No. 1169/2011) and only authorized nutrition and health-related claims on food packaging are allowed (EC Regulation 1924/2006). A recent study of the prevalence of nutrition and health claims across Europe found that one in four of the pre-packaged foods included in the study carried at least one nutrition and health claim (Hieke et al. 2016).

Food labelling was first regulated to protect consumers against fraud and health hazards. For example, in Norway early in the twentieth century food falsification was one of the food-related health concerns and food control legislation was passed in 1933 (Elvebakken 2001). The focus in the early legislation was on regulating how ingredients were declared on the food packages. Today food labelling, in addition to safety assurance, aims at also providing health advice. The rising rates of obesity and nutrition-related diseases are often referred to as a main cause behind the new awareness of food labelling. Current food labelling aims at protecting consumers' rights and providing consumers with information on nutritional quality so that they can make informed choices (Roos 2009). Food labelling regulations have longer traditions in Europe, North America and Australia, but now also countries in Asia, Africa, the Middle East and Latin America have adopted or have started discussing mandatory or voluntary nutrition labelling (Mandle et al. 2015, Perez and Edge 2014).

Health is increasingly viewed as a personal responsibility, and current nutrition policies focus on ways to promote healthy diets and helping consumers to make informed choices by giving consumers information, including health-related food labelling, as a strategy to improve their personal health (Kjærnes and Roos 2012). This chapter addresses health-related food labelling from a consumption perspective by focusing attention on two types of contemporary health-related food labelling: front-of-pack logos and food warnings. Front-of-pack logos are today debated and implemented worldwide (e.g., Julia and Hercberg 2016, Mandle et al. 2015, Nestle and Ludwig 2010, Perez and Edge 2014), and warning labels on food products are also discussed as potential regulatory measures to protect the population against food hazards and unhealthy foods (Boncinelli et al. 2017, Roos et al. 2010).

The chapter is organized into three main sections. The first section offers an overview of some of the research and literature on health-related food labelling and consumers, with a special focus on front-of-pack health logos. The second section presents mainly results from a Norwegian study of consumer views of food warnings, and the third section summarizes the main arguments and offers some policy implications and pointers for future research.

Health-Related Labelling and Consumers

Health-related labelling, and especially front-of-pack labelling, has increasingly been put forward as a policy tool for promoting healthy eating. Consequently, there is now also a large collection of studies on what labelling format consumers prefer, consumer understanding and potential impact of labelling, and a number of reviews have been conducted (e.g., Campos et al. 2011, Cecchini and Warin 2016, Cowburn and Stockley 2014, Grunert and Wills 2007, Hieke and Taylor 2012, Mandle et al. 2015, van Kleef and Dagevos 2015, Vyth et al. 2012). Some of the reviews have focused on consumer's understanding and use of food labelling schemes (Campos et al. 2011, Cowburn and Stockley 2014), impact on food choices and consumption (Campos et al. 2011, Cecchini and Warin 2016), and consumer perspective on front-of-pack labelling (van Kleef and Dagevos 2015).

Research has suggested that consumers are generally very positive to getting more information including food labelling (Grunert and Wills 2007). However, it seems that consumers when buying food mainly look at product information and less on nutrition and health-related information, such as fat or sugar content. Surveys conducted both in Europe and Australia suggest that consumers mainly check price, expiry dates and brand names when buying food (Hall and Osses 2013, Roos 2009, Roos et al. 2010). The use of health-related labels varies depending on the type of food product. Some studies have shown that consumer do not always read labels on familiar everyday products that they usually buy based on habit (Guthrie et al. 1995), and that the product category makes a difference (Malam et al. 2009). When people buy unhealthy "junk food" they are aware that these foods are not very healthy, and thus they do not feel a need to check the label. It has furthermore been suggested that many other factors, including routines, price, time and motivation, may influence reading of labels (Paterson et al. 2001). Earlier research has also dealt with identifying the consumers who use food labels (Campos et al. 2011, Cowburn and Stockley 2005, Garrett 2007, Grunert and Wills 2007). The use of nutrition labelling is more likely among middle-aged female consumers who have a high socio-economic status, high educational level, interest in nutrition, positive attitude to diet and knowledge of the link between diet and disease. However, research on the use of labels has not often included all types of consumers and, for example, poor population groups have rarely been studied (Mandle et al. 2015). It has been proposed that one of the advantages of front-of-pack labelling is that the use of a symbol increases comprehension of nutrition labels across sociodemographic and ethnic groups (Sinclair et al. 2013). Nevertheless, it is necessary to also study if they use food labels, understand what symbols mean and distinguish symbols from other information on the food packages. Additionally, although we have studies in which consumers report that they comprehend and use nutrition labels, we know little about what consumers do in actuality. A review of consumer understanding and use of nutrition labelling suggested that even though the reported use of nutrition labels is high, actual use at the point of food purchase may be much lower (Cowburn and Stockley 2005). Most consumers who report that they use nutrition labels seem to use nutrition labelling to avoid some nutrients (for example, fat or sugar) and calories, and the authors of the review concluded that labelling may have a limited but relevant role in promoting healthy diets. However, some studies suggest that consumers do not use labels actively when they are shopping. A qualitative study in the U.K. and Australia showed that consumers rarely used logos when shopping, and that shopping seemed to be based on routine (Rayner et al. 2001).

Recent literature has recognized three main front-of-pack labelling schemes: reference intakes (earlier guideline daily amounts), traffic light labelling and health logos (Perez and Edge 2014). This chapter focuses on front-of-pack health logos. In addition to assisting consumers in making healthier food choices, front-of-pack labelling schemes aim at motivating food producers to reformulate foods and produce healthier products. Thus, front-of-pack labelling may also have positive effects for consumer health without consumers having to read, use or comprehend the label. Logos for health and environment started appearing on packages in the 1980s. Sweden was one of the first countries that in 1989 introduced a voluntary health logo, a "keyhole" symbol, for identifying foods with low content of fat, sugar and salt and high content of dietary fibre within different product groups. In the same year the Australian Heart Foundation implemented the "tick", which a few years later also was introduced in New Zealand. Other examples are the Heart Symbol in Finland that was introduced in 2000 and the Choices Programme that today is implemented in the Netherlands, Belgium, Poland, Czech Republic, Argentina and Nigeria. The Swedish "keyhole" symbol was in 2009 further implemented in two other Nordic countries: Denmark and Norway (Kjærnes and Roos 2012, Roos 2009, http//:www.nokkelhullsmerket.no).

A survey among Norwegian consumers conducted before the implementation showed that consumers were in favour of simplified nutritional labelling and thought that making healthy choices would be easier if it was introduced (Roos 2009). However, it was concluded that although consumers were mostly positive, they had not been active requesting simplified nutrition labelling and it was still unclear whether they would really use the symbol or whether it would make it easier to make choices. Non-governmental organizations and food retailers took the first steps to support a health logo and lobbied the authorities to implement the "keyhole". The success of the "keyhole" seems to a large degree build upon the collaboration between "all interested parties"—the food sector, the industry, the retailers, the producers and consumer organizations, and others. All stakeholders were invited into the process early on. That the government stands firmly behind "keyhole" is also important for its success in Norway.

In addition to front-of-pack labelling on pre-packaged food products, this type of health-related food labelling has also been expanded to foods served in restaurants, fast food outlets and cafeterias. For example, the "keyhole" was introduced in lunch cafeterias in Sweden in 1993, four years after it had been launched on food products.

Calorie menu labelling especially in fast food restaurants has recently been proposed as a policy approach to address obesity in the U.S. (Swartz et al. 2011, Chaufan et al. 2011). However, research has indicated that awareness of calorie labels is low and that calorie menu labelling has no or limited effect on calorie consumption among consumers, whereas health-related labelling in lunch cafeterias at work places has been shown to be more effective (Swartz et al. 2011, Fernandes et al. 2016). One explanation for the limited effect may be that people usually eat in restaurants for pleasure and enjoyment and do not plan to eat healthy in this context. Therefore, in a restaurant setting, consumers are not interested in health-related information.

In sum, earlier literature on food labelling has mainly approached food labelling as information and implies that public health professionals view front-of-pack labelling as helpful for consumers to make informed choices. However, it is important to keep in mind that food choice is not limited to shopping but a complex process driven by experience, social processes and individual decision-making and that health information is only one of many factors consumers take into consideration (Kjærnes and Holm 2007). It is also important to consider the quality of existing studies when interpreting results. A review of the methodological quality of front-of-pack labelling research showed that the methodological quality was generally low (Vyth et al. 2012). Earlier studies have reported that consumers are generally positive to labelling and have described those who uses food labelling, but we know less about use of labelling in real life situations. Research in everyday life would help make visible the complexity of how both individual and socio-cultural factors (for example, norms, routines, household, time, price, knowledge, motivations, preferences, information and trust) are involved in consumers' negotiations related to selecting foods in the supermarket. Front-of-pack labelling has been highlighted by health policy makers as a promising strategy to help consumers making healthier food choices at the point of purchase, but this type of labelling may in the future increasingly have to compete for consumer's attention with other logos and information, including warning labels, which will be discussed next.

Consumer views on warning labels

Warning labels have been used on packages to alert consumers to different risks. Alcohol, tobacco and medication packages have a longer tradition with posting health-related warnings than food packages, and chemical and electrical products often have safety labels and warnings on how to use products safely. Food safety messages have generally been communicated to consumers by mass media, web-alerts, certification schemes and through warnings on food labels if foods contain certain ingredients that may be harmful for groups with specific health problems (e.g., allergies). Food safety regulation has traditionally restricted the contents of hazardous substances in food, but the possible use of new types of regulation has been discussed. For example, regulators have discussed the potential to use as food safety measures warnings that facilitate more informed consumer choices and give more responsibility to consumers.

The literature on warning labels on food is limited and has until recently mainly been restricted to allergens. For example, studies have shown that consumers with food allergies are not satisfied with the labelling and report problems with readability and visibility of the information (Cornelisse-Vermaat et al. 2007). Warning labels have some limitations, for example, one study concluded that warning labels might not be effective measures for food safety because attention to health warnings may be short-lived (Hughner et al. 2009). In addition to using warnings as food safety measures, there is currently an increased interest in exploring the potential use of warning labels on "junk" food as public health regulation of unhealthy food choices (Lacaniloa et al. 2011, Boncinelli et al. 2017, Effertz et al. 2014). Earlier studies show variation in how consumers notice and react to warning labels on food, and indicate that it is necessary to have more understanding of consumer views of warning labels on food before considering using warning labels on food as an alternative measure to banning hazardous substances. There is a need to know more about if, when and how warning labels on food may be appropriate food safety measures from the point of view of consumers.

A study of mandatory warning labels on food from the point of view of consumers was recently carried out in Norway (Roos et al. 2010). The aim of this study was to explore the expectations, understanding, and use of food warning labels in the Norwegian population. The study was based on both quantitative and qualitative data. To get an understanding of consumers' views, knowledge and use of warning labels, a web survey was conducted in 2009. The target was the adult population (15–87 years) and a sample consisting of 1001 responded. Consequently, two focus groups (N = 6 and N = 7; age 23–67 years) were in 2010 conducted in Oslo with consumers who have small children or have dietary restrictions. The following warning labels that were mandatory on food in Norway at the point of the study were included: "Do not refreeze after defrosting"; "Not recommended for children under the age of 3 years"; "May not be given to infants under 12 months"; "Excessive intake may have laxative effects"; "Contains liquorice—persons with high blood pressure should avoid large intakes" and; "Intake of plant sterols above 3 g per day should be avoided". In addition, two potential future warnings were included: "Color (E xxx) may have an adverse effect on activity and attention in children" and; "Dietary supplements containing beta carotene are not recommended for use among heavy smokers (more than 20 cigarettes a day)".

The web survey revealed that, except for a few warnings, there was mostly limited knowledge and awareness among Norwegian consumers about warning labels on food (Fig. 1). Many of the respondents stated that they had seen the more general warnings related to refreezing after defrosting and the two warnings related to children and infants, but the more specific warnings, especially the ones related to plant sterols and liquorice, seem to only have been seen by a very small number of the respondents. Interestingly a few even claimed that they had seen the two warnings related to beta-carotene and food color that were not in use. The results suggest that warnings on food are not a very well-known type of information, and maybe not usually distinguished from other information sources that consumers may have on foods. This was also evident from the focus group discussions, which suggested that consumers look for a whole range of information, are generally positive to more information and do not distinguish mandatory and voluntary information. Only some had noticed warnings, but they seemed positive to getting more information and thought that warning labels might be useful information for groups with specific health problems. However, it

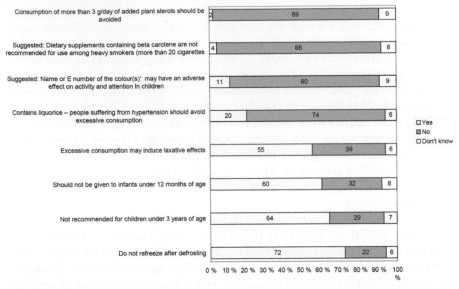

Fig. 1. Proportion of Norwegian consumers (N = 1001) who reported that they have seen the following warnings on food products. Percent. Adjusted for age, gender and geographical region. Source: Roos et al. 2010.

was also discussed that maybe not all read warnings and motivation is needed. For example, one participant pointed out that warnings are not easily accessible on the front of a package, but have to be read carefully on the back.

To get an understanding of consumer views on warning labels, responsibility and regulation, the web survey included a part with various statements related to warning labels on food (Fig. 2). Norwegian consumers do not seem to have a common view because on most of these statements the responses were distributed evenly on all response options. In addition, the high proportion of "don't know" responses on some of the statements may reflect that these were issues that consumers have not thought about nor have strong opinions about. However, one exception was the statement "warning labels on food products give useful information to consumers" where the majority agreed with the statement and very few disagreed. More than half also agreed with the statement "I think that Norwegian grocery stores can sell products that can be risky for some groups (for example, people with high blood pressure) if the food products have warning labels". About half of the consumers did not agree that the use of warning labels on food products could imply that the authorities turn down their responsibility for food safety, nor that warning labels could be alarming to consumers. This may be interpreted as an indication that many respondents being positive to warning labels did not see a contradiction between protective legislation and the use of warning labels.

The focus groups also indicated that consumers consider warning labels in line with other information as positive and useful, especially for potentially at-risk consumers with specific health conditions. Issues related to food and health were regarded as consumers' own responsibility, but they seemed to think that authorities

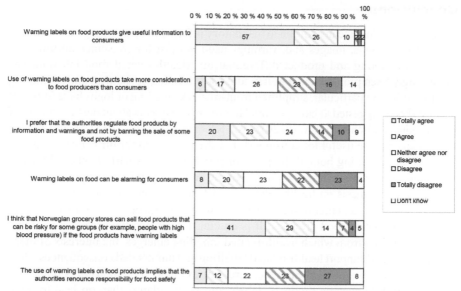

Fig. 2. The proportion of consumers (N = 1001) who reported that they agree or disagree with the following statements on warning labels on food. Percent. Adjusted for age, gender and geographical region. Source: Roos et al. 2010.

should label harmful things and provide information related to allergies, diabetes and other diseases. The participants in general supported banning dangerous and harmful ingredients if there are other alternatives available. However, they recognized that this might not always be in the interest of the producers. A difference between if warnings are used on everyday staple food products and foods that are consumed rarely was expressed. There seemed to be less interest in the use of warning labels on food products consumed on an everyday basis. Based on the findings of this study it was not possible to conclude if Norwegian consumers think that the use of warning labels on food is an appropriate alternative measure for food safety. The study suggests that consumers have limited understanding of differences between regulatory policies. Norwegian consumers do not seem to view warning labels on food as a new type of regulation or shift in responsibility but instead as information and an extension of what authorities already do. They expressed that warning labels provide better transparency and the possibility for special groups of consumers who need specific information to control what they buy. Very little scepticism was voiced among the participants maybe because warning labels have not received consumer mobilization and media attention in Norway. This study shows that considerable information and consumer education would be required if the use of warning labels on food products would be selected as a policy measure.

Based on this Norwegian study we can conclude that warning labels on food may have a potential to inform and help consumers in selecting foods, but it is not possible to conclude what views consumers have on the use of warning labels as an alternative food safety measure. Further studies both among consumers in different countries and contexts are necessary for being able to assess if warning labels may be an appropriate new type of regulation from the point of view of consumers.

Conclusions

Food packages include in addition to mandatory information also a range of voluntary front-of-pack logos, images and warnings used not just for consumer information, but also marketing and product differentiation. Health-related food labelling has increasingly been promoted as a policy tool for informing consumers and promoting healthy eating. In particular, simplified front-of-pack labelling or logos have in recent research been suggested to have potential to inform consumers about the healthiness of food products (Cechini and Warin 2016, van Kleef and Dagevos 2015). Health-related food labels may be useful for consumers, but there is a need for further research into actual use of labelling both at the point of purchase and as part of other phases of food consumption, including planning, cooking and eating. It is also important to bear in mind that emphasis on health-related food labelling can contribute to simplistic ideas that health is a matter of individual responsibility and determined by individual choices and specific food products. It is crucial to look into the broader social and political contexts from which health-related labelling emerges, the interests of those who propose and support health-related labelling and unintended consequences when considering this as a public health measure. For example, food industry is obviously more focused on specific products, labels and health claims than on healthy food patterns and diets in general.

In order for front-of-pack logos and food warnings to be effective and useful to consumers, several aspects, including what type of information consumers attend, understand, trust and take into account when they decide what to buy, have to be considered and the design and logo has to be understandable, relevant and useful for consumers. However, one limitation is that health logos seem to mainly have an effect on consumers that already are interested in health. In addition, food labelling competes with advertising elements, various logos and brand names, and consumers do not seem to differentiate between different types of information on food packages. Clear labelling and consistent format may support consumers in making healthier choices, but consumer choices are not just matter of individual responsibility. Policy measures that focus on motivating and educating people to make healthy choices seem to assume that choice is autonomous and that individual responsibility is possible. However, emphasis on individual motivation, education and information, including labelling, may have undesirable consequences and reproduce further social inequalities (Kjærnes and Roos 2012). On the other hand, one of the aims of front-of-pack labelling is to be an incentive for product reformulation and innovation and this might have health consequences that benefit all groups of consumers.

There is a recognized need for standardization and monitoring of health-related food labelling (Nestle and Ludwig 2010, Perez and Edge 2014, Rayner et al. 2013). For example, INFORMAS (The International Network of Food and Obesity/Non-communicable Diseases Research, Monitoring and Action Support) is a global network that has included a module that aims to monitor food labelling in retail setting globally (Rayner et al. 2013). The proposed indicators to be monitored cover the following components of food labelling: presence or absence of lists of ingredients, nutrient declarations, supplementary nutrition information, nutrition claims and health claims.

Warning labels on packages have been discussed as a policy option that balances the interests of companies and consumers. The warnings on food packages are supposed to raise awareness among consumers of the risky nature of a food product. A Norwegian study of warning labels indicated that consumers view warning labels as useful information especially for some groups with specific health conditions. However, it was also drawn attention to that as packages contain so much diverse information it can be difficult for consumers to identify the important "alerting" information a health warning is expected to convey. Thus, it was concluded that additional information and education is required if warning labels on foods are to be used as measures for food safety. However, there is a need for further research into food warnings before it is possible to conclude it they can be recommended as a policy option and effective way of trying to achieve food safety and health among consumers.

Finally, it is important to point out that health-related labelling is just one policy measure among others that governments can implement to promote health. Labelling itself is not sufficient if we wish to achieve all health policy goals. Actually, there is a potential risk that labelling may detract attention from other more effective strategies such as taxation and access to healthy foods. Therefore, there is a need for additional research into the effectiveness of other policy measures in relation to food labelling and the role of governments, industry and consumers. Industry sees food as a commodity and is mainly interested in using labelling for differentiating products, whereas for consumers food is much more than a product to choose from a shelf in the shop. For consumers food is a central part of everyday life and social relations. Consumers are when choosing food, in addition to health, influenced by a multitude of factors including routines, price, time constraints, taste, social relations, cultural norms, trust, knowledge and motivation, which makes food consumption a complex process.

References

Boncinelli, F., F. Gerini, G. Pagnotta and F. Alfnes. 2017. Warning labels on junk food: experimental evidence. Int. J. Consum. Studies 41: 46–53.

Campos, S., J. Doxey and D. Hammond. 2011. Nutrition labels on pre-packaged foods: A systematic review. Public Health Nutr. 14: 1496–1506.

Cecchini, M. and L. Warin. 2016. Impact of food labelling systems on food choices and eating behaviours; a systematic review and meta-analysis of randomized studies. Obes. Rev. 17: 201–210.

Chaufan, C., P. Fox and G.H. Hong. 2011. Food for thought: menu labelling as obesity prevention public health policy. Critical Public Health 21: 353–358.

Cornelisse-Vermaat, J.R., J. Voordouw, V. Yiakoumaki, G. Theodoridis and L. Frewer. 2007. Food-allergic consumers' labelling preferences: a cross-cultural comparison. Eur. J. Public Health 18: 115–120.

Cowburn, G. and L. Stockley. 2014. Consumer understanding and use of nutrition labelling: a systematic review. Public Health Nutr. 8: 21–28.

Effertz, T., M.-K. Franke and T. Teichert. 2014. Adolescents' assessments of advertisements for unhealthy food: an example of warning labels for soft drinks. J. Consum. Policy 37: 279–299.

Elvebakken, K.T. 2001. Næringsmiddelskontroll – mellom helse – og næringshensyn. Tidsskrift for Den Norske Lægeforening 121: 3613–3616.

Fernandes, A.C., R.C. Oliveira, R.P.C. Proença, C.C. Curioni, V.M. Rodrigues and G.M.R. Fiates. 2016. Influence of menu labelling on food choices in real-life settings: a systematic review. Nutrition Reviews 74: 534–548.

Garrett, S. 2007. Research literature relating to nutrition labeling and product selection at point of purchase: A short review and conceptual treatment. Food Standards Agency.

Grunert, K. and J. Wills. 2007. A review of European research on consumer responses to nutrition information on food labels. Journal of Public Health 15: 385–399.

Guthrie, J.F., J.J. Fox, L.E. Cleveland and S. Welsh. 1995. Who uses nutrition labelling, and what effects does label have on diet quality? Journal of Nutrition Education 27: 163–172.

Hall, C. and F. Osses. 2013. A review of inform understanding of the use of food safety messages on food labels. Int. J. Consum. Studies 37: 422–432.

Hieke, S. and C.R. Taylor. 2012. A critical review of the literature on nutritional labeling. Journal of Consumer Affairs 46: 120–156.

Hieke, S., N. Kuljanik, I. Pravst, K. Miklavec, A. Kaur, K.A. Brown, B.M. Egan, K. Pfeifer, A. Gracia and M. Rayner. 2016. Prevalence of nutrition and health-related claims on pre-packaged foods: a five-country study in Europe. Nutrients 8: 137.

Hughner, R.S., J.K. Maher, N.M. Childs and W.E. Nganje. 2009. Fish: friend or foe? Food policy and subpopulation warnings for consumers. Food Policy 34: 185–197.

Julia, C. and S. Hercberg. 2016. Research and lobbying conflicting on the issue of front-of-pack nutrition labelling in France. Archives of Public Health 74: 51.

Kjærnes, U. and L. Holm. 2007. Social factors and food choice: consumption as practice. *In*: Frewer, L. and H. van Trijp (eds.). Understanding Consumers of Food Products. Cambridge.

Kjærnes, U. and G. Roos. 2012. Food and welfare: Nordic nutrition policy and the current paradox of regulating obesity in Norway. pp. 43–63. *In*: Hellman, M., G. Roos and J. von Wright (eds.). A Welfare Policy Patchwork: Negotiating the Public Good in Times of Transition. Helsinki, Nordic Centre for Welfare and Social Issues.

Lacaniloa, R.D., S.B. Cash and W.L. Adamowicz. 2011. Heterogeneous consumer responses to snack food taxes and warning labels. The Journal of Consumer Affairs 45: 108–122.

Malam, S., S. Clegg, S. Kirwan, S. McGinigal and MBMRB Social Research. 2009. Comprehension and use of UK nutrition signposting labelling schemes. Prepared for Food Standards Agency.

Mandle, J., A. Tugendhaft, J. Michalow and K. Hofman. 2015. Nutrition labelling: a review of research on consumer and industry response in the global South. Glob. Health Action 8: 25912.

Nestle, M. and D.S. Ludwig. 2010. Front-of-package food labels: public health or propaganda? JAMA 303: 771–772.

Paterson, D., R. Zappelli and A. Chalmers. 2001. Qualitative research with consumers. Food labelling issues. Australian New Zealand Food Authority.

Perez, R. and M.S. Edge. 2014. Global nutrition labelling: moving towards standardization? Nutrition Today 49: 77–82.

Rayner, M., A. Boaz and C. Higginson. 2001. Consumer use of health-related endorsements on food labelling in the United Kingdom and Australia. Journal of Nutrition Education 33: 24–30.

Rayner, M., A. Wood, M. Lawrence, C.N. Mhurchu, J. Albert, S. Barquera, S. Friel, C. Hawkes, B. Kelly, S. Kumanyika, M. L'Abbé, A. Lee, T. Lobstein, J. Ma, J. Macmullan, S. Mohan, C. Monteiro, B. Neal, G. Sacks, D. Sanders, W. Snowdon, B. Swinburn, S. Vandevijvere and C. Walker. 2013. Monitoring the health-related labelling of foods and non-alcoholic beverages in retail settings. Obes. Rev. 14: Suppl. 1: 70–81.

Roos, G. 2009. Food labelling in the light of Norwegian nutrition policy. pp. 105–116. *In*: Oddy, D., P.J. Atkins and V. Amilien (eds.). The Rise of Obesity in Europe: A Twentieth Century Food History. Ashgate, Farnham.

Roos, G., U. Kjærnes and T. Ose. 2010. Warning labels on food from the point of view of consumers. SIFO Project report no. 6-2010, Oslo, National Institute for Consumer Research. (available at: www. hioa.no/extension/hioa/design/hioa/images/sifo/files/file77197_advarselsmerking_rapport_web.pdf).

Sinclair, S., D. Hammond and S. Goodman. 2013. Sociodemographic differences in the comprehension of nutritional labels on food products. Journal of Nutrition Education and Behavior 45: 767–772.

Swartz, J.J., D. Braxton and A.J. Viera. 2011. Calorie menu labelling on quick-service restaurant menus. An updated systematic review of the literature. Int. J. Behav. Nutr. Phys. Act. 8: 135.

van Kleef, E. and H. Dagevos. 2015. The growing role of front-of-pack nutrition profile labeling. A consumer perspective on key issues and controversies. Critical Reviews in Food Science and Nutrition 55: 291–303.

Vyth, E.L., I.H.M. Steenhuis, H.E. Brandt, A.J.C. Roodenburg, J. Brug and J.C. Seidell. 2012. Methodological quality of front-of-pack labeling studies: A review plus identification of research challenges. Nutrition Reviews 70: 709–720.

WHO. 2004. Global strategy on diet, physical activity and health. Resolution WHA 57.17. WHO, Geneva.

18

Genetically Modified Food Product Labeling Effects

How Dietary Restraint Impacts Consumer Cognition and Behavior

Anita G. Rodríguez[1],* and *Erin Baca Blaugrund*[2]

Introduction

Advanced technological developments and inundation of conflicting information throughout the media about diet and health, have led to consumer interest in high quality foods (Caswell and Mojduszka 1996). One important way consumers obtain information about food is through food labeling—such as country of origin, nutrition facts or ingredients (Borra 2006), and whether or not the product is genetically modified (Caswell 2000a). However, there are multiple terms indicating whether a product is genetically modified (GM), a genetically modified organism (GMO), or genetically engineered (GE). Previous research has referred more commonly to the food label as containing 'GM' or 'non-GM' as a way to inform consumers about these products. The first term reference was GE, but the United States Food and Drug Administration (US FDA) has not established a strict policy regarding this labeling procedure (FDA 2015). As a result, companies may utilize any of the terms. According to the FDA (2015), GE is the most accurate term, with GM and GE being used interchangeably.

[1] Interdisciplinary PhD Student, Marketing/Plant and Environmental Sciences, New Mexico State University, College of Business, PO Box 30001 MSC 5280, Department of Marketing, Las Cruces, NM 88003-8001.
[2] College Assistant Professor, New Mexico State University, College of Business, PO Box 30001 MSC 5280, Department of Marketing, Las Cruces, NM 88003-8001.
E-mail: erinbaca@nmsu.edu
* Corresponding author: anita@nmsu.edu

Effects of GM labeling on perceptions (Costa-Font et al. 2008) of a food's perceived healthfulness (Caswell 2000b, Crespi and Marette 2003) and subsequent consumption is an understudied area of examination (Lowe 1993). For example, with new GM labeling requirements being instituted in the United States (Lugo 2016, Baker and Burnham 2001), the impact of these requirements on consumer cognition and behavior are unknown. Understanding for whom (e.g., dietary restrained eaters) and for what foods (e.g., perceived whole food vs. processed food). For simplicity, we utilize a variation of Monteiro et al. (2010) food classification description to define whole food as an unprocessed or minimally processed food item, which entails "no processing", or refers to mainly a physical process that makes the whole food "more durable, accessible, convenient, palatable or safe". Overall, the inclusion of GM labeling may impact health perceptions and subsequent consumption and can help elucidate potential public policy options that could better meet the needs of government, manufacturers, and consumer stakeholders.

Previous research indicates wide heterogeneity amongst consumers regarding their response to GM foods. For example, food quality rankings for GM foods vary considerably. When deciding what to purchase, some consumers require products to indicate GM status, while others are indifferent (Caswell 2000b, Noussair et al. 2004, Klerck and Sweeney 2007). For consumers who prefer to know GM food status, their rationale varies for both refraining from or consuming GM foods based upon product attributes and labeling (Caswell 2000b, Noussair et al. 2004). Considering the aforementioned consumer response heterogeneity to GM foods, an existing consumer individual difference variable—such as dietary restraint or persistent regulation and monitoring of food consumption (Ward and Mann 2000)—may provide clear guidance regarding prediction of GM food labeling impacts on consumers. Furthermore, consumers may expect certain food types to either include or not include GM attributes. For example, GM labels on whole foods may be expected to be tied to particular attributes that benefit the consumer (e.g., increased size, better taste). When no attributes can be discerned from a GM labeled whole food, consumers may be fearful (Noussair et al. 2004, König et al. 2004) of how that food may affect their health. In both cases—i.e., dietary restraint and food type—perceptions of a food's healthfulness may be impacted, leading to more or less consumption.

In this chapter, consumer dietary restraint and GM food type are placed into a conceptual framework that describes how each influence a food's healthfulness and subsequent consumption. Specifically, restraint theory originally established by Herman and Mack (1975) explored how eating behavior was affected by psychological and non-psychological factors (Herman and Polivy 1975, Canetti et al. 2002). Several studies have utilized dietary restraint theory in terms of measuring dimensions of individual eating behavior utilizing modified questionnaires (Stunkard and Messick 1985), exploring dietary and disinhibition amongst young girls (Carper et al. 2000), and comparing validity of the three scales utilized to measure dietary restraint (Laessle et al. 1989). Therefore, we suggest that individuals who persistently monitor their food consumption may have increased sensitivity to GM food labeling. The presence of a GM food label may lead to a direct result of a decrease in healthfulness perception of the food item. No presence of a GM food label may have the opposite effect and in fact, increase the healthfulness perception. Downstream effects of these perceptions may result in greater inclinations to consume more of the perceived healthful food versus less healthful food for those who have dietary restraints (e.g., Payne et al. 2014). Furthermore, the proposed conceptual framework also suggests GM food labeling may have surprising effects on perceived food healthfulness depending on whether that food is perceived whole versus processed. For example, GM food labels in general, may result in decreased perceptions of healthfulness. When compared to the same food item that does not contain a GM food label, the result may be greater decreases in the perceived healthfulness of that food item. This is in comparison to processed foods when there is no discernible benefit from the genetic modification (i.e., "frankenfood effect"). Yet, when there is no GM food label, whole foods may be expected to be perceived as more healthful than a processed food. When the potential asymmetric negative evaluation on GM labeled whole foods is considered, consumers may be less willing to consume these foods.

The US recently passed GM food legislation requiring food manufacturers to use one of three label options: (1) USDA symbol indicating GMO presence; (2) print label using plain language, or; (3) QR code linked to ingredient information (Lugo 2016). Hence, this chapter is relevant and timely for consumers to have a better understanding of these new labels. Previously, the US and Canada national regulations included only voluntary labeling policies for GM foods, with a few exceptions such as in Canada, if there is potentially "a health or safety issue with a food" that may be diminished through labeling, then labeling is required (Blair and Regenstein 2015). More specifically, if the foods nutritional value or ingredients have changed, or if the food contains allergens then labelling is mandatory (Blair and Regenstein 2015). Europe, Australia/ New Zealand, South Korea, Japan, and China have already instituted "mandatory GM food labelling" policies, regardless of percentage or ingredients (Blair and Regenstein 2015). In all countries, however, consumer individual difference variables (e.g., dietary restraint) and product types thought to interact with GM labeling (e.g., whole vs. processed foods) have yet to be thoroughly examined. Examining the possible interactive effect of these variables with GM labeling on perceived food healthfulness and subsequent consumption may allow policy makers to more strategically balance consumers' "right to know" with potential unintended consequences for manufacturers.

This chapter focuses on the following: first, a brief overview of GM labeling is given for context. Next, a conceptual framework is introduced that examines how dietary restraint may interact with GM food labeling to result in greater or lesser perceptions of a food's healthfulness and subsequent consumption. Food type (i.e., whole vs. processed) is also introduced as an important boundary condition regarding the effect of GM food labeling on perceived healthfulness and consumption of a food. Finally, implications of the conceptual framework are discussed for public policy, manufacturers, and consumers.

Background

Genetically Modified Organisms (GMOs) were first introduced in the 1980s (König et al. 2004, Genetics Generation 2015). However, GMOs became prevalent in the marketplace in the 1990s (Phillips and Isaac 1998, König et al. 2004, Hartman 2014). Additional terminology used to reference GM have been "biotechnology", "genetically engineered", "bio-engineered" or "genetically altered" (Miller et al. 2003)—with some of the terms having more positive or negative connotations. The global market value of GM crops has continually increased to "US $15.7 billion representing 22% of the US $72.3 billion global crop protection market in 2013, and 35% of estimated US $45 billion global commercial seed market (James 2014)." The global organic food market has also steadily increased "at an average rate of 20% annually" (Pino et al. 2012). Since the inception of GM foods, there have been mixed views and research regarding the healthfulness of GM foods and/or GM ingredients in the marketplace (Harrison and Han 2005, Loureiro and Bugbee 2007, Costa-Font et al. 2008, Lee et al. 2012, Blair and Regenstein 2015). As a potential consequence, GM food labeling has been required in 64 nations (Blair and Regenstein 2015, Center for Food Safety 2016). However, not all food products require GM food labels. In fact, no country currently requires labels on products produced from animals fed GM animal feed, for example "meat, milk and eggs" (Blair and Regenstein 2015).

In the U.S., grocery trade associations frequently lobby against regulations requiring GM food labels on US food/food products under the assumption that GM labels may hurt manufacturers' brand image (Lee, Conroy and Motion 2012). However, in 2014, Vermont enacted a law—effective 1 July, 2016 (Grocery Manufacturing Association 2015, Pifer 2013, Kollipara 2015, Vermont Right to Know GMOs 2016), and became the first U.S. state to mandate GM labeling on foods sold throughout Vermont, with the exception of Vermont cheese, which is produced utilizing a GM ingredient (Strom 2016, Entine and Lim 2015). Following Vermont, the U.S. now requires GM foods that include more than 5% of their ingredients (by weight) to include a GM label (Blair and Regenstein 2015). In the U.S., GM labels are not required on products produced with GM technology, such as cheese produced utilizing "chymosin and other enzymes" containing GM microorganisms due to GM material being eliminated during processing (Blair and Regenstein 2015, Entine and Lim 2015). In 2013, newly established EU regulations required chymosin, "to be declared" on ingredient list, yet GM organisms utilized in "production process is not required to be explicitly labelled on the product" (Blair and Regenstein 2015).

Given the aforementioned GM food labeling policies, it is not well understood (1) how those potentially most sensitive to GM labeling with dietary restraints or other conditions respond to GM labeling, and (2) how GM labeling impacts perceptions of different types of foods such as whole or processed. This chapter introduces a framework to help answer these questions and provides potential guidance to government, manufacturers, and consumer stakeholders regarding balancing public and private interests.

Conceptual framework

The conceptual framework shown in Fig. 1, below suggests the impact of GM/non-GM food labels on perceived healthfulness of food depends both on those who have dietary restraints—low vs. high and the food type—whole vs. processed food is being evaluated. Specifically, dietary restraint and food type is expected to modify the impact of GM labels on a food's evaluation of perceived healthfulness and subsequent consumption. It is suggested that increased perceived healthfulness may increase

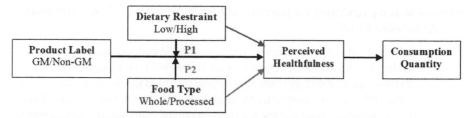

Fig. 1. GM labeling effects on perceived food healthfulness and consumption quantity.

subsequent consumption when healthfulness is a valued attribute for dietary restrained consumers. However, when a food type such as whole food is not perceived to have a discernible benefit from the genetic modification, the perceived healthfulness may decrease resulting in decreased consumption across consumers. These relationships are examined in the subsequent discussion.

GM food product labeling and dietary restraint

Dietary restrained individuals may be more conscientious of food attributes. Attributes of food products includes not only realization of a food's safety, but also value, packaging, as well as the process used to produce the food, and nutrition aspects (Lancaster 1971, Caswell and Mojduszka 1996, Hooker and Caswell 1996). Nutritional attributes are frequently inferred from food package labeling, which manufacturers largely control given commercial free speech rights (e.g., Payne and Niculescu 2012). However, the US Federal Nutrition Labeling and Education Act (NLEA) of 1990 gives the US Food and Drug Administration oversight over ingredient posting, percent juice labeling, and health claims (FDA-NLEA 1994, Nayga Jr. 1999, Burton and Andrews 1996). Packaging health claims, however—while correct and legal—can have boomerang effects on consumers. For example, "low-fat" claims in contrast with no claims on packaging can result in purchasing and subsequently eating more simply because perceptions of appropriate serving size increase while consumption guilt simultaneously decreases (Wansink and Chandon 2006). Considering those who have dietary restraints, these individuals may be the most sensitive to what is contained in their food and perceived nutrition gleaned from labels may affect them disproportionally compared to those who are less restrained. Indeed, restrained eaters are more likely to use nutrition labels to make decisions about food choices (Nayga Jr. 1999).

Legislation requiring GM labels may highlight nutritional attributes that are associated with GM foods more for dietary restrained eaters than others. For example, although "nutritional enhancements" are possible via GM technologies (Teisl et al. 2008), the negative information surrounding GM foods are most prevalent due to fear of unknown effects of food modifications, specifically genetic modifications, on overall health (Baker and Burnham 2001, König et al. 2004). Considering dietary restrained eaters' persistent regulation and monitoring of food consumption (Ward and Mann 2000), GM labels may have a significant effect on the perceived healthfulness of a food.

Role of dietary restraint on perceived healthfulness and consumption of GM labeled food

Dietary restrained individuals are more cognizant of food and subsequent food intake considering their heightened sense of health (Ruderman 1986). This heightened sense of health leads those who are dietary restrained to consider food labels when determining both the health and likely consumption of a food (Nayga Jr. 1999). For example, Payne et al. (2014) investigated how dietary restraint impacted individuals' "consumption intentions" when evaluating smaller portioned front-of-package "99 calorie or 100 calorie" snack packs that were objectively "less or more healthy". Results indicated that individuals who were dietary restrained intended to consume more of the healthier 99-calorie labeled food. This was thought to be because multiple cues of regulatory success (i.e., 99 calories, healthier food) led to restrained eaters being "licensed" to consume more; in effect, dietary regulatory control had been assigned to health cues instead of the self (e.g., Argo and White 2012, Provencher et al. 2009).

Considering dietary restrained individuals' emphasis on health may make them more knowledgeable about health attributes of food (Bublitz et al. 2010), GM labeled products may affect both their perception of a food's health and its subsequent consumption. This is in contrast to existing research that describes consumers actually purchasing and consuming more of GM labeled food. For example, a study conducted in a grocery store located in Hawaii found that papayas labeled as 'Hawaiian Grown GMO Papayas' or 'Hawaiian Grown Papayas-Genetically Modified Organism' and placed next to non-labeled GM papayas led consumers to prefer to purchase papayas with the GMO label than papayas without a GMO label (Shehata 2011). It is unclear if consumers purchased the papaya simply because there was a label associated with the product or if it was a lack of awareness regarding GMO terms. Additionally, in a similar study, Powell et al. (2003) found consumers preferred GE labeled sweetcorn over non-labeled ones. Curiously, consumers seem to be more prone to choosing a product containing a label. Consumers may perceive that labeled GMO products imply improved quality, or a consumers' curiosity to try a GMO fruit, or merely lack of knowledge regarding terms or acronym used for "GMO" (Shehata 2011). This would be in line with research that suggests that consumers, in general, do not comprehend nutrition information on food packages (Pelletier et al. 2004, Levy et al. 2000, Reid and Hendricks 1995). Considering the aforementioned discussion, the following propositions are offered:

P1: The perceived healthfulness of GM vs. Non-GM labeled foods is influenced by the individuals' dietary restraint such that:

P1a: High dietary restrained individuals are more likely than low dietary restrained individuals to decrease food consumption when the presence of a GM label has decreased the perceived healthfulness of that food item.

P1b: High dietary and low dietary restrained individuals are more likely to increase food consumption when the presence of a non-GM label has increased the perceived healthfulness of that food item.

GM labeling and food type

Depicted in Fig. 1 is an important moderating variable which involves the impact of GM labeling on consumers' evaluation of a food's perceived healthfulness and its subsequent consumption likelihood. In particular, the extent to which a food is perceived to be 'surprisingly GM' (Gaskell et al. 2004) may asymmetrically impact the perceived healthfulness of the food item and subsequent consumption. This is because the term GM (genetically modified)—may suggest some type of human intervention (Shaw 2002) in the production of a food that may be discernible from its outward appearance. In essence, when food is evaluated, people may believe that they know what the natural form the food should be resulting in potential surprise and/or disgust if a food deviates too far from expected natural forms of the food in question. These perceived "natural foods" may not be expected to have included human modification leading some consumers to consider the effect of the modification on their health and perhaps not choosing and/or consuming a particular food (Noussair et al. 2004). This may be particularly evident in contexts such as the grocery store where food decisions are frequently heuristically based and affect driven considering constraints on their time and motivation (Payne et al. 2014). While GM labeling and food type are significant considerations for consumers, authors are interested in relationship of GM labeled foods on individuals' perceived health and consumption.

Role of food type on perceived health and consumption of GM labeled food

Processed foods fall on a continuum of human intervention. Many snack foods, convenience foods, and packaged foods are expected to have had multiple points of human engineering including a foods' shape, size, taste, texture, smell, and ingredients found therein. Considering the level of human intervention, it may be the case that GM labels would not affect the perception of that food's healthfulness and subsequent consumption likelihood. Processed foods are generally perceived to be less healthy already and in fact consumers are making this conscious choice to consume the product. A GM label on processed foods may be an expected attribute of an already human engineered food. There is an expectation that there is a contrast between the perception of processed foods and non-processed or whole foods.

Whole foods such as baby carrots are those for which little or no human engineering are inferred, resulting in perceptions that the food is "natural". GM labels on whole foods may or may not affect perceptions of healthfulness. In general, consumers typically seek tangible benefits from foods they consume. A tangible benefit is identified as something that will be beneficial to consumers and/or something you can touch, feel, or see. For example, in the aforementioned example of GM labeled papayas (Shehata 2011), consumers actually purchased more papayas containing a label than when the same GM papayas were not labeled. It is unknown whether consumers associate the GM label to a particular visual attribute of the papaya for which consumers perceived some tangible benefit that falls outside the normal expectation of product benefits. In contrast, if consumers cannot associate a GM label to a tangible benefit of a whole food, they may become wary of the modification not knowing exactly how or if it benefits them—perhaps in a way that they would not feel comfortable because the expectation is not what is conventional. Considering the aforementioned discussion, the following propositions are offered:

P2: Food type changes how GM labels impact a foods perceived healthfulness such that:

P2a: Whole foods with GM labels, are more likely than processed foods with GM labels, to be perceived as less healthy when a tangible benefit of the GM label is not readily identifiable.

P2b: When a food is perceived to be less healthy because a tangible benefit of a GM label on a whole food is not readily identifiable, the food's consumption likelihood should decrease.

Discussion

In this chapter, the conceptual framework depicted in Fig. 1 introduces the concept that GM labels are most likely to impact perceive health of a food and subsequent consumption patterns for individuals with dietary restraint—low vs. high, and the type of food in question, whole vs. processed. To further understand this impact, a thorough review of NLEA guidelines may assist food manufacturers in developing GM food product labels, as it contains renowned public policy legislation for nutrition and food marketing information (Nyaga Jr. 1999). The Food and Agriculture Organization (FAO) of the United Nations indicates that regardless if food labels are mandatory or voluntary, some standards must be considered (FAO 2016). The implications of such labels may impact the consumer perception of food product quality, and potentially promote food market equity amongst different classes of individuals. Presence of GMOs listed on food package labels may impact global market sales, particularly in countries that have limited or in some cases, banned their use. The USDA organic label indicates that foods are void of bioengineering, such as "not bioengineered", "non-GMO", or other similar claims (USDA 2017). USDA Organic Standards require that organically produced food "prohibit use of genetic engineering" (Erikson 2000) and encapsulated in the Organic Foods Production Act (USDA 2017). Understanding how international trade may be affected depending upon government regulations and how consumers in those countries view GM food product labeling are important considerations. Given new GM food labeling requirements being instituted in the US (Lugo 2016), this conceptual framework provides timely information for GM labeling stakeholders. Specifically, this conceptual framework has implications for government, manufacturers, and consumers.

Government implications

Government implications resulting from recently mandated GM food product labeling (Lugo 2016, Baker and Burnham 2001) will impact trade and sales throughout the US and globally. The authors' conceptual framework is essential, as it will investigate how dietary restraint and food type moderate products labeled GM and the subsequent individuals' healthfulness perception of food. As previously stated, the increased perceived healthfulness of non-GM foods on dietary restrained individuals may in fact increase subsequent consumption.

When food type does not have a discernible GM benefit that is obvious on the label, for example, increased shelf life, enhanced nutritional value, or some texture/ flavor variation, consumers may have impressions about the food type that may impact their consumption pattern. In fact, the perceived health benefits of the food item given the label may decrease resulting in decreased consumption amongst dietary restrained consumers, and subsequent decrease purchases of the item. Therefore, information conveyed on GM food labels will also impact whether or not a consumer purchases the item, in particular whole food items. Overall, GM food labels are designed to help ensure consumers are informed (Harrison and Han 2005).

High dietary restrained individuals may evaluate and identify foods containing GM labels as less healthy, and therefore may decrease consumption of those foods, while simultaneously increasing consumption of similar non-GM labeled foods. GM labels on whole foods in contrast to processed foods may impact high dietary restrained individuals' health perceptions, as they may expect "processed foods" to contain GMs, but not necessarily whole foods, such as fruits or vegetables. Therefore, GM food labeling regulations will provide not only the everyday consumer, but also dietary restrained consumers with accurate information so they have the ultimate choice to consume or avoid these types of foods, if desired. The tangible benefits associated with GM labeling is unknown. Additionally, the perception of the tangible benefit being reduced because the food item is perceived to be unhealthy is an important part of the investigation for the ensuing research. For example, if high dietary restrained consumers are aware of GM presence in whole foods because of the label, consumption of those foods should decrease. However, low dietary restrained individuals' consumption of GM labeled whole foods may have adverse effect and actually increase their consumption given the simple presence of a label.

Manufacturer implications

Manufacturer implications of federally regulated GM food product labeling include corporate responsibility of increasing consumer awareness, obtaining accurate information, modifying labels accordingly, and potential cost implications. Based on sales, manufacturers' purchase of GM ingredients or food products may shift, as GM labels may appeal to different consumer segments. Additional research conducted by Baker and Burnham (2001) concluded that consumer group pressure resulted in some manufacturers taking initiative to assure consumers that their food products were GM-free. Manufacturers ensuring that foods are GM free, has created a scenario in which some food product recipes are altered, due to "swapping out ingredients" in search of GM-free ingredient options. This has also led to increased food prices involved in searching for replacement ingredients, as well as losing various nutrients, and even food product flavor being impacted (Biology Fortified 2016). These factors in turn can also affect high dietary restrained individuals' evaluations of GM foods, and may impact their consumption of GM labeled food products and influence manufacturers acquisition and distribution of GM food products, based on consumer demand and responding to consumer concerns/requests.

As previously noted, the tangible benefit of the GM label is unknown. What is common knowledge is whether there is a perceived difference between whole foods and processed foods. Shahidi (2009) indicated health conscious consumers are better served by consuming minimally processed food products. Thus manufacturers may benefit by increasing consumer knowledge of GM food attributes, as well as production processes, and/or potential environmental impact. Authors posit, when a food is perceived to be less healthy because a tangible benefit of a GM label on a whole food is unknown, that food's consumption likelihood should decrease for high dietary restrained individuals. This action would impact manufacturers that may already be producing non-GM foods, but have not yet implemented labeling policies that would take advantage of significant marketing opportunities. Not only would appropriate labeling be beneficial for manufacturers' ability to market non-GM foods, but this may also assist consumers in making purchasing decisions.

Consumer implications

Consumer implications of GM food product labeling will assist consumer understanding and influence "consumers' perceptions or beliefs" (Nayga Jr. 1999), so that consumers may make informed purchase decisions (Harrison and Han 2005). A US FDA survey revealed that "more than half of US consumers (54%) read labels" when purchasing a food product for the first time (FDA 2010). As aforementioned, manufacturers "swapping out ingredients" for GM free options may need to utilize less expensive or lower quality ingredient options. This action will impact consumers, specifically consumers with food allergies, or food sensitivities (Biology Fortified 2016). Consumer implications for high dietary restrained individuals may involve purchase avoidance of GM labeled foods (vs. non-GM labeled foods). Understanding how dietary restraint and product type modify product label and how it impacts consumers perceived health and consumption quantity are important considerations which have not been previously explored. This chapter emphasizes how high dietary restrained consumers are most impacted by GM food product labels, as they may evaluate GM foods as less healthy, and thus decrease consumption of that food. As suggested, this may actually cause an increase in the consumption of non-GM labeled foods.

Consumer implications of GM labels on whole foods in contrast with processed foods may be perceived as less healthy when a tangible benefit of the GM label is not known. High dietary restrained individuals will prefer to purchase non-GM whole foods, while low dietary restrained individuals may be indifferent and purchase either, depending upon availability, quality, and/or price.

Redesigning labels and including additional information to appeal to high dietary restrained individuals is imperative. Transparent information for GM labels on whole foods will help ensure consumers have pertinent information to make decisions on foods they choose to consume. Ultimately, recent federal regulations requiring GM food product labels will facilitate opportunity for information transparency. While decisions of foods to consume is ultimately an individuals' choice, full knowledge of product contents is necessary. The government is instrumental in enforcing regulations, and manufacturer compliance is essential for U.S. consumers to make informed decisions regarding whether 'to eat, or not to eat' GM foods.

Future implications

Future implications of GM food product labeling, dietary restraint, as well as food type are ample. Initially, the GM food debate was isolated to European countries, yet more recently U.S. consumer interest in GM food products has developed (Baker and Burnham 2001). Given existing research regarding receptivity of GM foods in the U.S. in comparison with other countries, specifically Europe, it will be interesting to investigate the impact on both. Mandatory food product labeling may veritably be a regulatory trade impediment (Crespi and Marette 2003). According to Liu et al. (2013), U.S. and European criteria regarding food-production technology regarding (GMO) differ, "U.S. food retailers implicitly accept" GMO, in contrast, "European food retailers explicitly oppose" GMO.

Teisl et al. (2008) research indicated consumers' perspectives concerning GM foods are susceptible to information about pros and cons identified with these foods, and Huffman et al. (2003, 2007) emphasized that GM labeled foods is a critical policy issue. US consumers insist food be free of "harmful substances" such as but not limited to "pesticides, chemical additives, hormones, and antibiotics" (Baker and Burnham 2001). Guido et al. (2010) researches concluded that consumers in France and Italy chose to purchase organic food based upon ethical perceptions as it avoided "detrimental consequences on human health and on the environment". More recently, increased consumer concern regarding GMOs revolve around their ability to withstand increased pesticide/herbicide applications. Allaying consumer "safety concerns" and prospective dangers with possible assets GM foods offer are governmental officials' priorities concerning GM food product labeling (Baker and Burnham 2001). Thus, incorporating other variables into the authors' conceptual model will provide numerous opportunities for future research. Government informing consumers of presence of GMOs in whole or processed foods is significant.

Labeling that is easy to process, easily noticeable, and provides quantity and quality information that consumers seek is key to understanding GM vs. non-GM foods. Consumer knowledge of transparent GM information via diverse labeling formats that doesn't necessarily dissuade them from purchasing the product, but simply permits informed consent will be beneficial to the educational process of GM foods. Given new GM food labeling mandates, the 'real-life' consumer effect may be investigated. GM food product labels may impact consumers differently depending upon whether consumer takes the opportunity to read and comprehend the information on food product packages.

References

Argo, J.J. and K. White. 2012. When do consumers eat more? The role of appearance self-esteem and food packaging cues. J. Mark. 76: 67–80.

Baker, G.A. and T.A. Burnham. 2001. The market for genetically modified foods: consumer characteristics and policy implications. Int. Food and Agribusiness Mark. Rev. 4: 351–360.

Biology Fortified. 2016. Six real consequences of GMO labeling—you may be shocked by #5! <https://www.biofortified.org/2016/04/six-consequences-of-gmo-labeling-you-may-be-shocked-by-5/> (access 14 April 2017).

Blair, R. and J.M. Regenstein. 2015. Genetic Modification and Food Quality: A Down to Earth Analysis. John Wiley and Sons.

Borra, S. 2006. Consumer perspectives on food labels. Am. J. Clin. N. 83: 1235S–1235S.

Bublitz, M.G., L.A. Peracchio and L.G. Block. 2010. Why did I eat that? Perspectives on food decision making and dietary restraint. J. Cons. Psych. 20: 239–258.

Burton, S. and J.C. Andrews. 1996. Age, product nutrition, and label format effects on consumer perceptions and product evaluations. J. Cons. Aff. 30: 68–89.

Canetti, L., E. Bachar and E.M. Berry. 2002. Food and emotion. Behav. Process 60: 157–164.

Carper, J.L., J.O. Fisher and L.L. Birch. 2000. Young girls' emerging dietary restraint and disinhibition are related to parental control in child feeding. Appetite 35: 121–129.

Caswell, J.A. and E.M. Mojduszka. 1996. Using informational labeling to influence the market for quality in food products. Am. J. Agricultural Econ. 78(5): 1248–1253.

Caswell, J.A. 2000a. Labeling policy for GMOs: To each his own? AgBioForum, J. Agrobiotechnology Manag. and Econ. 3: 53–57.

Caswell, J.A. 2000b. Analyzing quality and quality assurance (Including Labeling) for GMOs. AgBioForum, J. Agrobiotechnology Manag. and Econ. 3: 225–230.

Center for Food Safety. 2016. International Labeling Laws. <http://www.centerforfoodsafety.org/issues/976/ge-food-labeling/international-labeling-laws#> (access 29 July 29 2016).

Costa-Font, M., J.M. Gil and W.B. Traill. 2008. Consumer acceptance, valuation of and attitudes towards genetically modified food: Review and implications for food policy. Food Policy 33: 99–111.

Crespi, J.M. and S. Marette. 2003. "Does contain" vs."Does not contain": Does it matter which GMO label is used? Eur. J. Law and Econ. 16: 327–344.

Entine, J. and X. Lim. 2015. Cheese: The GMO food die-hard GMO opponents love (and oppose a label for), Genetic Literacy Project. <https://www.geneticliteracyproject.org/2015/05/15/cheese-gmo-food-die-hard-gmo-opponents-love-and-oppose-a-label-for/> (access 27 October 2016).

Erickson, B.E. 2000. Detecting genetically modified products in food. Anal. Chem. 72: 454A–459A.

Food and Agriculture Organization of the United Nations. 2016. Food labelling. <http://www.fao.org/ag/humannutrition/foodlabel/en/> (access 14 October 2016).

Food and Drug Administration. 1994. Nutrition Labeling and Education Act (NLEA) Requirements (8/94 – 2/95). <http://www.fda.gov/ICECI/Inspections/InspectionGuides/ucm074948.htm> (access 15 September 2016).

Food and Drug Administration. 2010. Survey Shows Gains in Food-Label Use, Health/Diet Awareness. <http://www.fda.gov/ForConsumers/ConsumerUpdates/ucm202611.htm> (access 22 July 2016).

Food and Drug Administration. 2015. Page last updated 10/19/2015. <https://www.fda.gov/Food/IngredientsPackagingLabeling/GEPlants/ucm461805.htm> (access 7 May 2017).

Gaskell, G., N. Allum, W. Wagner, N. Kronberger, H. Torgersen, J. Hampel and J. Bardes. 2004. GM foods and the misperception of risk perception. Risk Anal. 24: 185–194.

Genetics Generation. 2015. Introduction to Genetically Modified Organisms (GMOs). <http://knowgenetics.org/introduction-to-genetically-modified-organisms-gmos/> (access 14 October 2016).

Grocery Manufacturing Association. 2015. GMA Statement on Failure of Congress to Stop Patchwork of State GMO Labeling Laws in 2015. <http://www.gmaonline.org/news-events/newsroom/gma-statement-on-failure-of-congress-to-stop-patchwork-of-state-gmo-labelin/> (access 27 May 2016).

Guido, G., M.I. Prete, A.M. Peluso, R.C. Maloumby-Baka and C. Buffa. 2010. The role of ethics and product personality in the intention to purchase organic food products: A structural equation modeling approach. Int. Rev. Econ. 57: 79–102.

Harrison, R.W. and J.H. Han. 2005. The effects of urban consumer perception on attitudes for labeling of genetically modified foods. J. Food Distr. Res. 36: 29.

Hartman, K. 2014. GMO labeling: A case of asymmetric information and the "Nudge". Policy Perspect. 21: 48–59.

Herman, C.P. and D. Mack. 1975. Restrained and unrestrained eating. J. Pers. 43: 647–660.

Herman, C.P. and J. Polivy. 1975. Anxiety, restraint, and eating behavior. J. Abnorm. Psychol. 84: 666–672.

Huffman, W.E., J.F. Shogren, M. Rousu and A. Tegene. 2003. Consumer willingness to pay for genetically modified food labels in a market with diverse information: Evidence from experimental auctions. J. Agr. Resour. Econ. 28: 481–502.

Huffman, W.E., M. Rousu, J.F. Shogren and A. Tegene. 2007. The effects of prior beliefs and learning on consumers' acceptance of genetically modified foods. J. Econ. Behav. Organ. 63: 193–206.

Hooker, N.H. and J.A. Caswell. 1996. Regulatory targets and regimes for food safety: A comparison of North American and European approaches, Part One: Choosing Strategies for Health Risk Reduction. The Economics of Reducing Health Risk from Food proceedings of NE-165 Conference, June 6–7, 1995, Washington, D.C.

James, C. 2014. Global Status of Commercialized Biotech/GM Crops: 2014. ISAAA Briefs No. 49. ISAAA: Ithaca, NY.

Klerck, D. and J.C. Sweeney. 2007. The effect of knowledge types on consumer-perceived risk and adoption of genetically modified foods. Psychol. and Mark. 24: 171–193.

Kollipara, P. 2015. U.S. House moves to block labeling of GM foods, Science. <http://www.sciencemag.org/news/2015/07/us-house-moves-block-labeling-gm-foods> (access 18 July 2016).

König, A., A. Cockburn, R.W. Crevel, E. Debruyne, R. Grafstroem, U. Hammerling, I. Kimber, I. Knudsen, H.A. Kuiper, A.A. Peijnenburg, A.H. Penninks, M. Poulsen, M. Schauzu and J.M. Wal. 2004. Assessment of the safety of foods derived from genetically modified (GM) crops. Food and Chem. Toxic. 42: 1047–1088.

Laessle, R.G., R.J. Tuschl, B.C. Kotthaus and K.M. Prike. 1989. A comparison of the validity of three scales for the assessment of dietary restraint. J. Abnorm. Psychol. 98: 504–507.

Lancaster, K. 1971. Consumer Demand: A New Approach. Columbia University Press.

Lee, M.S.W., D. Conroy and J. Motion. 2012. Brand avoidance, genetic modification, and brandlessness. Aust. Mark. J. 20: 297–302.

Levy, L., R.E. Patterson, A.R. Kristal and S.S. Li. 2000. How well do consumers understand percentage daily value on food labels? Amer. J. Health Promo. 14: 157–160.

Liu, A.H., M. Bui and M. Leach. 2013. Considering technological impacts when selecting food suppliers: Comparing retailers' buying behavior in the United States and Europe. J. Bus.-Bus. Mark. 20: 81–98.

Loureiro, M.L. and M. Bugbee. 2007. Trading off risks and benefits when valuing new GM technologies: a consumer perspective. Int. J. Agr. Resour. Governance and Ecol. 6(1): 111–123.

Lowe, M.R. 1993. The effects of dieting on eating behavior: a three-factor model. Psychol. Bull. 114(1): 100–121.

Lugo, D. 2016. U.S. Senate passes GM food labeling bill. Science. <http://www.sciencemag.org/news/2016/07/us-senate-passes-gm-food-labeling-bill> (access 18 July 2016).

Miller, J.D., M. Annou and E.J. Wailes. 2003. Communicating biotechnology: Relationships between tone, issues, and terminology in US print media coverage. J. Appl. Commun. 87: 29–40.

Monteiro, C.A., R.B. Levy, R.M. Claro, I.R.R. de Castro and G. Cannon. 2010. Increasing consumption of ultra-processed foods and likely impact on human health: evidence from Brazil. Public Health N. 14: 5–13.

Nayga Jr., R.M. 1999. Toward an understanding of consumers' perceptions of food labels. Int. Food Agribusiness Manage. Rev. 2: 29–45.

Noussair, C., S. Robin and B. Ruffieux. 2004. Do consumers really refuse to buy genetically modified food? Econ. J. 114: 102–120.

Payne, C.R. and M. Niculescu. 2012. Social meaning in supermarkets as a direct route to improve parents' fruit and vegetable purchases. Agr. Resour. Econ Rev. 41: 124.

Payne, C.R., M. Niculescu and C.E. Barney. 2014. Consumer consumption intentions of smaller packaged snack variants. Int. J. Cons. Stud. 38: 238–242.

Payne, C.R., M. Niculescu, D.R. Just and M.P. Kelly. 2014. Shopper marketing nutrition interventions. Physiol. Behav. 136: 111–120.

Pelletier, A.L., W.W. Chang, J.E. Delzell and J.W. McCall. 2004. Patients' understanding and use of snack food package nutrition labels. J. Am. Board. Fam. Pract. 17: 319–323.

Phillips, P.W.B. and G. Isaac. 1998. GMO labeling: Threat or opportunity? AgBioForum, J. Agrobiotechnology Man. Econ. 1: 25–30.

Pifer, R.H. 2013. Mandatory labeling laws: What do recent state enactments portend for the future of GMOs. Penn St. Law Rev. 118: 789–814.

Pino, G., A.M. Peluso and G. Guido. 2012. Determinants of regular and occasional consumers' intentions to buy organic food. J. Cons. Aff. 46: 157–169.

Polivy, J. 1998. The effects of behavioral inhibition: Integrating internal cues, cognition, behavior, and affect. Psychol. Inq. 9: 181–204.

Powell, D.A., K. Blaine, S. Morris and J. Wilson. 2003. Agronomic and consumer considerations for Bt and conventional sweet-corn. Brit. Food J. 105: 700–713.

Provencher, V., J. Polivy and C.P. Herman. 2009. Perceived healthiness of food. If it's healthy, you can eat more! Appetite 52: 340–344.

Reid, D.J., S.A. Conrad and S.M. Hendricks. 1995. Tracking nutrition trends, 1989–1994: an update on Canadians' attitudes, knowledge and reported actions. Can. J. Pub. Health = Revue canadienne de sante publique 87: 113–118.

Ruderman, A.J. 1986. Dietary restraint: A theoretical and empirical review. Psychol. Bull. 99: 247.

Shahidi, F. 2009. Nutraceuticals and functional foods: whole versus processed foods. Trends in Food Sci. Technol. 20: 376–387.

Shaw, A. 2002. It just goes against the grain. Public understandings of genetically modified (GM) food in the UK. Pub. Understanding Sci. 11: 273–291.

Shehata, S. 2011. Market testing of labeled and unlabeled GMO papaya fruits in Honolulu chain stores. J. Food Distr. Res. 42: 107.

Strom, S. 2016. GMOs in Food? Vermonters Will Know. *N Y Times*. <http://www.nytimes.com/2016/07/01/business/gmo-labels-vermont-law.html?_r=1> (access 18 July 2016).

Stunkard, A.J. and S. Messick. 1985. The three-factor eating questionnaire to measure dietary restraint, disinhibition and hunger. J. Psychosom. Res. 29: 71–83.

Teisl, M.F., S. Radas and B. Roe. 2008. Struggles in optimal labelling: how different consumers react to various labels for genetically modified foods. Int. J. Cons. Stud. 32: 447–456.

USDA Organic Standards. 7 USC Ch. 94: ORGANIC CERTIFICATION From Title 7- AGRICULTURE, 6524. Organically produced food. <http://uscode.house.gov/view.xhtml?path=/prelim@title7/chapter94&edition=prelim> (access 12 May 2017).

Vermont Right to Know GMOs. 2016. Vermont's GMO labeling law is live! <http://www.vtrighttoknowgmos.org/vermonts-gmo-labeling-law-live/> (access 22 July 2016).

Wansink, B. and P. Chandon. 2006. Can "low-fat" nutrition labels lead to obesity? J. Mark. Res. 43(4): 605–617.

Ward, A. and T. Mann. 2000. Don't mind if I do: disinhibited eating under cognitive load. J. Pers. Soc. Psychol. 78: 753–763.

Waxman, H.A. 1989–1990. H.R.3562—Nutrition Labeling and Education Act of 1990. Congress.gov. < https://www.congress.gov/bill/101st-congress/house-bill/3562> (access 22 July 2016).

Williams, P. 2005. Consumer understanding and use of health claims for foods. Nutr. Rev. 63: 256–264.

19

Sustainability, Certification Programs, and the Legacy of the Tokyo 2020 Olympics

*Yoshiko Naiki** and *Isao Sakaguchi*#

Introduction

Tokyo was announced as the host of the 2020 Olympic and Paralympic Games (hereinafter, Olympics or Olympic Games) in September 2013 in Buenos Aires. At that time, a Japanese 2020 Olympic delegation member explained that *"omotenashi"*, the spirit of Japanese hospitality,[1] was the traditional Japanese way in which Tokyo would welcome foreigners. However, as the 2020 Olympics gets closer, the current situation in Japan has cast doubt on the feasibility of delivering *omotenashi* for a sustainable Olympics.[2]

Since the London 2012 Olympics, sustainability has become common practice when preparing for and holding the Olympic Games, which means that all goods and services provided for the Olympic venues—such as timber used for the construction and the food provided at the Olympic village—must meet certain sustainability

* Osaka School of International Public Policy, Osaka University; 1-31 Machikaneyama, Toyonaka, Osaka, Japan, 560-0043.
# Department of Law, Gakushuin University, 1-5-1 Mejiro, Toshimaku, Tokyo, Japan, 171-8588.
  E-mail: isao.sakaguchi@gakushuin.ac.jp
* Corresponding author: ynaiki@osipp.osaka-u.ac.jp

[1] Japan Times, Omotenashi: The Spirit of Selfless Hospitality, October 19, 2013.
[2] Mainichi Shinbun, Otemoride 'Omotenenashi'? , July 4, 2016 (Evening Edition) (in Japanese).

standards. The Tokyo Organizing Committee of the Olympic and Paralympic Games (hereinafter, the Organizing Committee) drafted and published Sustainable Sourcing Codes. However, two questions arise: first, and most importantly, what are the sustainability conditions, and secondly, do we have enough products in the Japanese market that meet the Olympic sustainability conditions? Therefore, the feasibility of a "sustainable" *omotenashi* is doubtful. This chapter focuses in particular on the sustainable agricultural and fishery products to be provided at the Olympic venues.[3]

Currently there are many mechanisms available to evaluate the sustainability of goods and services for the Olympic Games. In this chapter, we focus on the use of certification programs (Auld 2015), and, in particular, the non-state and voluntary mechanisms that are operational and referred to as "global certification programs". While most of these programs were originally started in Europe, their scope is broader and has a global reach, including developing countries. Some programs are widely recognized, such as the GLOBALG.A.P. for agricultural products and the Marine Stewardship Council (MSC) and the Aquaculture Stewardship Council (ASC) for fishery products.

These global certification programs usually focus on two main tasks: establishing sustainability standards and verifying whether users (i.e., applicants) are compliant with these standards. Their sustainability standards usually include environmental, economic, and social/labor criteria. The verification task, or the so-called "certification", is an important function of global certification programs. While there are many variations for certifying products in practice, global certification programs usually adopt a third-party auditing system that involves conformity assessment bodies (CABs), which are auditing bodies independent of the global certification programs' commercial interests and political pressures: CABs certify whether the applicants' products meet the sustainability standards set down by the global certification programs. These CABs also have to be accredited by independent accreditation bodies (ABs). Adopting a third-party certification system strengthens the accountability and credibility of non-state voluntary certification programs. Producers and retailers attempt to obtain certification to improve their brand image and business reputation, thereby increasing sales or gaining new market access.

Another feature of these global certification programs is the need to engage with a broad range of stakeholders (i.e., producers, industry, retailers, and NGOs) to establish and revise a program's sustainability standards and decision-making procedures. Multi-stakeholder governance is an important feature of global certification programs, even though they may have been initially developed by an NGO or a specific industry.

What we address in this chapter is not research into the rise or effects of such global certification programs as there is already a large body of international relations (IR) research available that explains certification initiatives as a part of wider studies on "private regulation" or "private governance" (Vogel 2008). Instead, the focus in this chapter is on the emergence of new local certification programs as competitors to the existing global certification programs. In IR literature, the emergence of competing certification programs has already been examined as well. In this chapter, therefore,

---

[3] In this chapter, we focus on the Sustainable Sourcing Codes for agricultural and fishery products, not addressing the Sourcing Codes for timber and livestock products due to space limitations.

we specifically examine the emergence and expansion of competing local certification programs in Japan against the background of the Tokyo 2020 Olympics. The question to be examined is why local certification programs are currently being promoted in Japan when global certification programs were widely used at the last two Olympic Games in London and Rio de Janeiro.

The remainder of this chapter is organized as follows. Section 2 reviews the literature that addresses the relationship with and competition between global certification and local certification programs. Building on this literature review, we introduce three perspectives on the emergence of the new certification programs. Section 3 examines how Japanese local certification programs for agricultural and fishery products have emerged to rival GLOBALG.A.P. and MSC/ASC. In Section 4, we assess the emergence of Japanese local certification programs, utilizing the insights and arguments in the literature review in Section 2. The final section presents policy implications and discusses ways in which these multiple coexisting programs (global and local programs) in Japan can be improved.

Literature Review: Global Certification Programs vs. Local Certification Programs

This section reviews the relevant literature that helps us to understand why the new local certification programs have been developed to compete with the existing global certification programs as IR scholarship on "private regulation" and "private governance" can provide important insights for our research. An early famous case of the emergence of competing programs was the studies in forestry where the industry created new competing programs (Cashore et al. 2004). Here, we examine three insights and arguments drawn from the literature. We are of the view that these three insights are interrelated.

The first insight extracted from the literature is focused on "fragmentation"; i.e., when multiple certification programs co-exist in a sector and there is no dominant certification program in the sector. Regulatory fragmentation occurs with the emergence of new "local" certification programs to rival "global" programs. Different sectors (e.g., coffee, fisheries, forestry or clothing) have been found to have different patterns with some sectors being more fragmented than others. Research has identified several elements to explain this fragmentation—when and how new programs emerge.

Fransen and Conselmann (2015) described the conditions in which an early certification initiative can dominate a sector and prevent any further entries that can cause fragmentation; when a sector has a high industrial concentration and the certification program is an industry-led (instead of by NGOs) organization that prefers to establish more lenient standards with lower participation costs.

On the other hand, Auld (2014) argued that if the demand for certified products and consumer interests in sustainability increase in a market, a new program development is likely because a program developed later could find it easier to enter the certification-business market because of the business benefits garnered from earlier programs. Foley and Hébert (2013) offer a complementary argument that certification helps to foster a market for governance, and they illustrate this by exploring the back-and-forth decisions of the Alaska salmon fishery to participate in the MSC certification program.

Once a market for sustainable seafood had been established, the salmon fishery had an interest in supporting its own labeling initiative as a vehicle to access this market. At the same time, Auld (2014) also suggests that the relevant actors' motivations for creating a new program could be affected by the "inclusive opportunities" provided in an existing program; in other words, when the actors feel that there are "limited possibilities for voice" in the existing program, they may seek to develop a new program.

The second insight is concerned with the research question as to how (or to what extent) global certification programs are received and put in practice in the producing countries, and, in particular, in developing countries. We view that the emergence of new local certification programs to rival global programs is to some extent influenced by the success or failure of the existing global certification programs; that is, whether local producers have adopted the sustainability standards set by the global certification programs, thereby having obtained certification.

Several case studies have explained the growth factors for global program certification in developing countries. Bartley (2010) identified two dominant factors: export demand in countries where consumers are conscious of credibly certified sustainable products and the existence of public scrutiny or transnational pressure on firms and producers to supply sustainable products. However, Bartley (2010) also claimed that internal factors such as the domestic regulatory environment were equally important, and provided insights as to how global certification programs "are filtered, renegotiated, or compromised" when they are applied to domestic settings. Similarly, Schleifer (2017) gave three domestic conditions for the acceptance of global certification programs in developing countries: (1) adjustment of standard stringency to local conditions; (2) industry-wide support; and (3) concurrence with the national regulatory environment.

The third insight is related to "public-private interactions". Even in studies on "private regulation" or "private governance", the role of the state is still pertinent. The large body of research has examined many different types of public-private interactions (Green 2017) and there have also been many different analytical perspectives on how to conceptualize such interactions. However, for the purpose of this chapter, it is important to focus on the "state responses" to the emergence of new local programs. Gulbrandsen (2014) examined the government support for certification programs in the forestry and fishery sectors that led to a growth in such programs; for example, a state may recognize a new certification program so as to enforce state regulations or may recognize certain certification programs in its public procurement to give these programs legitimacy.

More specifically, Foley and Havice (2016) examined several cases of emerging local certification programs for fisheries that were actually supported by (local) governments. They claimed that these new local programs could be viewed as positive efforts that attempted to accommodate "territorial attributes" such as local production features, industry interests, and state regulations. At the same time, such new programs also attempted to comply with international sustainability norms. In this respect, Foley and Havice (2016) argued that such local certification programs would play a constructive role of achieving "territorial sustainability governance" (in contrast to "transnational sustainability governance").

As state responses are important, we examine how the Japanese government has engaged in the emergence of Japanese local certification programs despite the existence of respected global certification programs. It is also important to analyze how state responses have influenced performance in the Japanese local certification programs.

The Emergence and Development of Local Certification Programs in Japan

Uptake of global certification programs in Japan and the Tokyo 2020 olympics sustainable sourcing codes

This subsection first examines the extent to which global certification programs—such as the GLOBALG.A.P., MSC, and ASC—have been adopted in Japan: how many Japanese producers are interested in obtaining global program certification? As discussed in the following, global certification program implementation in Japan has not been very successful to date.

The GLOBALG.A.P. (formerly named "EurepGAP") was originally developed in 1997 by European retailers and has its head office in Cologne, Germany. G.A.P. stands for "good agricultural practices (GAP)"—practices that are "addressing environmental, economic and social sustainability for on-farm production and post-production processes resulting in safe and healthy food and non-food agricultural products" (FAO 2003). Therefore, the GAP standards cover agricultural product food safety, environmental protection, worker health, and traceability, taking a holistic integrated approach to all farming stages. For instance, in the current GLOBALG.A.P., there are 218 fruit and vegetable standards, of which 99 are focused on food safety, 69 on the environment, 22 on traceability, and 28 on worker health (GLOBALG.A.P. 2016).

In Japan, the GLOBALG.A.P. Association of Japan was founded in 2010, but a local formal office was only opened in 2015. To date (as of October 2017), around 450 local Japanese producers have obtained GLOBALG.A.P. (fruit and vegetables) certification (in contrast, there are around 183,900 GLOBALG.A.P. (fruit and vegetables) certification issued world-wide).[4]

MSC commenced in 1997 with its main office being in London. ASC started later in 2010 and its main office is in Utrecht, the Netherlands. MSC's local office in Japan was opened in 2007 and ASC local office was opened in 2017. At the time of writing, in Japan, there are only three MSC-certified fisheries and only one ASC-certified farm.[5] In contrast, there are 296 MSC-certified fisheries (as of December 2016) and 539 ASC-certified farms (as of November 2017) in the world.[6]

---

[4] See, GLOBALG.A.P. Japan's news on 18 October, 2017 <https://www.ggap.jp/?p=120> (accessed on 9 December, 2017) (in Japanese).

[5] See, for the MSC <https://www.msc.org/track-a-fishery-ja/certified-ja>; for the ASC < https://www.asc-aqua.org/ja/> (accessed on 9 December, 2017) (in Japanese).

[6] See, for the MSC < https://www.msc.org/documents/environmental-benefits/global-impacts/msc-global-impacts-report-2017>; for the ASC <https://www.asc-aqua.org/news/certification-update/> (accessed on 9 December, 2017).

It was against this backdrop that the Tokyo 2020 Olympics Sustainable Sourcing Codes for agricultural and fishery products were developed by the Organizing Committee.[7]

The two Sustainable Sourcing Codes for agricultural and fishery products have a similar, simple structure.[8] In terms of agricultural products,[9] the Sustainable Sourcing Code Annex 2-2, Section 2 establishes three sustainability conditions for procured agricultural products: (1) to ensure the safety of the products with reference to relevant Japanese laws; (2) to ensure a harmonious balance between agricultural production activities, the surrounding environment, and ecosystems with reference to relevant Japanese laws; and (3) to ensure the safety of workers with reference to relevant Japanese laws. Therefore, these three conditions very briefly cover food safety, environmental protection, and worker safety, with the baselines for safety and protection being current Japanese laws.

Appendix 2-2, Section 3 states that "Agricultural products certified under the JGAP Advance or GLOBALG.A.P. scheme are accepted as ones that satisfy conditions (1) to (3) of Section 2." Note that "JGAP Advance"—a Japanese local certification program, which is explained in more detail below—was established in parallel with the GLOBALG.A.P.

Similarly, in terms of fishery products, the Sustainable Sourcing Code, Appendix 2-4, Section 2 establishes four sustainability conditions for procured fishery products: (1) to ensure that fishery products are caught or raised in an appropriate manner with reference to the FAO Code of Conduct for Responsible Fisheries, laws, ordinances, etc. relevant to the fishery industry; (2) to ensure that wild-caught fishery products are managed in a planned manner, and that conservation of the ecosystem is taken into consideration based on scientific data; (3) to ensure that farmed fishery products are being raised by aquaculture that takes into consideration the conservation of the ecosystem through planned maintenance and improvements in the farm environment, and is conducted with appropriate measures to ensure food safety based on scientific data; and (4) to ensure the safety of workers with reference to relevant laws, ordinances, etc.

Requirement (2) addresses wild-caught fishery products, while requirement (3) is focused on farmed fishery products. Requirements (1) and (4) apply to both wild-caught and farmed fishery products; one meets the FAO Code of Conduct for Responsible Fisheries, and the other ensures the safety of workers. In terms of worker safety, note that baseline for safety is "relevant laws, ordinances, etc." and, different from the Sourcing Code for agricultural products, is not limited to Japanese laws. The reason Japanese laws are not specifically mentioned in this fishery product Sourcing Code is

---

[7] The Tokyo Organizing Committee, Tokyo 2020 Olympic and Paralympic Games Sustainable Sourcing Code (1st edition), < https://tokyo2020.jp/en/games/sustainability/ >.

[8] Since this chapter's focus is on certification programs, we only address cases where suppliers use certification programs to prove that their products meet sustainability conditions. However, the Tokyo 2020 Olympic Sustainability Sourcing Codes actually provide another way for suppliers to prove sustainability without using certification programs. Thus, the scope of products to be accepted as sustainable goods for the Olympics will be broader. See, the Sustainable Sourcing Code, Appendix 2-2, Section 4 and Appendix 2–4, Section 4.

[9] As suggested in footnote 3, agricultural products do not include livestock products—the latter products are subject to a separate Sourcing Code.

probably because of the FAO Code of Conduct for Responsible Fisheries. While this Code is non-binding, it was unanimously adopted at the FAO conference in 1995, and therefore perceived as the global principles and standards for fisheries (FAO 1995). This FAO Code includes several provisions related to worker safety.[10]

Then, Appendix 2-4, Section 3 states that "fishery products with MEL, MSC, AEL, or ASC certification are accepted as ones that satisfy conditions (1) to (4) of Section 2." Note that "MEL (Marine Eco-label Japan)" and "AEL (Aquaculture Eco-label)"—Japanese local certification programs for wild caught and farmed fishery products—are considered in parallel with the MSC and the ASC.

In the next two sub-sections, we examine how these Japanese local certification programs—such as the JGAP Advance, MEL, and AEL—have developed and evolved in the context of the Tokyo 2020 Olympics.

JGAP advance: Japanese local certification program for agricultural products

Currently there are around 4000 farms certified under the JGAP standards (as of March 2017).[11] The local Japanese GAP program—JGAP—was originally initiated in November 2006 by 30 farm producers producing EurepGAP-certified products.[12] At that time, the JGAP was also seeking GLOBALG.A.P. benchmarked status and in 2007, JGAP's "version 2.1" standards achieved GLOBALG.A.P. "equivalent" status.

In 2010, in a regular version update, the JGAP adopted a dual-standard system: "JGAP's export standards" for Japanese producers who wanted to export to foreign markets, and "JGAP's basic standards" for Japanese producers not intending to export their products. At that time, the JGAP 2010 standards were no longer benchmarked against the GLOBALG.A.P. (Naiki 2014).

In 2016, when JGAP's standards were updated to "JGAP 2016", the JGAP Advance—which was referred to the Tokyo 2020 Sustainable Sourcing Code for agricultural products—was developed. The JGAP 2016 consists of two types of standards: a JGAP Basic and the JGAP Advance,[13] which was similar to concepts and structures in the JGAP 2010. However, at that time, a Japanese government subsidy was given to develop the JGAP Advance. In 2015, the MAFF (Ministry of Agriculture, Forest, and Fishery) announced specific grants for projects focused on developing

---

[10] For instance, the FAO Code, Article 8.2.5 provides as follows: "Flag States should ensure compliance with appropriate safety requirements for fishing vessels and fishers in accordance with international conventions, internationally agreed codes of practice and voluntary guidelines. States should adopt appropriate safety requirements for all small vessels not covered by such international conventions, codes of practice or voluntary guidelines."

[11] See, the Japan GAP Foundation's webpage, <http://jgap.jp/JGAP_Assoc/pamph_house.pdf> (accessed on 9 December, 2017) (in Japanese).

[12] The group was originally established as "Japan Good Agricultural Initiative (JGAI)" in 2005.

[13] More precisely, the JGAP Basic has been renamed simply as "JGAP", while the JGAP Advance has been renamed as "ASIAGAP" from August 2017. See, the Japan GAP Foundation's press release on 6 July, 2017, <http://jgap.jp/JGAP_News/NewsRelease20170706-jgap_advance-asiagap.pdf> (accessed on 9 December, 2017) (in Japanese). However, this chapter retains the former names (JGAP Basic and JGAP Advance) because it is easier to conceive of the concepts and structure of the JGAP certification standards.

GAP standards for promoting the export of Japanese agricultural products (MAFF 2015). The JGAP Foundation was given this subsidy and developed the subsequent JGAP Advance standards.

However, there was a condition attached to this grant, namely, that the GAP standards for exports should be in line with the latest version of the GFSI (Global Food Safety Initiatives) Benchmarking Requirements (GFSI 2017). The GFSI is the food-industry led initiative that commenced in 2000. When a certification program satisfies the GFSI Benchmarking Requirements, it becomes a GFSI-recognized program, which is understood "as a stamp of approval and a signal of strong food safety standards" and "has come to be required by many buying companies as a prerequisite to doing business."[14] Currently, 13 programs have been recognized by the GFSI—one of which is GLOBALG.A.P. Therefore, to be regarded as an internationally accepted program, such as GLOBALG.A.P., JGAP Advance needs to achieve GFSI recognition. As of writing, JGAP Advance is in GFSI's benchmarking process.[15]

Looking back, MAFF has been gradually involved in the promotion of the JGAP Advance certification standards. Initially, MAFF calmly observed the developments in the private and local GAP programs in Japan. Apart from the JGAP initiative, various different GAP programs with quite simple and basic standards have been developed in different prefectures in Japan. Because of these multiple local GAP programs, in 2010, the MAFF attempted to unify GAP programs by developing the "guidelines on common standards for GAP" (which was updated in 2012) (MAFF 2012). However, at that time, the MAFF did not intend to harmonize the multiple existing GAP programs; rather, the guidelines had 40 standards for each of nine different product categories (i.e., vegetables, rice, wheat, fruits, and tea) that simply show how each standard matched existing Japanese laws and ordinances. Therefore, it appears that the government was only interested in regulating GAP programs so that they would fall within the scope of the government's laws and requirements.

However, in April 2016, the MAFF announced an "Action plan based on the guidelines on common standards for GAP," which included three actions: (1) to ensure local GAP programs met with the "guidelines on common standards for GAP"; (2) to encourage producers to obtain GLOBALG.A.P. certification; and (3) to promote Japan's own GAP program for international transactions. Action (3) specifically referred to the development of the JGAP Advance standards.

This new stance taken by MAFF seems to be because of two reasons. First, as the Tokyo 2020 Olympics Organizing Committee had started to draft the Sustainability Sourcing Codes, the government was concerned about whether there would be a sufficient volume of certified foods available given the small number of GLOBALG.A.P.-certified products in the Japanese market. The second reason was because of the finalization of the TPP (Trans-Pacific Partnership) Agreement (signed in February 2016), as this Agreement would have required the government to

---

[14] See, GFSI's webpage on "What are GFSI-recognised Certification Programmes?," <http://www.mygfsi.com/certification/recognised-certification-programmes.html > (accessed on 9 December, 2017).

[15] See, Japan GAP Foundation's press release on 29 November, 2017, <http://jgap.jp/JGAP_News/NewsRelease20171129-asiagap-gfsi-shinsei.pdf> (accessed on 9 December, 2017) (in Japanese). See also, the MAFF's explanations on this, <http://www.maff.go.jp/j/seisan/gizyutu/gap/asiagap_gfsi.html> (accessed on 9 December, 2017) (in Japanese).

remove tariff protections over domestic agricultural products—to respond to domestic producers' concerns, the Japanese government needed to develop new strategies to promote agricultural production and expand exports. On this point, the government was seeking to promote the export of Japanese agricultural products that had obtained GAP certification. It is against this backdrop that the MAFF began to engage in local GAP programs and, in particular, promote the JGAP Advance certification standards.

Related to this MAFF movement, in May 2017, the Liberal Democratic Party of Japan announced the "Strategy for Standards and Certification" in the context of agriculture, forests, and fisheries (MAFF 2017), in which it was proposed that three times as many as the present certification number needed to gain GAP certification in Japan by the end of 2019 to ensure the supply of certified foods to the Tokyo 2020 Olympics. It was also stated that producers seeking GAP certification must either meet the JGAP Advance or the GLOBALG.A.P., which are regarded as internationally accepted certification standards.

MEL and AEL: Japanese local certification programs for fishery products

The MEL was founded in 2007 by the Japanese Fisheries Association and the AEL was founded in 2014 by the Japanese Foodist Association.[16] Currently there are 28 fisheries certified by the MEL and 19 aquaculture farms certified by the AEL (as of May, 2017).[17] The number of MEL certified-fisheries and AEL certified-farms has been increasing. As previously noted, there are only three MSC certified fisheries and only one ASC certified farm in Japan.[18] To further promote the local programs, the Fisheries Agency in the MAFF granted subsidies to cover MEL and AEL assessment costs under the "Promotion Project for Japanese Fisheries Certification Schemes of International Acceptance" (Fisheries Agency 2016).

One of the main problems faced by Japanese fishermens and farmers when applying for the MSC or the ASC is their voluntary fishery/farm management plans promoted by the Fisheries Agency. The MSC and the ASC certification standards clearly require that each fishery/farm implement an effective fishery/farm management system for responsible and sustainable fishing/farming. However, in Japan, such management plans are basically developed and implemented voluntarily by the fishery/farm associations, with the government only approving the plans. Fishery cooperatives and associations follow the Fisheries Agency guidelines when developing their management plans (Makino 2011); however, these management plans are not strictly required to be science-based under the current guidelines.

In fact, many of the management plans are simply based on past landing volumes for wild-capture fisheries rather than sustainable yield levels, and past seed fish/shell numbers for aquaculture rather than the environmental carrying capacity of the farming

---

[16] In 2016, MEL became a legal entity as a general incorporated association.
[17] See, for the MEL, <http://www.fish-jfrca.jp/04/mel_4.html>; for the AEL, <http://www.fish-jfrca.jp/04/a_eco_1.html> (accessed on 9 December, 2017) (in Japanese).
[18] However, note that the number of fisheries/farms certified does not tell us very much as some fisheries/farms might have very large volumes of production and others very small. In addition, on the market side, we need to consider whether retailers and seafood companies hold the chain of custody certification.

grounds (Sakaguchi 2017). Under such weak management planning, it is difficult for Japanese fishermens/farmers to meet the robust MSC and ASC criteria. In contrast, MEL and AEL do not require any substantial transformation in their current practices.

MEL and AEL also have another problem. As previously noted, credible global certification programs usually adopt a third-party assessment system by designating independent CABs. The FAO guidelines specifically require ISO/IEC 17065 for CABs and ISO/IEC 17011 for ABs to verify whether a CAB meets the standard set in ISO/IEC 17065 (FAO 2009, FAO 2011), which is intended to guarantee that third-party assessment is truly independent of commercial interests and political pressures. This is a prerequisite for credibility for any sustainability certification programs.

On this point, both MEL and AEL have designated the Japan Fisheries Resources Conservation Association (JFRCA) as the sole current CAB;[19] however, the JFRCA does not have ISO/IEC 17065 status. Further, MEL and AEL have also respectively designated the Japan Fisheries Science and Technology Association (JFSTA)[20] and NPO Support Center for Fishery Resources Improvement and Administration (FRIA Support) as the AB;[21] however, neither JFSTA nor FRIA Support hold the ISO/IEC 17011 status. In short, MEL and AEL certification assessments are not independent third-party assessment systems and do not have the appropriate external quality control required by FAO guidelines.

This combination of weak criteria and poor assessment quality control is the reason behind a number of highly doubtful "certified" fisheries/farms. For example, MEL certified the purse seine fishery for endangered Pacific Bluefin tuna in 2015 (JFRCA 2015) and the AEL certified the Pacific Bluefin tuna farming for wild seed fish in 2017 (JFRCA 2017). The pursuit of both the purse seine fishery and farming were the main reason for the stock collapse.

To respond to the growing criticisms concerning transparency and assessment, MEL has drafted new certification standards for wild capture fisheries, which became effective in February 2018.[22] MEL also drafted a new set of certification standards for aquaculture separately from the AEL standards, which became effective in March 2018.[23] This latter action is intended that MEL is planning to merge AEL in the future. MEL announced that it will apply for the Global Seafood Sustainability Initiative (GSSI) recognition for these new certification standards. The GSSI was established in 2013 through a partnership of major retailers, seafood processors, and the FAO to provide a system by which to verify conformity with the FAO guidelines. The GSSI released its Global Benchmarking Tool in 2015, and currently, MSC, the Alaska Responsible Fisheries Management (RFM) Certification, the Iceland Responsible Fisheries Management (IRFM) Certification, and the Best Aquaculture Practices

---

[19] Japan Fisheries Resources Conservation Association, <http://www.fish-jfrca.jp/index.html> (accessed on 9 December, 2017) (in Japanese).

[20] Japan Fisheries Science and Technology Association, <http://www.jfsta.or.jp/> (accessed on 9 December, 2017) (in Japanese).

[21] NPO Support Center for Fishery Resources Improvement and Administration, <http://www.fria-support.jp/> (accessed on 9 December, 2017) (in Japanese).

[22] See, Marine Eco-Label Japan's news on 1 February, 2018, <http://www.melj.jp/> (accessed on 9 April, 2018) (in Japanese).

[23] See, Marine Eco-Label Japan's news on 9 March, 2018, <http://www.melj.jp/> (accessed on 9 April, 2018) (in Japanese)..

(BAP) of Global Aquaculture Alliance have all been recognized.[24] GSSI recognition has become a vital condition for the MEL to be accepted as a credible certification program in Japanese and global seafood markets. The Fisheries Agency in the MAFF also expects the MEL to achieve GSSI recognition.[25]

Analysis of Japan's Local Context

Looking at Japan's cases related to the JGAP, MEL, and AEL, there is some support in previous literature research (in Section 2) for the emergence of local certification programs.

The "fragmentation" perspective suggests the existence of an early certification program before the emergence of a new competing program. Interestingly, in Japan's case, we can claim that the first-mover programs in Japan were the Japanese local certification programs (JGAP, MEL, and AEL) rather than the global certification programs (GLOBALG.A.P., MSC, and ASC) because the global certification program presence in Japan has been so weak, especially on the producer side. Therefore, if Japanese stakeholders develop local certification programs with lenient standards based on local perspectives and if these local certification programs expand, it is not easy for global certification programs to compete with them as the global certification programs have stringent standards. At the same time, it is also true that Japanese local programs are currently facing pressure to improve their certification standards by achieving GFSI or GSSI recognition. If such pressure really works, competition dynamics may change: both local and global programs can compete on a level playing field.

In terms of market demand for and consumer interests in certified products, Japan is still experiencing limited growth. However, the recent further diffusion of Japanese local programs among producers is coincident with the growing "ad-hoc" demand for certified products for the Tokyo 2020 Olympics. While the global certification programs have always existed as an option for Japanese producers/fishers/farmers, it was not considered a better option because they feel that their specific local situations are not well considered and there are no "inclusive opportunities" within the global certification program decision-making where a very few number of Japanese producers and retailers were involved. This is also due to the weak presence of global certification programs in Japan.

In addition, the two external conditions—export demand in foreign countries and transnational pressure from the NGOs—for the uptake of global certification programs do not exist in Japan; therefore, there have been limited opportunities for global certification program acceptance in Japan. Similarly, the domestic conditions for global certification programs to be well received have not been met in Japan. Global certification programs usually do not allow for adjustment of their stringent

---

[24] See, GSSI's webpage, "GSSI-recognized Seafood Certification Schemes," <http://www.ourgssi.org/benchmarking/recognized-schemes/> (accessed on 9 December, 2017).

[25] See, the House of Representatives, Japan, Committee on Agriculture, Forestry, and Fisheries, No.12, Committee's meeting minutes on 17 May, 2017, <http://www.shugiin.go.jp/internet/itdb_kaigiroku.nsf/html/kaigiroku/000919320170517012.htm> (accessed on 9 December, 2017) (in Japanese).

universal standards. While there is some flexibility permitted in these global certification standards,[26] these are not always seen as sufficient for local operations. Normally, rather than local producers who want to avoid the high costs of complying with stringent global standards, retail and supermarkets can become key supporters of global certification programs. In Japan's case, the largest supermarket in Japan, AEON, is a strong supporter of GLOBALG.A.P, MSC, and ASC; however, due to Japan's low industrial concentration in retail and supermarkets, AEON's leadership alone in supporting the global programs is not strong enough to generate an industry-wide uptake.

Furthermore, the relationships between global certification programs and national regulatory environment in Japan need to be viewed in the context of "public-private interactions". There have been several state responses in Japan's case. The Japanese government provided support for the local certification programs by granting subsidies to upgrade existing local standards (JGAP), covering assessment costs (MEL and AEL), and through the promotion of the JGAP, MEL, and AEL as qualified certification programs to be included in the Tokyo 2020 Olympics Sustainable Sourcing Codes in line with GLOBALG.A.P., MSC, and ASC. The Sourcing Codes were drafted and published by the Organizing Committee, which is seen as a private body rather than a public one; however, it is misleading to regard the Organizing Committee as a purely private body. It can be presumed that the government's intentions were reflected in the Sourcing Codes. To ensure that Japanese local programs were considered in the Sourcing Codes, the government, and, in particular, the MAFF, carefully formulated the national regulatory environment; for example, in the way the MAFF has developed a GAP policy that was then channeled into the Sourcing Code formulation.

As the literature review suggests, government support for certification programs normally assists these certification programs in building credibility and increasing their reputation. However, are the Japanese local certification programs sufficiently credible as global certification programs? In other words, the question is whether Japanese government support for local certification programs is successfully promoting "territorial sustainability governance" (Foley and Havice 2016). As stated, the Japanese local programs are currently facing pressure to raise their certification standards by achieving GFSI and GSSI recognition. GFSI and GSSI recognition are prerequisite for the Japanese local certification programs to be accepted in the world market.[27]

Policy Implications for Sustainability Governance

Research on the existence of multiple certification programs is not new. Many studies have focused on certain sectors (such as forestry, fisheries, or clothing) and drawn

---

[26] For example, in terms of GLOBALG.A.P., it is possible to set up "National Technical Working Groups (NTWGs)" whose work is "to identify specific local adaptation and implementation challenges and develop guidelines." See, The GLOBALG.A.P. National Technical Working Group, <http://www.globalgap.org/uk_en/who-we-are/ntwgs/> (accessed on 9 December, 2017). However, the work of the NTWGs is not to lower the GLOBALG.A.P.'s universal standards at a local level.

[27] In terms of the JGAP Advance, it appears to target the Asian market instead of the European or US markets, as the JGAP Advance has been renamed as "ASIAGAP" from August 2017. See, *supra* note 13.

comparison across sectors. Moreover, there has been increased attention on country-specific analyses when there are multiple certification programs operating at the same time. In particular, when a new local certification program emerges in developing countries as a competitor to global programs, it is regarded as a new trend of "Northern versus Southern certification standards" (Schouten and Bitzer 2015). This chapter, therefore, offers a new example of how local certification programs act as competitors to existing global certification programs within one country. We explored the local factors in Japan that have driven the development of alternative local programs rather than supporting existing global programs.

As suggested, in the fields of "private regulation" and "private governance", it is impossible to exclude the entry of new programs if there is demand for certification (Auld 2014, Foley and Hebert 2013). In Japan, there is growing "ad-hoc" demand for certified products for the Tokyo 2020 Olympics. Therefore, even though there is a credible global certification program as an option, the expansion of local programs may not be avoided. A situation similar to that in Japan can occur anywhere in the world—a trend that can be observed now in several Asian countries.

While this chapter attempted to explain the reasons for the emergence of these new programs, we did not fully address how this situation of multiple certification programs can be adequately handled. Although answering this question is beyond the scope of this chapter, we would like to make several suggestions that could lead to a more coordinated relationship between the multiple certification programs in Japan.

First, while multiple certification programs cause "regulatory fragmentation" and "competition", does this only result in disadvantages to sustainability governance? Some studies have argued that fragmentation and competition have both advantages and disadvantages (Fransen and Conzelmann 2015). As has been argued, fragmentation can lead to consumer confusion regarding the different eco-labelling of certified products and increase assessment costs for producers when retailers require them to be certified under the different certification programs (Fransen 2015). What is more concerning is that fragmentation could lead to a race to the bottom with each new program developing more lenient standards (Fransen 2012). If multiple certification programs in Japan have different certification standards in terms of the substance and procedures, granting equal recognition to the Japanese local certification programs and the global certification programs in the Olympic Sustainability Sourcing Codes could foster "a culture of mistrust and rivalry" between the certification programs (Fransen 2012).

However, as has also been pointed out, "Many analysts…disagree whether it produces a race to the bottom or the top" (Eberlein et al. 2014). In this regard, the Japanese local programs are currently facing pressure to raise their certification standards by achieving GFSI and GSSI recognition (and in fact, the JGAP Advance has entered into GFSI's benchmarking process). However, only the Japanese government is promoting GFSI and GSSI recognition and no other actors seem to be strongly interested in the local Japanese programs' efforts to achieve GFSI and GSSI recognition at this time. In order to drive changes in the local Japanese programs and push them to achieve GFSI and GSSI recognition, comparative assessments between the local and global programs should be carried out. Publication of such comparative assessments by a neutral research institute will release information on differences between certification programs and may increase attention by the public and relevant actors (Overdevest 2005). Such an information release may have the effects of encouraging the local Japanese programs to further improve certification standards by achieving GFSI and GSSI recognition.

Indeed, "rule development [by certification programs] is not static" and it is important "to understand the *processes* through which change occurs" (Cashore et al. 2003). Currently, Japanese local programs seem to be in the process of building institutionalization robustness and if Japanese programs really struggle to achieve GFSI and GSSI recognition, it may lead to a race to the top among all existing certification programs in Japanese market. However, it is still early to make a full analysis of the development processes of Japanese local certifications at this moment.

What then are the advantages of the presence of multiple global and local certification programs in Japan? As the market demand for certified products and consumer interests in sustainable production and consumption are still limited in Japan, it is crucial for local actors to engage in sustainability certification activities. Therefore, the presence of multiple certification programs could be seen as a transitional period or a "learning" phase in the longer sustainability certification process, which could diffuse "sustainability norms" at a local level. In particular, the existence of local certification programs could raise stakeholder awareness and serve as the initial learning phase for development of long-term sustainability norms. A transitional period may also be necessary to develop a sustainability community at a local level. For instance, it may be an important period to educate and raise "professionals" in a sustainability community—such a community includes not only producers, retailers, and consumers, but also the assessors/auditors who conduct the verification tasks in CABs. As bringing in foreign assessors/auditors to Japan raises costs, the demand for local assessors/auditors is increasing. The Japanese government should also collaborate with these private actors to craft state sustainability policies. It remains to be seen whether current government involvement could lead to "learning" and the fostering of a truly sustainable community in Japan.

Acknowledgements

Earlier versions of this chapter were presented at Aoyamagakuin University, Gakushuin University, and the Institute of Developing Economies. The authors would like to thank the participants in these events. Special thanks also go to Graeme Auld for his valuable comments. This research is partly supported by JSPS KAKENHI Grant Numbers 15KT0118 and 16KT0093.

References

Auld, G. 2014. Constructing Private Governance: The Rise and Evolution of Forest, Coffee, and Fisheries Certification. Yale University Press, New Haven, USA.

Auld, G. 2015. Policy making: Certification as governance. pp. 2610–2617. *In*: Bearfield, D.A. and M.J. Dubnick (eds.). Encyclopedia of Public Administration and Public Policy, 3rd edition. Taylor & Francis, New York, USA.

Bartley, T. 2010. Transnational private regulation in practice: The limits of forest and labor standard certification in Indonesia. Business and Politics 12: 1–34.

Cashore, B., G. Auld and D. Newsom. 2003. The United States' race to certify sustainable forestry: Non-state environmental governance and the competition for policy-making authority. Business and Politics 5: 219–259.

Cashore, B., G. Auld and D. Newsom. 2004. Governing Through Markets: Forest Certification and the Emergence of Non-State Authority. Yale University Press, New Haven, USA.

Eberlein, B., W.K. Abbott, J. Black, E. Meidinger and S. Wood. 2014. Transnational business governance interactions: Conceptualization and framework for analysis. Regul. Gov. 8: 1–21.

Foley, P. and K. Hébert. 2013. Alternative regimes of transnational environmental certification: Governance, marketization, and place in Alaska's salmon fisheries. Environ. Plann. A. 45: 2734–2751.

Foley, P. and H. Havice. 2016. The rise of territorial eco-certifications: New politics of transnational sustainability governance in the fishery sector. Geoforum. 69: 24–33.

Food and Agriculture Organization of the United Nations. 1995. The FAO Code of Conduct for Responsible Fisheries. <http://www.fao.org/docrep/005/v9878e/v9878e00.htm> (accessed on 14 October 2017).

Food and Agriculture Organization of the United Nations. 2003. Development of a Framework for Good Agricultural Practices, COAG/2003/6. <ftp://ftp.fao.org/docrep/fao/meeting/006/y8704e.pdf> (accessed on 14 October 2017).

Food and Agriculture Organization of the United Nations. 2009. Guidelines for the Ecolabelling of Fish and Fishery Products from Marine capture Fisheries (Revision 1). < http://www.fao.org/docrep/012/i1119t/i1119t.pdf > (accessed on 19 October 2017).

Food and Agriculture Organization of the United Nations. 2011. Technical Guidelines on Aquaculture Certification. < http://www.fao.org/3/a-i2296t.pdf> (accessed on 19 October 2017).

Fransen, L. 2012. Corporate Social Responsibility and Global Labor Standards: Firms and Activists in the Making of Private Regulation. Routledge, New York, USA.

Fransen, L. 2015. The politics of meta-governance in transnational private sustainability governance. Policy. Sci. 48: 293–317.

Fransen, L. and T. Conzelmann. 2015. Fragmented or cohesive transnational private regulation of sustainability standards? A comparative study. Regul. Gov. 9: 259–275.

Fisheries Agency (Japan). 2016. Public offering for promoting projects regarding fishery ecolabels originally developed in Japan. <http://www.jfa.maff.go.jp/j/gyosei/supply/hozyo/attach/pdf/160908_s1-1.pdf> (accessed on 14 October 2017) (in Japanese).

Global Food Safety Initiative. 2017. The GFSI Benchmarking Requirements, Version 7. <http://www.mygfsi.com/certification/benchmarking/gfsi-guidance-document.html> (accessed on 14 October 2017).

GLOBALG.A.P. 2016. GLOBALG.A.P. fruit & vegetables certification: The first choice for retailers & producers around the world. <http://www.globalgap.org/.content/.galleries/documents/160506_Fruit_and_Vegetables_Booklet_en.pdf> (accessed on 14 October 2017).

Green, J.F. 2017. Blurred lines: Public-private interactions in carbon regulations. Int. Interact. 43: 103–128.

Gulbrandsen, L.H. 2014. Dynamic governance interactions: Evolutionary effects of state responses to non-state certification programs. Regul. Gov. 8: 74–92.

Japan Fisheries Resource Conservation Association. 2015. Purse seine fisheries at Japan sea. <http://www.fish-jfrca.jp/04/pdf/mel/wajima_maki.pdf> (accessed on 14 October 2017) (in Japanese).

Japan Fisheries Resource Conservation Association. 2017. Bluefin tuna aquaculture in Ehime prefecture. <http://www.fish-jfrca.jp/04/pdf/ael/JFRCA151703A.pdf> (accessed on 14 October 2017) (in Japanese).

Ministry of Agriculture, Forestry, and Fisheries. 2012. Guidelines for common standards for GAP. <http://www.maff.go.jp/j/seisan/gizyutu/gap/guideline/pdf/guide_line_120306.pdf> (accessed on 14 October 2017) (in Japanese).

Ministry of Agriculture, Forestry, and Fisheries. 2015. Public offering for promoting projects regarding GAP for export goods. <http://www.maff.go.jp/j/supply/hozyo/seisan/pdf/koubo_150220_3.pdf> (accessed on 14 October 2017) (in Japanese).

Ministry of Agriculture, Forestry, and Fisheries. 2016. Action plan regarding the proliferation/expansion of GAP based on the GAP guidelines. <http://www.maff.go.jp/j/seisan/gizyutu/gap/g_kaigi/280428/pdf/2.pdf> (accessed on 14 October 2017) (in Japanese).

Ministry of Agriculture, Forestry, and Fisheries. 2017. Proposal for strategies of standards and certifications. <http://www.maff.go.jp/j/seisan/gizyutu/gap/g_kaigi/290529/pdf/siryou1.pdf> (accessed on 14 October 2017) (in Japanese).

Makino, M. 2011. Fisheries Management in Japan: Its Institutional Features and Case Studies. Springer, New York, USA.

Naiki, Y. 2014. The dynamics of private food safety standards: A case study on the regulatory diffusion of GLOBALG.A.P. Int. Comp. Law. Q. 63: 137–166.

Overdevest, C. 2005. Treadmill politics, information politics, and public policy: Toward a political economy of information. Organ. Environ. 18: 72–90.

Sakaguchi, I. 2017. Sustainable seafood campaign and sustainable fisheries (a series of newspaper articles). Minato Shinbun. 14th April 2017 and 24th March 2017 (in Japanese).

Schleifer, P. 2017. Private regulation and global economic change: The drivers of sustainable agriculture in Brazil. Governance 30: 687–703.

Schouten, G. and V. Bitzer. 2015. The emergence of Southern standards in agricultural value chains: A new trend in sustainability governance? Ecol. Econ. 120: 175–184.

Vogel, D. 2008. Private global business regulation. Annu. Rev. Polit. Sci. 11: 261–282.

20

Conclusion

Shigeru Matsumoto[1],* and *Tsunehiro Otsuki*[2]

Consumers are highly interested in food credence attributes and spend considerable on them while shopping. This book has three objectives: (1) to understand how consumers value various food credence attributes, (2) to compare consumers' valuation of specific food credence attributes across countries, and (3) to propose agro-food policies to reflect consumers' demand for food credence attributes. Since the problem of food credence attributes has been studied in many research fields, we invited scholars of different academic disciplines to exchange research ideas in this book volume.

Among various food attributes, food safety remains consumers' greatest concern. Part I began the discussion in Chapter 1 on food safety issues. Chapter 2 discussed consumers' opinion about the use of irradiation, genetic modification, and nanotechnologies. The chapter further discussed consumers' concern about chemical and hormone use in agricultural production. Chapters 3, 4, and 5 discussed animal diseases, GMO technologies, and radiation contamination, respectively. These chapters reported that consumers' reaction toward these food risks widely varies between regions as well as between individuals. In addition, it was confirmed that policy selection in the event of food contamination has a large impact on the consumers' confidence in the aftermath. Although many findings have been reported in this book, it is important to accumulate knowledge of consumers' response toward food risks as well as effects of countermeasures toward contamination.

Assurance of safety and quality of food has driven challenges to balance the cost and benefit to consumers and suppliers as markets today have liberalized globally. Part II focused on the role of regulations or voluntary programs in food attributes on

[1] Department of Economics, Aoyama Gakuin University, Room 828, Building 8, 4-4-25 Shibuya, Shibuya, Tokyo, Japan, 150-8366.
[2] Department of Economics, Osaka School of International Public Policy, Osaka University, 1-31 Machikaneyama, Toyonaka, Osaka, 560-0043 Japan.
    E-mail: otsuki@osipp.osaka-u.ac.jp
* Corresponding author: shmatsumoto@aoyamagakuin.jp

international trade and cross-border supply-chain management. Chapter 6 provided the economic model to analyze the welfare impact of the policies to regulate food credence attributes in an open economy context. It proposed schemes to test protectionism that such policies sometimes present. Chapter 7 provided empirical analyses to examine the effect of food safety regulations on the international food trade. It demonstrated in case studies of poultry meat trade that higher food safety is likely to increase import demand, while it is likely to reduce export supply by imposing trade cost. Chapter 8 addressed the role of organic farming and geographic indication in cross-border value-chain management in the case studies of countries in the Mekong region. It discussed that careful management of cross-border value chains is critical for a region's win–win outcome. Chapter 9 addressed agricultural biodiversity as a diversification of farm production while providing environmental, economic, and sociocultural benefits to the local and global society. Empirical analysis of the chapter identified household and market characteristics that are critical to promote agricultural biodiversity.

Consumers are concerned about not only the physical aspects of foods but also their cultural and ethical aspects. Part III discussed consumers' expectation about agricultural production. Although many consumers think that agricultural products need to be produced in sustainable and ethical manner, they have very limited knowledge about agricultural production. One solution for bridging this gap is to provide consumers with information on agricultural production so that they can make an informed choice. Therefore, a policy maker often tries to standardize the production system and create an index to help consumers make an informed choice. However, Chapters 10 and 11 cautioned that such policy measures can lead to unintended consequences since consumers' food choice is a very complex task. It is very important to recognize the demerits of standardization and simplification of the food policy. Although many consumers state that they are concerned about animal welfare in agricultural production, it is unclear whether they are actually considering it in food selection. Chapter 12 summarized the challenges remained to reflect people's concern about animal welfare. Under time constraints, consumers often use production area information when choosing foods. While Chapter 13 analyzed the use of country of origin information, Chapter 14 analyzed the use of regional information. Both chapters demonstrated the way of information provision that greatly affects consumers' use of production region information. Production area information can conflict with other information on food attribute. Chapter 15 explained the problems that consumers would face when production information is provided with organic product information.

Part IV studies food marketing and regulation. Chapter 16 showed that consumers in modern societies rely much more on extrinsic cues than on intrinsic cues when making a food selection. Although governments are attempting to influence consumer diet by responding to this change, such attempts have not been much successful at the present stage. Chapter 17 found out the conditions that makes health-related labeling effective and useful to consumers. Only consumers who are interested in their health utilize health-related labeling; therefore, labeling strategies may not be effective to consumers who need to modify dietary behavior. If consumers' dietary restriction is related to another food attribute, then the effect of information provision varies between consumers with different dietary restrictions. Chapter 18 analyzed how GMO information affected consumers with different dietary restrictions. Chapter

19 discussed the difference and competition between Japanese local certification programs such as the Tokyo 2020 Olympics Sustainable Sourcing Codes and the existing global certification programs such as GLOBALG.A.P. It demonstrated that the Japanese local certification programs are growing even though they compete with the existing global programs.

As mentioned earlier, this book was written by researchers from various academic disciplines. However, irrespective of their academic disciplines, several common policy recommendations have been proposed, which are summarized here.

As food attribute information becomes more complex, consumer cognition becomes important in food selection. Food policy needs to be designed under cognitive constraints of consumers. However, it is not well understood how consumers with restricted cognitive abilities choose food at the present stage. When the cognitive ability of consumers is limited, we need to be very cautious when promoting standardization and simplification in food management. More research is warranted to describe consumer behavior on actual shopping occasion.

Modern agricultural production is fragile and unsustainable in both developing and developing countries. Although the programs to support more sustainable agricultural production have been developed in recent years, their impacts are still marginal. Many consumers in developed countries are also willing to support such sustainable agricultural practices. However, the marketing and regulation systems have not yet been developed appropriately to promote them. Further research is required for finding effective marketing and regulating strategies.

When designing food policy, it is necessary to strongly recognize that our diet is closely tied to ethics and culture. As we need to respect each other's culture, we need to respect each other's dietary habits as well. When designing food policies, it is not sufficient to just evaluate food attributes from a scientific point of view. We need to humbly listen to consumers' expectation about their food. The policies introduced without consultation are neither endorsed nor trusted by consumers. Hence, they are not effective at all.

Obesity and sodium, fat, or sugar intake are problems in many countries. Various programs have been introduced to improve consumer dietary behavior. However, they have not successfully altered people's behavior who need to change diet behavior. Although the government can influence consumers' diet and nutrition, classical approaches relying on regulations and/or price incentive programs alone are not so effective. A combination of classical approaches and the new program appealing to consumer's ethic or value will be necessary to change consumer behavior.

Index

Printed and bound by CPI Group (UK) Ltd, Croydon, CR0 4YY

01/11/2024

01782623-0006